Cell Biology Protocols

Cell Biology Protocols

Editors

J. Robin Harris
Institute of Zoology,
Johannes Gutenberg-Universität, Mainz, Germany

John Graham
JG Research Consultancy, Upton, Wirral, UK

David Rickwood
Department of Biological Sciences, University of Essex, Colchester, UK

John Wiley & Sons, Ltd

Other Wiley Editorial Offices

John Wiley & Sons Inc., 111 River Street, Hoboken, NJ 07030, USA

Jossey-Bass, 989 Market Street, San Francisco, CA 94103-1741, USA

Wiley-VCH Verlag GmbH, Boschstr. 12, D-69469 Weinheim, Germany

John Wiley & Sons Australia Ltd, 42 McDougall Street, Milton, Queensland 4064, Australia

John Wiley & Sons (Asia) Pte Ltd, 2 Clementi Loop #02-01, Jin Xing Distripark, Singapore 129809

John Wiley & Sons Canada Ltd, 22 Worcester Road, Etobicoke, Ontario, Canada M9W 1L1

Wiley also publishes its books in a variety of electronic formats. Some content that appears
in print may not be available in electronic books.

Library of Congress Cataloging-in-Publication Data:

Cell biology protocols / editors, J. Robin Harris, John Graham, David Rickwood.
        p. cm.
    ISBN-13: 978-0-470-84758-9
    ISBN-10: 0-470-84758-1
    1. Cytology – Laboratory manuals.   I. Harris, James R.   II. Graham, J. M. (John M.),
1943- III. Rickwood, D. (David)
    QH583.2.C47 2005
    571.6078 – dc22

                                                                                    2005029713

British Library Cataloguing in Publication Data

A catalogue record for this book is available from the British Library

ISBN-13: 978-0-470-84758-9 (H/B)
ISBN-10: 0-470-84758-1 (H/B)

Typeset in 10.5/12.5pt Times by Laserwords Private Limited, Chennai, India
Printed and bound in Great Britain by Antony Rowe Ltd, Chippenham, Wiltshire
This book is printed on acid-free paper responsibly manufactured from sustainable forestry
in which at least two trees are planted for each one used for paper production.

Contents

Preface   **xi**

List of Contributors   **xiii**

1   Basic Light Microscopy   **1**
   Minnie O'Farrell

   Introduction   1
   Key components of the compound microscope   2
   Techniques of microscopy   6

   Protocols
   1.1   Setting up the microscope for bright field microscopy   7
   1.2   Setting Köhler illumination   8
   1.3   Focusing procedure   9
   1.4   Setting up the microscope for phase contrast microscopy   11
   1.5   Setting up the microscope for epifluorescence   14
   1.6   Poly-L-lysine coating   18
   References   19

2   Basic Electron Microscopy   **21**
   J. Robin Harris

   Introduction   21
   EM methods available   22

   Protocols
   2.1   Preparation of carbon-formvar, continuous carbon and holey carbon support films   25
   2.2   The 'droplet' negative staining procedure (using continuous carbon, formvar–carbon
         and holey carbon support films)   27
   2.3   Immunonegative staining   29
   2.4   The negative staining-carbon film technique: cell and organelle cleavage   31
   2.5   Preparation of unstained and negatively stained vitrified specimens   33
   2.6   Metal shadowing of biological specimens   35
   2.7   A routine schedule for tissue processing and resin embedding   37
   2.8   Agarose encapsulation for cell and organelle suspensions   39
   2.9   Routine staining of thin sections for electron microscopy   40
   2.10 Post-embedding indirect immunolabelling of thin sections   42
   2.11 Imaging the nuclear matrix and cytoskeleton by embedment-free electron microscopy   44
         Jeffrey A. Nickerson and Jean Underwood
   References   50

3 Cell Culture   51
Anne Wilson and John Graham

Cells: isolation and analysis   51
Anne Wilson
Mechanical disaggregation of tissue   52

Protocols

3.1   Tissue disaggregation by mechanical mincing or chopping   54
3.2   Tissue disaggregation by warm trypsinization   56
3.3   Cold trypsinization   58
3.4   Disaggregation using collagenase or dispase   60
      Anne Wilson
3.5   Recovery of cells from effusions   63
      Anne Wilson
3.6   Removal of red blood cells by snap lysis   64
3.7   Removal of red blood cells and dead cells using isopycnic centrifugation 65
      Anne Wilson
3.8   Quantitation of cell counts and viability   67
      Anne Wilson
3.9   Recovery of cells from monolayer cultures   71
      Anne Wilson
3.10 Freezing cells   74
3.11 Thawing cells   76
      John Graham
3.12 Purification of human PBMCs on a density barrier   80
3.13 Purification of human PBMCs using a mixer technique   82
3.14 Purification of human PBMCs using a barrier flotation technique   83
References   84

4 Isolation and Functional Analysis of Organelles   87
John Graham

Introduction   88
Homogenization   88
Differential centrifugation   90
Density gradient centrifugation   91
Nuclei and nuclear components   92
Mitochondria   93
Lysosomes   94
Peroxisomes   94
Rough and smooth endoplasmic reticulum (ER)   95
Golgi membranes   96
Plasma membrane   96
Chloroplasts   97
Protocols
4.1   Isolation of nuclei from mammalian liver in an iodixanol gradient (with notes on
      cultured cells)   98
4.2   Isolation of metaphase chromosomes   100

4.3	Isolation of the nuclear envelope	102
4.4	Nuclear pore complex isolation	104
	J. Robin Harris	
4.5	Preparation of nuclear matrix	106
4.6	Preparation of nucleoli	107
4.7	Isolation of a heavy mitochondrial fraction from rat liver by differential centrifugation	108
4.8	Preparation of a light mitochondrial fraction from tissues and cultured cells	110
4.9	Purification of yeast mitochondria in a discontinuous Nycodenz® gradient	112
4.10	Purification of mitochondria from mammalian liver or cultured cells in a median-loaded discontinuous Nycodenz® gradient	114
4.11	Succinate–INT reductase assay	116
4.12	Isolation of lysosomes in a discontinuous Nycodenz® gradient	117
4.13	β-Galactosidase (spectrophotometric assay)	119
4.14	β-Galactosidase (fluorometric assay)	120
4.15	Isolation of mammalian peroxisomes in an iodixanol gradient	121
4.16	Catalase assay	123
4.17	Analysis of major organelles in a preformed iodixanol gradient	124
4.18	Separation of smooth and rough ER in preformed sucrose gradients	127
4.19	Separation of smooth and rough ER in a self-generated iodixanol gradient	129
4.20	NADPH-cytochrome c reductase assay	131
4.21	Glucose-6-phosphatase assay	132
4.22	RNA analysis	133
4.23	Isolation of Golgi membranes from liver	134
4.24	Assay of UDP-galactose galactosyl transferase	136
4.25	Purification of human erythrocyte 'ghosts'	137
4.26	Isolation of plasma membrane sheets from rat liver	139
4.27	Assay for 5′-nucleotidase	141
4.28	Assay for alkaline phosphodiesterase	143
4.29	Assay for ouabain-sensitive Na^+/K^+-ATPase	144
4.30	Isolation of chloroplasts from green leaves or pea seedlings	145
4.31	Measurement of chloroplast chlorophyll	147
4.32	Assessment of chloroplast integrity	148

5 Fractionation of Subcellular Membranes in Studies on Membrane Trafficking and Cell Signalling

153

John Graham

Introduction	154
Methods available	154
Plasma membrane domains	155
Analysis of membrane compartments in the endoplasmic reticulum–Golgi–plasma membrane pathway	156
Separation of membrane vesicles from cytosolic proteins	157
Endocytosis	158

Protocols
5.1	Separation of basolateral and bile canalicular plasma membrane domains from mammalian liver in sucrose gradients	160

5.2	Isolation of rat liver sinusoidal domain using antibody-bound beads	162
5.3	Fractionation of apical and basolateral domains from Caco-2 cells in a sucrose gradient	163
5.4	Fractionation of apical and basolateral domains from MDCK cells in an iodixanol gradient	165
5.5	Isolation of lipid rafts	167
5.6	Isolation of caveolae	170
5.7	Analysis of Golgi and ER subfractions from cultured cells using discontinuous sucrose–D_2O density gradients	172
5.8	Analysis of Golgi, ER, ERGIC and other membrane compartments from cultured cells using continuous iodixanol density gradients	174
5.9	Analysis of Golgi, ER, TGN and other membrane compartments in sedimentation velocity iodixanol density gradients (continuous or discontinuous)	177
5.10	SDS–PAGE of membrane proteins	180
5.11	Semi-dry blotting	182
5.12	Detection of blotted proteins by enhanced chemiluminescence (ECL)	183
5.13	Separation of membranes and cytosolic fractions from (a) mammalian cells and (b) bacteria	185
5.14	Analysis of early and recycling endosomes in preformed iodixanol gradients; endocytosis of transferrin in transfected MDCK cells	188
5.15	Analysis of clathrin-coated vesicle processing in self-generated iodixanol gradients; endocytosis of asialoglycoprotein by rat liver	191
5.16	Polysucrose–Nycodenz® gradients for the analysis of dense endosome–lysosome events in mammalian liver	194
	References	196

6 *In Vitro* Techniques   **201**
Edited by J. Robin Harris

	Introduction	203
	Protocols	
	Nuclear components	
6.1	Nucleosome assembly coupled to DNA repair synthesis using a human cell free system *Geneviève Almouzni and Doris Kirschner*	204
6.2	Single labelling of nascent DNA with halogenated thymidine analogues *Daniela Dimitrova*	210
6.3	Double labelling of DNA with different halogenated thymidine analogues	214
6.4	Simultaneous immunostaining of proteins and halogen-dU-substituted DNA	217
6.5	Uncovering the nuclear matrix in cultured cells *Jeffrey A. Nickerson, Jean Underwood and Stefan Wagner*	220
6.6	Nuclear matrix–lamin interactions: *in vitro* blot overlay assay *Barbara Korbei and Roland Foisner*	228
6.7	Nuclear matrix–lamin interactions: *in vitro* nuclear reassembly assay	230
6.8	Preparation of *Xenopus laevis* egg extracts and immunodepletion *Tobias C. Walther*	234
6.9	Nuclear assembly *in vitro* and immunofluorescence *Martin Hetzer*	237
6.10	Nucleocytoplasmic transport measurements using isolated *Xenopus* oocyte nuclei *Reiner Peters*	240
6.11	Transport measurements in microarrays of nuclear envelope patches by optical single transporter recording *Reiner Peters*	244

Cells and membrane systems

6.12 Cell permeabilization with Streptolysin O   248
      Ivan Walev

6.13 Nanocapsules: a new vehicle for intracellular delivery of drugs   250
      Anton I. P. M. de Kroon, Rutger W. H. M. Staffhorst, Ben de Kruijff and Koert N. J. Burger

6.14 A rapid screen for determination of the protective role of antioxidant proteins in
      yeast   255
      Luis Eduardo Soares Netto

6.15 *In vitro* assessment of neuronal apoptosis   259
      Eric Bertrand

6.16 The mitochondrial permeability transition: PT and $\Delta\Psi$m loss determined in cells or
      isolated mitochondria with confocal laser imaging   265
      Judie B. Alimonti and Arnold H. Greenberg

6.17 The mitochondrial permeability transition: measuring PT and $\Delta\Psi$m loss in isolated
      mitochondria with Rh123 in a fluorometer   268
      Judie B. Alimonti and Arnold H. Greenberg

6.18 The mitochondrial permeability transition: measuring PT and $\Delta\Psi$m loss in cells and
      isolated mitochondria on the FACS   270
      Judie B. Alimonti and Arnold H. Greenberg

6.19 Measuring cytochrome *c* release in isolated mitochondria by Western blot analysis   271
      Judie B. Alimonti and Arnold H. Greenberg

6.20 Protein import into isolated mitochondria   272
      Judie B. Alimonti and Arnold H. Greenberg

6.21 Formation of ternary SNARE complexes *in vitro*   274
      Jinnan Xiao, Anuradha Pradhan and Yuechueng Liu

6.22 *In vitro* reconstitution of liver endoplasmic reticulum   277
      Jacques Paiement and Robin Young

6.23 Asymmetric incorporation of glycolipids into membranes and detection of lipid
      flip-flop movement   280
      Félix M. Goñi, Ana-Victoria Villar, F.-Xabier Contreras and Alicia Alonso

6.24 Purification of clathrin-coated vesicles from rat brains   286
      Brian J. Peter and Ian G. Mills

6.25 Reconstitution of endocytic intermediates on a lipid monolayer   288
      Brian J. Peter and Matthew K. Higgins

6.26 Golgi membrane tubule formation   293
      William J. Brown, K. Chambers and A. Doody

6.27 Tight junction assembly   296
      C. Yan Cheng and Dolores D. Mruk

6.28 Reconstitution of the major light-harvesting chlorophyll *a/b* complex into liposomes   300
      Chunhong Yang, Helmut Kirchhoff, Winfried Haase, Stephanie Boggasch and Harald Paulsen

6.29 Reconstitution of photosystem 2 into liposomes   305
      Julie Benesova, Sven-T. Liffers and Matthias Rögner

6.30 Golgi–vimentin interaction *in vitro* and *in vivo*   307
      Ya-sheng Gao and Elizabeth Sztul

Cytoskeletal and fibrillar systems

6.31 Microtubule peroxisome interaction   313
      Meinolf Thiemann and H. Dariush Fahimi

6.32 Detection of cytomatrix proteins by immunogold embedment-free electron
      microscopy   317
      Robert Gniadecki and Barbara Gajkowska

6.33 Tubulin assembly induced by taxol and other microtubule assembly promoters   326
      Susan L. Bane
6.34 Vimentin production, purification, assembly and study by EPR   331
      John F. Hess, John C. Voss and Paul G. FitzGerald
6.35 Neurofilament assembly   337
      Shin-ichi Hisanaga and Takahiro Sasaki
6.36 α-Synuclein fibril formation induced by tubulin   342
      Kenji Uéda and Shin-ichi Hisanaga
6.37 Amyloid-β fibril formation *in vitro*   345
      J. Robin Harris
6.38 Soluble Aβ$_{1-42}$ peptide induces tau hyperphosphorylation *in vitro*   348
      Terrence Town and Jun Tan
6.39 Anti-sense peptides   353
      Nathaniel G. N. Milton
6.40 Interactions between amyloid-β and enzymes   359
      Nathaniel G. N. Milton
6.41 Amyloid-β phosphorylation   364
      Nathaniel G. N. Milton
6.42 Smitin–myosin II coassembly arrays *in vitro*   369
      Richard Chi and Thomas C. S. Keller III
6.43 Assembly/disassembly of myosin filaments in the presence of EF-hand
      calcium-binding protein S100A4 *in vitro*   372
      Marina Kriajevska, Igor Bronstein and Eugene Lukanidin
6.44 Collagen fibril assembly *in vitro*   375
      David F. Holmes and Karl E. Kadler

7   **Selected Reference Data for Cell and Molecular Biology**   379
    David Rickwood

    Chemical safety information   379
    Centrifugation data   386
    Radioisotope data   388

    Index   391

Preface

Cell biology is a rapidly expanding discipline that is dependent upon continual technical development. We have attempted to compile an exciting and broadly useful cell biology techniques book, containing tried-and-tested procedures as well as newly established ones. Thus, this book contains an extensive series of routine and up-to-date protocols of value for those studying diverse aspects of present-day cell biology. The book commences with the presentation of several essential light microscopical procedures and leads on to the basic procedures required for producing a range of different cellular, subcellular and macromolecular specimens for transmission electron microscopical study. Then follows a chapter dealing with cell culture and cell separation procedures that are widely used to provide starting material for cellular research. The numerous techniques needed to study subcellular organelles and isolated cellular membranes are presented in the next two chapters, thereby providing the main thrust of the book. A series of 44 more specialist techniques used for *in vitro* studies and reassembly approaches in cell biology appear in the next chapter, each contributed by authors knowledgeable and experienced in their field of study. Finally, a reference chapter contains useful information on chemical hazard/safety aspects, centrifugation and radioisotopes. The book has a strong practical content and is directed to those at all levels who perform research in cell biology.

J. Robin Harris
John Graham
David Rickwood

June, 2005

List of Contributors

Judie B. Alimonti
Special Pathogens Program, National Microbiology Laboratory, H2380, 1015 Arlington Ave. Winnipeg, Manitoba, Canada, R3E 3R2

Geneviéve Almouzni
Institut Curie, CNRS, UMR 218, Section Recherche, 26 rue Ulm, F-75248 Paris 05, France

Alicia Alonso
Universidad del Pais Vasco, EHU, CSIC, Unidad Biofis, Aptdo 644, E-48080, Spain

Susan L. Bane
Department of Chemistry, SUNY, Binghamton, NY 13902-6016, USA

Julie Benesova
Lehrstuhl für Biochemie der Pflanzen, Ruhr-Universität Bochum, D-447780 Bochum, Germany

Eric Bertrand
Novartis Pharma AG, CH-4002, Switzerland

Stephanie Boggasch
Institut für Allgemeine Botanik des Johannes-Gutenberg-Universität, Müllerweg 6, D-55099 Mainz, Germany

Igor Bronstein
BBSRC Institute for Animal Health, High Street, Crompton RG20 7NN, UK

William J. Brown
Biochemistry, Molecular and Cell Biology Sections, Cornell University, Ithaca, NY 14853, USA

Koert N.J. Burger
Department of Biochemical Physiology, Institute of Biomembranes, Room W210, Padualaan 8, 3584 CH Utrecht, The Netherlands

K. Chambers
Biochemistry, Molecular and Cell Biology Sections, Cornell University, Ithaca, NY 14853, USA

C. Yan Cheng
Population Council, Center of Biomedical Research, 1230 York Avenue, New York NY 10021, USA

Richard Chi
Department of Biological Science, Florida State University, Tallahassee, FL 32306-4370, USA

Anton I.P.M. de Kroon
Department Biochemistry of Membranes, Centre for Biomembranes and Lipid Enzymology, Institute of Biomembranes, Padualaan 8, 3584 CH Utrecht, The Netherlands

Ben de Kruijff
Department Biochemistry of Membranes, Centre for Biomembranes and Lipid Enzymology, Institute of Biomembranes, Padualaan 8, 3584 CH Utrecht, The Netherlands

Daniela S. Dimitrova
Center for Single Molecule Biophysics and Department of Microbiology, 304 Sherman Hall, SUNY at Buffalo, Buffalo, NY 14214, USA

A. Doody
Biochemistry, Molecular and Cell
Biology Sections, Cornell University,
Ithaca, NY 14853, USA

H. Dariush Fahimi
Institute of Anatomy and Cell Biology,
University of Heidelberg, INF 307,
Neuenheimer Feld 307, D-69120
Heidelberg, Germany

Paul G. Fitzgerald
Department of Cell Biology and Human
Anatomy, School of Medicine, 1 Shields
Avenue, Davis, CA 95616-8643, USA

Roland Foisner
Department of Molecular Cell Biology,
Institute of Medical Biochemistry, Vienna
Biocenter, University of Vienna, Dr. Bohr
Gasse 9/3, A-1030 Vienna, Austria

Barbara Gajkowska
The Laboratory of Cell Ultrastructure,
Polish Academy of Sciences, Warsaw,
Poland

Ya-sheng Gao
Department of Pathology, Duke
University Medical Center, Box No.
3020, Rm 225, Jones Bldg, Durham, NC
27, USA

Robert Gniadecki
University of Copenhagen, Bispebjerg
Hospital, Department of Dermatology
D92, Bispebjerg Bakke 23, DK-2400
Copenhagen NV, Denmark

Félix M. Goñi
Universidad del País Vasco, EHU, CSIC,
Unidad Biofis, Aptdo 644, E-48080, Spain

John Graham
JG Research Consultancy, 34 Meadway,
Upton Wirral CH49 6IQ, UK

Arnold H. Greenberg[†]
University of Manitoba, Department of
Medical Microbiology, 539-730 William
Avenue, Winnipeg, MB, R3E OV9,
Canada

†deceased

J. Robin Harris
Institute of Zoology, University of Mainz,
D-55099 Mainz, Germany

John F. Hess
Department of Cell Biology and Human
Anatomy, School of Medicine, 1 Shields
Avenue, Davis, CA 95616-8643, USA

Martin Hetzer
Molecular and Cell Biology Laboratory,
The Salk Institute for Biological Studies,
10010 North Torrey Pines Road, La Jolla,
CA 92037, USA

Matthew K. Higgins
MRC Laboratory of Molecular Biology,
Hills Road, Cambridge CB2 2QH, UK

Shin-ichi Hisanaga
Department of Biology, Tokyo
Metropolitan University, Graduate School
of Science, Hachioji, Tokyo 1920397,
Japan

David F. Holmes
Wellcome Trust Centre for Cell Matrix
Research, School of Biological Sciences,
University of Manchester, Stopford
Building, Oxford Road, Manchester MI3
9PT, UK

Karl E. Kadler
Wellcome Trust Centre for Cell Matrix
Research, School of Biological Sciences,
University of Manchester, Stopford
Building, Oxford Road, Manchester MI3
9PT, UK

Thomas C.S. Keller
Department of Biological Sciences,
Florida State University, Tallahassee, FL
32306-4370, USA

Helmut Kirchhoff
Institut für Botanik, Westfälische
Wilhelms-Universität Münster,
Schlossplatz 2, D-48149 Münster,
Germany

Doris Kirschner
Institut Carie, 26 rue d'Ulm, 75248 Paris
Cedex 05, France

Barbara Korbei
Department of Molecular Cell Biology,
Institute of Medical Biochemistry, Vienna
Biocenter, University of Vienna, Dr. Bohr
Gasse 9/3, A-1030 Vienna, Austria

Marina Kriajevska
University of Leicester, Clinical Sciences
Unit, Leicester General Hospital,
Gwendolen Road, Leicester LE5
4PW, UK

Sven-T. Liffers
Lehrstuhl für Biochemie der Pflanzen,
Ruhr-Universität Bochum, D-447780
Bochum, Germany

Yuechueng Liu
Department of Pathology, University of
Oklahoma Health Services Center,
Oklahoma City, OK 73104, USA

Eugene Lukanidin
Danish Center Society, Institute of Cancer
Biology, Department of Molecular Cancer
Biology, Strandblvd 49, 4-3, DK-2100
Copenhagen, Denmark

Ian G. Mills
Dept. of Neurobiology, MRC Laboratory
of Molecular Biology, Hills Road,
Cambridge CB2 2QH, UK

Nathaniel G.N. Milton
Department of Molecular Pathology &
Clinical Biochemistry, Royal Free
Hospital Campus, Rowland Hill Street,
London NW3 2PF, UK

Dolores D. Mruk
Population Council, Center of Biomedical
Research, 1230 York Avenue, New York,
NY10021, USA

Luis Eduardo Soares Netto
Departamento de Biologia, Instituto de
Biociências, Universidade de São Paulo,
Rua do Matão, 277; Sala 327, Cidade
Universitária, CEP 05508-900, São
Paulo-SP, Brazil

Jeffrey A. Nickerson
Department of Cell Biology, School of
Medicine, University of Massachusetts,
55 Lake Avenue N., Worcester, MA
01655, USA

Minnie O'Farrell
Department of Biological Sciences,
University of Essex, Wivenhoe Park,
Colchester CO4 3SQ, UK

Jacques Paiement
Département de Pathologie et Biologie
Cellulaire, Université de Montréal N-813,
Pavilion Principal, 2900
Edouard-Montpetit, Montréal, Québec
H3T 1J4, Canada

Harald Paulsen
Institut für Allgemeine Botanik der
Johannes-Gutenberg-Universität,
Müllerweg 6, D-55099 Mainz, Germany

Brian J. Peter
McMahon Laboratory, Neurobiology
Division, MRC-LMB, Hills Road,
Cambridge CB2 2QH, UK

Reiner Peters
Institut für Medizinische Physik und
Biophysik, Universität Münster,
Robert-Koch-Straße 31, D-48149
Münster, Germany

Anuradha Pradhan
Department of Pathology, University of
Oklahoma Health Services Center,
Oklahoma City, OK 73104, USA

David Rickwood
Department of Biological Sciences,
University of Essex, Colchester, UK

Matthias Rögner
Lehrstuhl für Biochemie der Pflanzen,
Ruhr-Universität Bochum, D-447780
Bochum, Germany

T. Sasaki
Department of Biology, Tokyo
Metropolitan University, Graduate School
of Science, Hachioji, Tokyo 1920364,
Japan

Rutger W.H.M. Staffhorst
Department Biochemistry of Membranes,
Centre for Biomembranes and Lipid
Enzymology, Institute of Biomembranes,
Padualaan 8, 3584 CH Utrecht, The
Netherlands

Elizabeth Sztul
Department of Cell Biology, University
of Alabama, McCullum Bldg, Rm 668,
1530 S. 3rd Avenue, Birmingham, AL
35294, USA

Jun Tan
The Roskamp Institute, University of
South Florida, 3515 E. Fletcher Avenue,
Tampa, FL 33613, USA

Meinolf Thiemann
Graffinity Pharmaceuticals AG, Im
Neuenheimer Feld 518-519, D-69120
Heidelberg, Germany

Terrence Town
Yale University School of Medicine and
Howard Hughes Medical Institute, 310
Cedar St., PO Box 208011, New Haven,
CT 06520-8011, USA

Kenji Uéda
Department of Biology, Tokyo
Metropolitan University, Graduate School
of Science, Hachioji, Tokyo 1920364,
Japan

Jean Underwood
Department of Cell Biology, University of
Massachusetts Medical School, 55 Lake
Avenue, Worcester, MA 01655, USA

Ana-Victoria Villar
Universidad del País Vasco, EHU, CSIC,
Unidad Biofis, Aptdo 644, E-48080, Spain

John C. Voss
Department of Biological Chemistry,
School of Medicine, 1 Shields Avenue,
Davis, CA 95616-8643, USA

Stefan Wagner
Department of Cell Biology, University of
Massachusetts Medical School, 55 Lake
Avenue, Worcester, MA 01655, USA

Ivan Walev
Institute for Medical Microbiology and
Hygiene, University of Mainz, Hochhaus
Augustusplatz, D-55131 Mainz, Germany

Tobias C. Walther
EMBL, Gene Expression Programme,
Meyerhofstrasse 1, 69117 Heidelberg,
Germany

Anne Wilson
Woodbine Terrace, Stanton, Ashbourne
Derbyshire DE6 2DA

F.-Xabier Contreras
Universidad del Pais Vasco, EHU, CSIC,
Unidad Biofis, Aptdo 644,
E-48080, Spain

Jinnan Xiao
Department of Pathology, University of
Oklahoma Health Services Center,
Oklahoma City, OK 73104, USA

Chunhong Yang
Institut für Allgemeine Botanik der
Johannes-Gutenberg-Universität,
Müllerweg 6, D-55099 Mainz, Germany

Robin Young
Département de Pathologie et Biologie
Cellulaire, Université de Montréal N-813,
Pavilion Principal, 2900
Edouard-Montpetit, Montréal, Québec
H3T 1J4, Canada

1

Basic Light Microscopy

Minnie O'Farrell

Protocol 1.1	Setting up the microscope for bright field microscopy	7
Protocol 1.2	Setting Köhler illumination	8
Protocol 1.3	Focusing procedure	9
Protocol 1.4	Setting up the microscope for phase contrast microscopy	11
Protocol 1.5	Setting up the microscope for epifluorescence	14
Protocol 1.6	Poly-L-lysine coating	18

Introduction

Light microscopy is an indispensable technique for cell and molecular biologists to study cellular structures and biological processes in both living and fixed cells. This chapter provides an overview of light microscopy, describes the important parts of the microscope and goes on to explain how to set up a standard research microscope for bright field and phase contrast microscopy. There is also a short section on confocal microscopy. More comprehensive descriptions of the different forms of light microscopy are found elsewhere [1–4].

Microscopes are instruments that produce an enlarged image of a specimen. The eye-pieces and the objectives are the main components of the magnification system of the microscope, the product of the magnification of the objective lens and the ocular lens give the total magnification of the microscope. The visibility of the magnified specimen depends on contrast and resolution. Contrast is the difference in light intensity between an object and its background. Some biological samples contain coloured compounds, for example pigmented animal cells and chlorophyll-containing chloroplasts in plant cells, but most biological samples are colourless and have to be fixed and stained before observation [5]. Such stained specimens are observed using bright field microscopy. Other kinds of microscope systems are available to enhance contrast in living samples; these include phase contrast, dark field, differential interference contrast (DIC) and fluorescence microscopy (Table 1.1). The flow chart in Figure 1.1 will help in the selection of the appropriate microscopic observation method.

Cell Biology Protocols. Edited by J. Robin Harris, John Graham, David Rickwood
© 2006 John Wiley & Sons, Ltd

Table 1.1 Techniques for producing contrast in light microscopy

Type	Mechanism	Requirements	Fixed cells	Live cells	Appearance
Bright field	Absorption of visible light following staining of specimen	Any light microscope; range of histochemical stains	Yes	No	Coloured image depending on stains
Phase contrast	Variations in refractive index within specimen	Phase objective and phase condenser	Yes	Yes	Many shades of grey
Dark field	Scattered light	Dark field stop in condenser	Yes	Yes	Bright objects against dark background
Differential interference contrast	Gradient of refractive index	Special objective lens	Yes	Yes	3D effect
Fluorescence	Excitation and emission of light by fluorophore	An excitation light source and appropriate filters for emission; range of fluorescent probes including naturally fluorescent proteins	Yes	Yes	Bright colours against a dark background

The resolution of the optical system, that is the ability to distinguish objects separated by small distances, determines the degree of detail observable. The limit of resolution of the light microscope is about 0.2 μm. Enlarging the image too much results in 'empty magnification' and the quality of the image deteriorates. The limits of resolution are determined by the quality of the objective and the condenser.

Key components of the compound microscope

The eyepieces, body tube, nosepiece and objectives are part of the magnification system of the microscope. The condenser, condenser-iris diaphragm, filters, field iris diaphragm and light source are the parts that compose the illumination system of the microscope. To use a microscope properly, and to get the most out of it, it is important to understand the purpose and function of each of the microscope's components (Figure 1.2).

The body and lamp

The binocular body, the arm and the base form the frame of the microscope. This provides the stability and holds the optical and other components rigid and in place. The lamp is in the base of the body; its brightness is controlled by an on/off switch and a rheostat control knob. Just above the lamp is a collector lens with a field diaphragm to control the area of illumination. The field diaphragm also aids focusing and centring of the illumination.

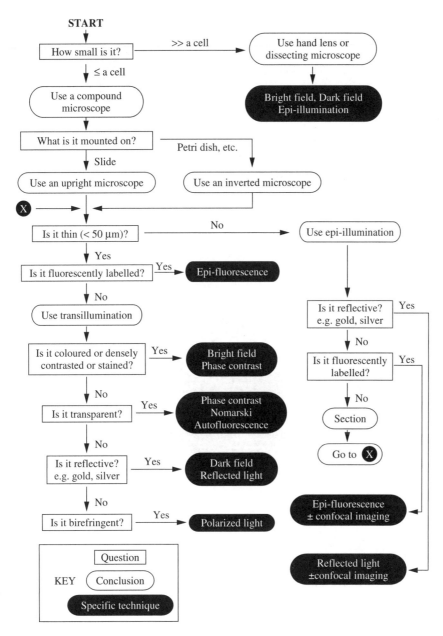

Figure 1.1 Flow chart for selection of observation methods. Reproduced from Rawlins (1992) *Light Microscopy*, Fig 1.1 Bios Scientific Publishers, Oxford

The condenser

The condenser provides a bright, even illumination of the specimen for a wide range of magnifications. The condenser can be focused and light transmission regulated by the condenser-iris diaphragm; correctly used these will optimize resolution, contrast and depth of field. Modified condensers are required for contrasting techniques such as phase contrast and differential interference contrast.

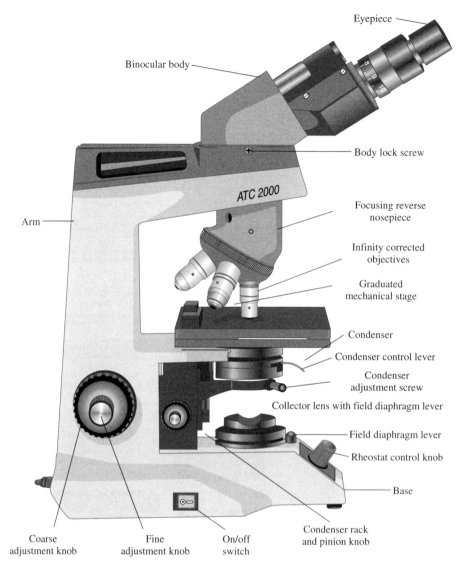

Eyepiece

Binocular body

Body lock screw

ATC 2000

Focusing reverse nosepiece

Infinity corrected objectives

Graduated mechanical stage

Arm

Condenser

Condenser control lever

Condenser adjustment screw

Collector lens with field diaphragm lever

Field diaphragm lever

Rheostat control knob

Base

Condenser rack and pinion knob

Coarse adjustment knob

Fine adjustment knob

On/off switch

Figure 1.2   The parts of the compound light microscope

The stage and focus mechanism

The specimen, usually on a slide, is held in place by a sprung arm on the mechanical stage. The stage can be moved in the x and y planes and mounted vernier scales can be used to locate sites of interest on the coverslip/slide. The course (outer) and fine (inner) focus adjustment knobs alter the level of the stage with respect to the objective.

The objective

The objectives lenses are mounted on a revolving nosepiece which allows for easy changes between magnification and also facilitates the maintenance of focus when the

Table 1.2 Properties of some objective lenses

Magnification of objective	Focal length (mm)	NA	Working distance (mm)	Diameter of field (mm)	Depth of field (μm)
10	16	0.20–0.30	4–8	1–2	c. 10
40	4	0.65–0.85	0.2–0.6	0.25–0.50	1–2
100 (oil)	2	1.20–1.30	0.11–0.16	0.1–0.2	0.5

different objectives are moved into position, giving parfocality. The objective lens of the microscope is the major component responsible for the magnification and resolution of the image; it is perhaps the single most important element of the microscope.

Basically, an objective consists of a set of lens elements that form an image of an object at a distinct distance beyond the objective; it collects light from every specimen point and forms a real intermediate image in the eyepiece focal plane. Besides collecting light from the specimen and 'magnifying' the latter, the objective contains lenses that correct the aberrations created as light passes through the collecting lens system. The ability to collect light and, therefore, to resolve detail is termed the numerical aperture (NA) of the objective. The limit of resolution is determined by the wavelength of light used (λ) and the NA, the light-gathering capacity of the objective:

$$\text{Resolution} = 0.61 \times \frac{\text{wavelength of light source}(\lambda)}{\text{numerical aperture (NA)}}$$

A dry objective cannot have an NA greater than 1 but an immersion medium, for example oil, can increase the NA beyond 1 (Table 1.2)

In selecting an objective for a given purpose it is useful to know certain figures. These are (1) the magnification, (2) the focal length, (3) numerical aperture, (4) the working distance, (5) the diameter of the field of view and (6) the depth of field. Average values for commonly used objectives are shown in Table 1.2

The working distance is the clearance between the lowest point of the objective and the upper surface of the coverslip. The depth of field is the range of distances over which objects can still appear reasonably sharp. The most important factor in deciding this quantity is the NA.

The objective lenses bear a number of inscriptions including the type, the magnification and the NA (Figure 1.3). Achromat, Plan Achromat and Plan Apochromat are the names of objectives of increasing quality. The Achromat lenses are colour corrected for two wavelengths (red, 656 nm and blue, 486 nm) and are corrected for spherical aberration in the green (546 nm). The Apochromats have been further corrected to give the best colour reproducibility. The Plan designation refers to correction for flatness of field across the whole image. The inscription on the objective lens shown in Figure 1.3 is, for example, 40/0.65 and 160/0.17. These figures indicate the initial magnification ×40, numerical aperture 0.65, for use with microscopes with a mechanical tube length of 160 mm and with a coverslip 0.17 mm thick.

The eyepieces

The real, intermediate image formed by the objective is observed and further magnified by means of an eyepiece. They usually have a magnifying power of 10× but can range

Figure 1.3   An objective lens showing the specifications engraved on the metal body tube

from 4× to 30×. Eyepieces over 12.5×, however, depending on the objective used, may result in 'empty magnification'. Apart from its magnification, an eyepiece is characterized by its field of view number. With the aid of this number it is possible to calculate the diameter of the field covered in the specimen plane. The field of view number of the eyepiece divided by the magnification of the objective gives the diameter of the actual field of view in millimetres.

There are also eyepieces specially designed for spectacle wearers. They are usually marked with a diagram of a pair of glasses. The interpupillary distance can be altered in most binocular microscopes.

Techniques of microscopy

Bright field microscopy

Bright field microscopy is probably the most widely used form of microscopy and is used mainly for fixed and stained specimens. For optimal resolution the microscope should be aligned correctly and one of the most important alignments is setting up Köhler illumination. This provides bright and even illumination over the specimen and allows for the control of contrast and depth of field.

PROTOCOL 1.1

Setting up the microscope for bright field microscopy

1. Set up the microscope on a clean, dry area of bench as vibration-free as possible.

2. Plug in and switch on the illuminator.

3. Centre and focus the light source according to the manufacturer's instructions. The light source on many microscopes is precentred.

4. Bring the ×10 objective into position on the nosepiece and ensure the condenser is in the bright field position, usually indicated by a zero (0).

5. Fully open the field diaphragm on the light source and the aperture diaphragm of the condenser.

6. Adjust the brightness of the light from the illuminator to a comfortable level using the rheostat.

7. Place a clean, prepared slide on the stage. Use a stained preparation or section. Bring the objective to within a few millimetres of the coverslip on the slide. Make sure that the specimen is in line with the objective.

8. Looking through the eyepieces gently lower the stage by rotating the course adjustment until the specimen appears in focus. Move through the focal position: continue rotating the knob until the specimen has appeared in focus, and is just beginning to become blurred again.

9. Now use the fine adjustment to make the image clear.

10. Adjust the interpupillary distance between the eyepieces so that it is comfortable for you.

Setting Köhler illumination

To focus the condenser and set up Köhler illumination:

1. Gently rack the condenser upwards to the top of its travel.

2. Close down the field iris diaphragm of the illuminator. Focus the image by lowering the condenser. The condenser is focused when both the specimen and the diaphragm edge are sharply defined.

3. Centre the field diaphragm image into the centre of the field of view, using the centring screws on the condenser, then open the field diaphragm to just fill the field of view.

4. Remove one eyepiece and, looking into the tube, adjust the condenser aperture diaphragm so that it covers two-thirds to three-quarters of the circular illuminated area. This controls the angle of the cone of light illuminating the object and helps avoid flare.

5. Replace the eyepiece; Köhler illumination is now set up. The brightness of the field should be altered by adjusting the light intensity of the illuminator. Do not alter light by adjusting the condenser or field diaphragm.

6. The alignment should be correct for all objectives. The nosepiece can be rotated to bring other objectives into position. Most microscopes are parfocal so only the fine focus should be used to make final adjustments to obtain clear images.

Use of oil immersion objectives

Oil immersion objectives are objectives that are used with immersion oil (instead of air) in the space between the front lens of the objective and the coverslip. As explained previously, this method leads to a gain in NA, and hence a gain in the resolution, brightness and clarity of the image. Immersion oil has a refractive index of 1.515 and glass has a refractive index of 1.51. The principles underlying the use of the $\times 100$ objective are similar to those used in setting up the lower power objectives. The only differences, in practice, are the introduction of an oil film between the top of the coverslip and the objective lens and the need for very great caution in focusing because of the very short working distance of $\times 100$ objectives.

When examining extremely small specimens, such as bacteria and individual eukaryote cells, locating the sample image can be difficult. It is easiest to focus on the sample first using the $\times 40$ objective. Then, using the fact that the objectives are parfocal, move to the $\times 100$ objective.

1. Focus on the specimen using the ×40 objective (in some cases, you may wish to start with an even lower power objective first).

2. Move the ×40 objective away from the slide, but do not yet move the revolving nosepiece fully round to bring the ×100 objective into position.

3. Leaving the slide in position on the stage, apply a small drop of immersion oil onto the illuminated spot on the coverslip. Do not use too much oil, it can run off the slide and on to the microscope.

4. Carefully move the ×100 objective into position, checking that there is space for it. The objective should make contact with the oil, but not the slide.

5. You should be approximately in focus already so only use the fine focus knob.

Oil immersion objectives and slides which have been used with oil should *always* be cleaned with lens tissue immediately after use. If left uncleaned the oil tends to dry up and may form a hard film on the objective and slide.

Phase contrast microscopy

If a typical living cell is observed under well-aligned bright field microscopy, very little detail can be observed. However, small differences in the refractive indices or the thickness within specimen structures can be converted into differences of light and dark. This is what is known as phase contrast microscopy and it is extensively used for observing living and unstained specimens.

In the phase contrast optics there is a phase ring in the rear focal plane of the objective and a phase annulus in the front focal plane of the condenser (Figure 1.4A). The ring and the annulus alter the amplitude and phase relationships of the light diffracted by structures in the cell compared to the non-diffracted light to generate interference, thereby enhancing contrast. Phase contrast produces an image of many dark and light grey gradations. Positive phase contrast areas with higher refractive indices, like mitochondria in the cytoplasm and nucleoli against the background of the nucleoplasm, appear darker.

Phase contrast microscopy is limited to thin specimens, single cells or very thin cell layers. The resolution is also limited by the phase ring and phase annulus.

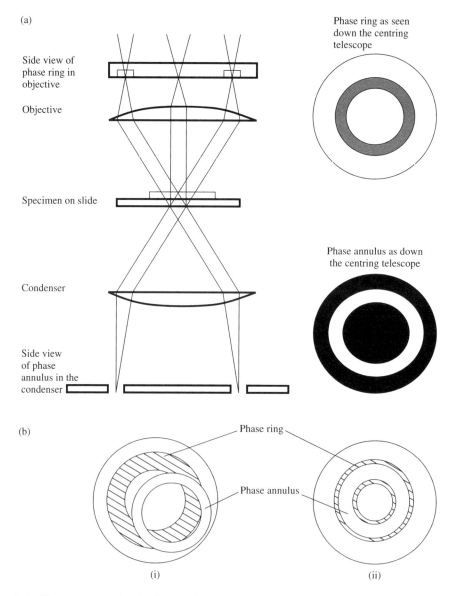

Figure 1.4   Phase contrast. (a) the light path in a phase contrast microscope showing the phase ring in the objective and the phase annulus in the condenser. (b) Incorrect (i) and correct (ii) alignment of the phase ring and phase annulus

Setting up the microscope for phase contrast microscopy

1. The objectives for phase microscopy are different from those used for bright-field only. They include a phase ring and are usually inscribed Ph. Phase contrast objectives can also be used for bright field work. The condenser is also different; it contains a series of phase annulae.

2. Köhler illumination is set up as described previously

3. To centre the phase contrast annular diaphragm, start with the objective ×10 Ph 1. Set the condenser turret to the appropriate position.

4. Move the specimen slide to a position where no object is seen, yet the cover-slip is still present.

5. Remove the eyepiece and in its place insert the centring telescope (phase telescope). Rotate the top part of the telescope to bring the image of the phase plate in the objective into sharp focus. The image of the condenser annulus will also be visible.

6. The position of the objective phase ring is fixed but the phase annulus in the condenser can be moved by manipulating the adjusting screws on the condenser. The phase annulus needs to be superimposed, concentrically, over the phase ring (see Figure 4B).

With many microscopes the phase annulae in the condenser turret are adjusted taking ×10 Ph 1 as the standard so it may not be necessary to perform the centring procedure at each magnification. However, for ideal resolution and for photomicrography it is advisable to check the centring at each magnification.

Differential interference contrast microscopy

This is another type of interference system that detects very small gradients or differences in thickness or refractive index. It is based on illumination of the specimen with two beams of light polarized at 90° to each other and with a lateral displacement (shear) of approximately the resolution limit of the objective. By altering the special beam-splitting prisms one can obtain an almost continuous change from bright field to dark field microscopy resembling degrees of highly oblique illumination and giving a three-dimensional appearance. Like phase contrast microscopy, this method is used to reproduce unstained transparent objects at high contrast. The image obtained gives a characteristic relief effect and offers some advantage over ordinary phase contrast optics. It is particularly useful for the study of dynamic events *in vivo*, for example cells in mitosis, and for studying the three-dimensional structure of cells and embryos.

Dark field microscopy

Dark field microscopy can be employed in the study of living micro-organisms such as spirochetes, protozoa and yeast and unstained tissues and cells.

In bright field microscopy stains are used to produce sufficient contrast in the specimen so that diffraction can occur, thus rendering the specimen detail visible. Staining, however, can produce artifactual changes in some biological specimens; also fixation prior to staining kills the cell. Therefore, dark field microscopy, like phase contrast, is very useful for the study of living specimens. With dark field microscopy sufficient contrast can be introduced between a transparent object and the surrounding medium to render the object visible. This is done by illuminating the object at such an oblique angle that intense rays of light that normally pass through the specimen and travel directly into the objective (illuminating the entire field of view) are directed past the objective. This produces a dark field of view but the contrast between the dark background and the diffracted rays scattered off the specimen renders the object visible, bright and shiny against the dark background.

Although bright field condensers can be adjusted to give an approximation of a dark field, special condensers produce the required hollow cone of light more effectively.

Of great importance when using dark field techniques is the extreme cleanliness of slides, coverslips and immersion substances. Air bubbles, dirt or any extraneous material will be refractile and therefore interfere with the image from the specimen.

Fluorescence microscopy

Fluorescence microscopy is a sensitive technique very widely used in both research and clinical laboratories. Some materials for microscopic examination are autofluorescent, e.g. lipids, vitamins, porphyrins, drugs and drug-containing tissues and carcinogenic compounds. In addition fluorescent dyes are used to stain cells and tissues. There are now a wide range of fluorescent labels for a large number of physiologically important biological molecules. Secondary fluorescence is also of major importance. This is obtained by such methods as immunofluorescence, whereby a biological molecule is located by using an antibody joined to a fluorochrome dye or when fluorescent probes attached to oligonucleotides are used to identify genes on chromosomes.

An atom or molecule, when struck by a quantum of light, undergoes a change in the arrangement of the electrons about its nucleus. The electron (or electrons) is displaced to a higher energy level in the form of heat and light. Since some of the energy is lost to heat, the light energy emitted is of a lower energy level and thus of a longer wavelength than the light that energized it. Therefore, when light of a short wavelength is directed towards a fluorochrome dye conjugate, the dye will emit radiation of a longer wavelength plus heat. By blocking out the exciting light ray, the emitted light rays can be isolated and detected by the eye through the microscope. The only difference between basic bright field microscopy and fluorescence microscopy is the use of an illumination system which produces only light of a shorter wavelength and a receiving system which receives only light of a longer wavelength.

In addition to the light source (usually a mercury or xenon lamp), a fluorescence microscope requires excitation and barrier filters. The excitation or primary filter will transmit shorter wavelength radiation only, while the barrier or secondary filter will transmit the longer wavelength emitted fluorescent light only. In order to obtain the best possible contrast, with bright specimen portions where fluorescence occurs and the darkest possible background where there is no fluorescence, the most commonly used condenser system is a dark field condenser.

Most contemporary fluorescent microscopes use epifluorescence or incident-light fluorescence. In these microscopes the objective also works as a condenser carrying the exciting beam and also collecting the emitted light. The assembly of filters in the filter block include the excitation filter and the barrier filter (interference filters) and a chromatic beam splitter (Figure 1.5). The microscope is relatively easy to align and it allows the use of thicker specimens and also the combination of the incident-light fluorescence excitation with a transmitted-light technique such as phase contrast.

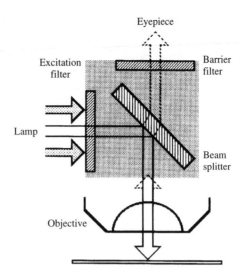

Figure 1.5   Filter and dichroic mirror block in an epifluorescence microscope

Setting up the microscope for epifluorescence

The alignment of the light source to achieve an evenly illuminated field is critically important in epifluorescent microscopy:

1. Turn on the lamp and allow to warm up.

2. Insert the filter block to be used.

3. Place a white piece of paper on the stage and adjust the light intensity to a suitable level with the slider. Open the condenser diaphragm and the field diaphragm.

4. Adjust the lamp collector focusing until a sharply focused image of the lamp is seen. Centre the image with the lamp centring screws.

5. Focus on a specimen slide. Close down the field diaphragm until it can be seen, focus the condenser and centre it using the centring screws on the condenser. When focused and centred open up the iris to the edge of the field of view.

Confocal microscopy

The resolution of fluorescence microscopy is limited by light emitted from above and below the focal plane leading to considerable blurring of the image. Confocal microscopy is intended to achieve a high axial resolution, that is structures in the centre of a cell can be distinguished from those at the top or the bottom. Contrary to conventional microscopy, it relies on point illumination, rather than field illumination. A schematic diagram of the corresponding configurations is shown in Figure 1.6. The specimen is illuminated by a point source, consisting of a laser beam focused through a small aperture. Hence, the intensity reaching out-of-focus points on the specimen is lower than using conventional, field illumination. In turn, fluorescent light leaving the specimen is focused on the small detector aperture. As can be seen in Figure 1.6, the combined effects of point illumination, together with point detection, mean that most of the out-of-focus light is excluded from the final image, considerably increasing the contrast and therefore the fine detail of the sample.

The chromatic beam splitter reflects the excitation light towards the specimen, while allowing the emission light to reach the detector. The optical path from the source aperture to the objective lens is essentially the same as that from the objective lens to the detector aperture. Therefore, both the illuminating and detector apertures are focused on the illuminated point in the specimen and this is the reason why the system is called *confocal*. The detector is a photomultiplier (PMT), because the signal to be detected is usually very weak.

In a confocal system only a single point is viewed at a time. Therefore, in order to build an image, that point has to be scanned over the sample and the measured intensities recorded. This is usually achieved by scanning the beam using two mirrors, giving rise to the name confocal laser scanning microscopy (CLSM). Measured intensities are recorded and displayed by a host computer, which also

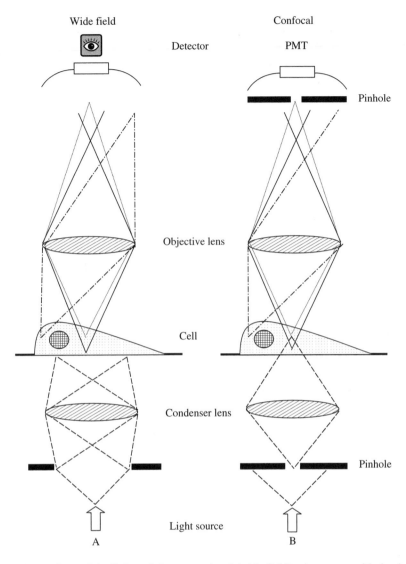

Wide field

Confocal

Detector

PMT

Pinhole

Objective lens

Cell

Condenser lens

Pinhole

Light source

A

B

Figure 1.6 Comparison of the light path in conventional (wide field) microscopy with that in confocal microscopy

drives the scanning system. All the thin sections along the axis obtained are largely devoid of out-of-focus light, this effect being called *optical sectioning*, and repetition of the $X-Y$ scanning along the Z axis allows the construction of a 3-D image of the object being analysed. Projection of all Z sections over a single plane (maximum projection) results in a 2-D image which, unlike conventional microscopy, contains only in-focus information; thus a large depth of field is achieved by adding together very thin, in-focus sections.

For optimum image acquisition, three elements have to be combined: the laser intensity, the size of the confocal aperture and the gain of the amplifier. Any increase of these parameters produces a bright image, but the effect may be detrimental on the image quality by reducing the resolution or increasing image noise or photobleaching. Table 1.3 summarizes

Table 1.3 Factors affecting image collection in confocal microscopy

Effect on:	Laser intensity	Aperture	Gain
Resolution	Not affected, but high intensity allows for smaller aperture	Closing the aperture increases the resolution and the Z-sectioning effect	Not affected directly but through the control of noise
Image noise	Not directly affected, but high intensity allows for lower gain	Not directly affected, but large aperture allows for lower gain	Raising the gain introduces more noise
Photobleaching	Higher intensity produces higher photobleaching	Not affected	Not affected

the effects of these components on the collected image.

Common problems and microscope care and maintenance

Problems in light microscopy usually arise from misalignment or dirt on the optical surfaces. Dark image and uneven illumination usually mean that the condenser is not correctly centred or the condenser diaphragm or field diaphragm are closed down too much or off-centre. The opposite can also occur with the image visible, but pale and undefined; this usually means that too much light is coming through and causing flare. The field or condenser diaphragms need to be closed down and the condenser adjusted. Dirt on the eyepieces or the objective will cause blurring and loss of definition. Dust on the lenses should be blown away with dry air and the lens cleaned with lens tissue moistened with distilled water. Stubborn grease can be removed with xylol; alcohols may dissolve the lens cement. Immersion oil should be wiped off at the end of every session of microscope use with lens tissue; never let the immersion oil dry to form a hardened film. Immersion oil should not be allowed to come into contact with any lenses except the oil immersion objective, but it is surprising how often a film of oil is found on low and medium power objectives. This should be cleaned off as for the oil immersion objectives. Care should be taken not to scratch the lens surfaces.

Care should also be taken to ensure that the tubes of the microscope are closed either by an eyepiece or a dust plug. Similarly all positions on the revolving nosepiece should also be closed with a dust plug.

The rackwork and other moving parts should be treated with great care; it is essential not to force anything if it is apparently jammed. Do not apply grease of an unspecified type to the sliding surfaces of the course-focusing adjustment or the gliding stage.

At all times the instrument should handled carefully; for example, when carrying the microscope hold the base with one hand and the body with the other. When not in use the microscope should always be protected by a vinyl dust cover or kept in a cabinet.

Preparation and staining of specimens

Attachment

When working with tissue culture cells or cell smears, the cells to be stained need to be attached to a solid support, usually a microscope slide or coverslip, to facilitate handling. This can be done in a number of ways. Adherent cells can be grown on coverslips in Petri dishes or in special slide chambers. Poorly adherent cells can be encouraged to attach to glass or plastic in culture by precoating the coverslips or slides with 1% (w/v) gelatin in distilled water or 500 µg/ml poly-L-lysine in distilled water. For easier handling suspension cells can be bound to slides using chemical linkers or by centrifuging onto slides using a cytocentrifuge [3].

Poly-L-lysine is frequently used to coat slides to facilitate attachment of suspension cells for fixing and staining. This positively charged polymer binds to glass sides through the charged lysine side groups; cells which have an overall negative charge bind to the positively charged polymer through non-covalent interactions.

1. Prepare a stock solution of poly-L-lysine (MW > 150 000 D) at a concentration of 1 mg/ml in distilled water. Do not store for long periods; make fresh each week.

2. Coat clean glass coverslips or slides by incubating in the poly-L-lysine solution for 10 min at room temperature.

3. Wash in several changes of water and air-dry.

4. Add a drop of cell suspension in PBS onto the coverslip. Incubate for 10 min at room temperature. After this the cells are ready for fixing.

Fixation

The fixation process preserves the specimen and stabilizes the cellular structure while permeabilizing the cell to allow access of stains or antibodies, in the case of immunostaining [5, 6]. Fixatives can be divided into two broad categories, organic solvents and cross-linking reagents. The organic solvents such as acetone and alcohols extract lipids and dehydrate the cells; macromolecules are precipitated. Cross-linking agents, for example formaldehyde or glutaraldehyde, form cross-links mainly between amino groups, stabilizing molecular structures. The choice of fixative depends on the sample and staining techniques to be used; some common fixatives are listed in Table 1.4. The fixation process

Table 1.4 Some commonly used fixatives

Organic fixatives	Suggested conditions
Ethanol 70–100%	5 min, room temperature. Air-dry
Methanol 70–100%	5 min, room temperature. Air-dry
Acetone : methanol (1 : 1 v/v)	5 min, room temperature. Air-dry
Glacial acetic acid : Methanol (1 : 3 v/v)	10 min, room temperature. Air-dry (for immunocytochemistry remove solvent and wash several times with PBS)
Cross-linking agents	
4% (w/v) Paraformaldehyde in PBS	10 minutes, room temperature. Wash cells several times with PBS (for immunocytochemistry cells/tissue should be permeabilized with 0.2% (v/v) Triton X100 or NP40 for 2 min at room temperature)
0.5% (v/v) Glutaraldehyde in PBS	30 min to several hours. Wash cells several times with PBS

Table 1.5 Some commonly used stains

Stain	Use
Giemsa (dilute stock in Hepes buffer, pH 6.8)	General staining, nuclei stain purple, cytoplasm blue. Also used for blood smears and bone marrow preparations
Hematoxylin (Mayer's) and Eosin	General staining, nuclei stain dark blue and cytoplasm pink-red
Papanicolaou's	General staining, nuclei blue, nucleoli blue or red and cytoplasm pink or blue depending on cell type
Sudan Black	Stains neutral fats green/brown, other lipids green/black; cytoplasm unstained

may cause artefacts and it is advisable to test several fixatives to determine the most effective in a particular situation. Also the procedures and conditions should be carefully controlled for consistency.

Preparation of tissue sections

Two commonly used methods used to prepare tissue for staining are sectioning of paraformaldehyde-fixed, paraffin-embedded tissues and sectioning of frozen tissues [6, 7]. Frozen sectioning is a relatively gentle way to prepare tissue samples with the advantage that the tissue is unfixed. A disadvantage is that specialist sectioning equipment in the form of a cryostat is required. Most histological studies are carried out on paraformaldehyde-fixed, paraffin-embedded tissue samples. The fixing and embedding processes are quite harsh and may not be suitable for immuno-histochemistry where particular care has to be taken with fixation.

Staining

Fixed cultured cells, cell smears or tissues sections need to be stained to give the contrast required for observation with bright field microscopy. There are many different stains that can be used to differentiate different cell types and different subcellular structures [3, 5, 6]. A selection of commonly used stains is given in Table 1.5. Most of these stains are commercially available, prepared as stable solutions.

References

1. Bradbury, S. and Bracegirdle, B. (1998) *Introduction to Light Microscopy*, 2nd edn. Bios Scientific Publishers, Oxford.
2. Rubbi, C. (ed.) (1994) *Light Microscopy: Essential Data*. John Wiley, London.
3. Spector, D., Goldman, R. and Leinwand, L. (1998) *Cells: A Laboratory Manual*. Cold Spring Harbor Laboratory Press, New York.
4. Wright, S. and Wright, D. (2002) in *Methods in Cell Biology*, Vol. 70: *Cell Biological Applications of Confocal Microscopy* (B. Matsumoto, ed.), pp. 1–85. Academic Press, New York.
5. Boon, M. and Drijver, J. (1986) *Routine Cytological Staining Techniques: Theoretical Background and Practice*. Macmillan, London.
6. Dealtry, G. and Rickwood, D. (eds) (1992) *Cell Biology LabFax*. Bios Scientific Publishers, Oxford.
7. Kiernan, J. (1990) *Histological and Histochemical Methods*. Pergamon Press, Oxford.

2

Basic Electron Microscopy

J. Robin Harris

Protocol 2.1 Preparation of carbon-formvar, continuous carbon and holey carbon support films   25

Protocol 2.2 The 'droplet' negative staining procedure (using continuous carbon, formvar–carbon and holey carbon support films)   27

Protocol 2.3 Immunonegative staining   29

Protocol 2.4 The negative staining-carbon film technique: cell and organelle cleavage   31

Protocol 2.5 Preparation of unstained and negatively stained vitrified specimens   33

Protocol 2.6 Metal shadowing of biological specimens   35

Protocol 2.7 A routine schedule for tissue processing and resin embedding   37

Protocol 2.8 Agarose encapsulation for cell and organelle suspensions   39

Protocol 2.9 Routine staining of thin sections for electron microscopy   40

Protocol 2.10 Post-embedding indirect immunolabelling of thin sections   42

Protocol 2.11 Imaging the nuclear matrix and cytoskeleton by embedment-free electron microscopy   44

Introduction

Electron microscopy (EM) is an essential tool for most cell biologists. When used appropriately, EM is able to provide direct visual evidence for the organization of biological structures at the subcellular and even molecular level. This chapter aims to provide cell biologists with some basic knowledge of the available EM specimen preparation techniques which will allow them to carry out the more straightforward analyses of cellular, subcellular and macromolecular samples. For more detailed methodologies the reader should consult one of the several available texts that is dedicated to EM [1–5].

Cell Biology Protocols. Edited by J. Robin Harris, John Graham, David Rickwood
© 2006 John Wiley & Sons, Ltd

EM methods available

The techniques used for biological specimen preparation fall into two main categories, those utilizing resin embedding followed by thin sectioning and positive staining or immunolabelling, and those using a thinly spread layer of particulate material, followed by metal shadowing, air-dry negative staining or rapid freezing/vitrification without or with negative staining. The technique of freeze-fracture, although widely used in the past, is somewhat less popular today although it remains useful for a number of situations [1, 5]; it is not easy for the inexperienced to perform as a routine procedure and is not included in this chapter. Thin sectioning of resin embedded cell suspensions, monolayers and organelles (see *Protocol 2.7*) essentially follows the established fixation and staining procedures that are widely used for tissues. Suspended cells and isolated organelles can be pelleted, followed by dispersal in low melting temperature agarose (see *Protocol 2.8*). Agarose can also be added to monolayers *in situ*, for direct processing on the plastic cell culture flask, miniature culture system or Petri dish. This agarose encapsulation approach is considered to be especially convenient, since small pieces of gelled agarose, containing the cellular material of interest, can be processed throughout the specimen preparation stages far more easily than by repeated centrifugal pelleting and resuspension, prior to embedding of the fixed and dehydrated material in resin. A specialist technique for embedment-free electron microscopy is given in *Protocol 2.11*.

Negative staining

Negative staining (see *Protocols 2.2 and 2.3*), by surrounding thinly-spread particulate biological materials with an amorphous coating of dried heavy metal salt, cannot usually be applied to structures as large and thick as intact cells or indeed cell nuclei, although some success has been achieved with blood platelets and cells that have been extracted with neutral surfactant or split open/wet-cleaved during the specimen preparation. When it comes to isolated organelles and their subfractions, such as mitochondria, chloroplasts, plasma membrane fractions such as cell junctions, rough and smooth ER, caveosomes, Golgi, nuclear envelope, cytoskeletal and fibrillar proteins, negative staining has considerable potential. The same applies to the use of negative staining for the study of oligomeric proteins, enzymes and macromolecular assemblies such as the 20S and 26S proteasome, ribosomes and the isolated nuclear pore complex [3]. A recently introduced improvement to the technique provides a standardized procedure for spreading biological particulates, supported by negative stain alone, across the holes of holey carbon support films [6].

Vitrification

The technique of vitrification of unstained biological particles suspended in a thin aqueous film, followed by cryoelectron microscopy, generally provides a superior structural approach than negative staining (see *Protocol 2.5*). Cryoelectron microscopy brings with it some technical difficulties, but these have largely been overcome in recent years [7]. It should, however, be borne in mind that unstained vitrified/frozen-hydrated specimens have to be studied under strict low temperature and low electron dose conditions, and that very often digital image processing is required in order to recover the structural information from the electron micrographs, because of the inherently low image contrast. Air-dried negatively stained specimens, adsorbed to carbon or spread across holes, can also be studied at low temperatures in the presence of glucose or trehalose which provide considerable protection of the biological material, increased sample mobility/reduced adsorption by the carbon support and reduced sample flattening. This low temperature negative staining approach is somewhat more easily performed than the study of unstained vitrified specimens and can yield a resolution in the order of 1.5 nm, but is likely to be inferior to the recently introduced procedure of cryonegative staining [8], which combines the benefits of both vitrification and negative staining.

Metal shadowing and freeze-fracture

Both negatively stained and vitrified unstained specimen preparation is usually performed with unfixed samples. On the other hand, platinum–carbon or tungsten–iridium metal shadowing often requires prior fixation of the biological material, together with total removal of fixative and buffer salts by prior dialysis against distilled water, washing after attachment to a mica or carbon substrate, or suspension in a buffer composed of volatile salts such as ammonium acetate or ammonium bicarbonate usually together with glycerol (see *Protocol 2.6*). The resolution obtained from metal shadowing used following freeze-fracture [5] is often somewhat inferior to negative staining and cryomicroscopy of unstained vitrified material, but excellent results have been achieved from freeze-dried and freeze-cleaved samples, where the granularity of the metal evaporated *in vacuo* is very fine [9].

Immunolabelling

Of considerable significance in modern cell biological studies is the ability to perform immunolabelling of antigens located within cellular structures and isolated macromolecules. Immunolabelling can be applied to biological material before or after processing for resin embedding (pre-/post-embedding labelling; see *Protocol 2.10*) and can also be successfully combined with negative staining [10, 11] (see *Protocol 2.3*), vitrification and metal shadowing techniques. Colloidal gold is the most widely used electron-dense marker for conjugation with antibody (IgG/Fab/Fab′) or other ligand such as protein A/protein G, avidin/streptavidin (for biotinylated proteins) or lectin. Colloidal gold particles ranging from 1 to 20 nm diameter are available commercially, which also enable double labelling procedures to be performed on the same tissue, using antibody–gold probes of different size. The preservation of antigenicity is a major consideration when post-embedding is to be performed. An underlying difficulty of the post-embedding labelling procedures in the fact that the antigenic epitopes under investigation may not withstand the high concentrations of fixatives (e.g. 3% glutaraldehyde) normally employed for tissue stabilization prior to dehydration and embedding. Consequently 2 percent (*para*)formaldehyde–0.5% glutaraldehyde is often used for fixation, with retention of cellular protein antigenicity, but often with inferior structural preservation. Recently 1.4 nm gold cluster (Nanogold™) labels and 0.8 nm undecagold have become available commercially. These probes can be chemically linked to antibodies and streptavidin and are likely to make an increasing impact within the areas of high-resolution cellular and macromolecular labelling in the future.

Specialized techniques

It is beyond the scope of the present book to deal with the use of vacuum coating, ultramicrotomy, cryoultramicrotomy, high pressure freezing, freeze-substitution, freeze-fracture and plunge freezing. These skilled procedures are thoroughly documented elsewhere, but are usually best learnt directly from technicians/scientists within the EM laboratory. Very often such staff may provide a service role, available to all users of the laboratory, dependent upon local collaborative arrangements and funding.

Equipment and reagent hazards

The EM preparative equipment available in different laboratories will vary somewhat with respect to the larger items, such as vacuum coating units, glow-discharge equipment, ultramicrotomes, cryoultramicrotomes, rapid freezing apparatus and cryostorage systems. All these items need to be used carefully according to the manufacturer's instructions. The smaller cheaper items of equipment tend to be widely available in all laboratories, having been supplied through the international network of well-established suppliers of EM equipment and consumables. Some of the reagents used for electron microscopy are hazardous. In particular, osmic acid should be used with care and always within a fume extraction hood. Osmic acid solutions should be kept in a sealed desiccator at 4 °C. Glutaraldehyde is also dangerous and should be handled in a fume hood; stock 25% (v/v) glutaraldehyde should be stored at 4 °C. All contaminated organic solvent waste should be disposed of in bulk via environmentally acceptable procedures. Uranyl acetate has a low level of natural radioactivity. Accordingly, waste solution and contaminated filter paper and tissues should be disposed of using specified, approved routes. Other waste heavy metal staining salts should be disposed of in accordance with local regulations.

Preparation of carbon-formvar, continuous carbon and holey carbon support films

Reagents

Chloroform

Formvar

0.25% (w/v) solution of formvar in chloroform ①

Glycerol

Sodium dodecyl sulphate (SDS)

Equipment

Carbon rods or carbon fibre

Carbon rod sharpener

EM grids, 400 mesh

Filter paper (e.g. Whatman No. 1)

Floating-off dish (with stop-tap control on outflow)

Glass microscope slides (ethanol cleaned)

Mica

One-sided razor blade

Petri dishes

Vacuum coating apparatus (with carbon rod, carbon fibre or electron gun source)

Procedure ②

1. Immerse a clean dry microscope slide into the formvar solution. Allow slide to drain vertically onto a filter paper and then dry.

2. Score three edges of one side of the slide with a single-sided razor blade and float off the formvar film onto a water surface (i.e. in the 'floating-off' dish).

3. Place EM grids shiny side up on the floating formvar and remove the formvar + grids from the water surface with a piece of stainless steel gauze, a piece of perspex or a filter paper. ③

4. Allow to dry and then carbon-coat to an optimal thickness (e.g. 10–15 nm). ④ Carbon–formvar films are then immediately ready for use as negative staining supports, with or without glow discharge treatment.

5. For production of support films of carbon alone, dissolve the formvar and wash away by immersing individual grids vertically into chloroform.

Alternatively for carbon support films:

1. Carbon-coat pieces of freshly cleaved mica with a layer of carbon (see step 4, above). ●

2. Position EM grids, dull-side-up, on a piece of stainless steel gauze under water in the 'floating-off' dish. ⑤

3. Float off the carbon onto the water surface, as step 3, above, position over the grids and lower the water level to bring the carbon onto the grids. Remove carefully and allow to dry beneath an angle lamp before use. ⑥

Notes

This procedure will take approximately 1−2 h.

① Allow several hours or overnight for the formvar to completely dissolve in the chloroform before using. Store in a well-stoppered bottle and avoid contamination with airborne dust.

② This protocol presents a combination of possibilities for the production of formvar, formvar-carbon and carbon support films. It can also be easily modified for the production of 'holey' carbon support films, if a glycerol−SDS−water suspension (0.5% v/v, 0.1% w/v, 0.35% v/v) in 0.15% (w/v) formvar−chloroform is used, or if the drying chloroform-formvar is subjected to microdroplets of water by breathing onto the microscope slide or placing the slide on a cooled metal block (see refs 2, 3, 6).

③ This manoeuvre will require a little practice.

④ The thickness of the evaporated carbon can be estimated in a reasonable manner by simply placing a piece of white paper alongside the grids. Alternatively, a calibrated quartz crystal, carbon thickness monitor may be available.

⑤ Alternatively, a piece of filter paper can be placed on the stainless steel gauze, before carefully positioning the grids under water.

⑥ Support films of carbon alone will be found to be much more fragile than carbon−formvar films. The former are, however, more often used for negative staining, but the latter may be found to be desirable for immunonegative staining (using nickel or gold EM grids) where an increased number of incubations and washing stages are necessary. Extremely thin carbon films may be supported across a thicker holey carbon film, as can samples mixed with negative stain + trehalose (see ref. 6).

Pause point

● Carbon-coated mica can be stored under dust-free conditions until required.

The 'droplet' negative staining procedure (using continuous carbon, formvar–carbon and holey carbon support films)

Reagents

2% Ammonium molybdate in distilled water (to pH 7.0 with NaOH)

5% Ammonium molybdate + 0.1% (w/v) trehalose in distilled water (to pH 7.0 with NaOH) ①

Aqueous sample suspension (e.g. protein, virus, organelle, membrane, lipid), at ~0.1–0.5 mg/ml

0.1% (w/v) trehalose in distilled water

2% Uranyl acetate in distilled water

4% Uranyl acetate + 0.1% trehalose (w/v) in distilled water

Equipment

Carbon-, formvar-carbon- or holey carbon-coated EM grids ② (see *Protocol 2.1*)

Filter paper and filter paper wedges (e.g. Whatman No. 1)

Fine curved forceps with rubber closing ring (or use reverse action forceps)

Glow discharge apparatus

Grid storage boxes

Needle

Parafilm™, or equivalent

Petri dishes

Pipettes (20 and 10 μl) and tips

Procedure [1, 3, 6, 7, 12]

1. Cut off an appropriate length of parafilm™. With the paper backing still in place, attach the paraffin wax layer loosely to the work bench by running a blunt object (e.g. curved-ended scissors) in straight lines around the edges using a ruler, and then produce a number of straight lines across the parafilm, depending upon the number of samples. ③ Then remove the paper overlay.

2. Position three or four 20 μl droplets of water spaced along the parafilm lines, with a 20 μl droplet of negative stain solution at the back and a 10 or 20 μl droplet of sample suspension at the front.

3. Take individual (briefly glow-discharge treated ④) carbon support films carefully by the edge in fine forceps and touch to the sample droplet.

4. After a period of time for sample adsorption, ranging from 5 to 60 s (depending upon sample concentration), remove the fluid on the grid by touching to the edge of a filter paper wedge.

5. Wash the grid by touching successively to the water droplets, and remove fluid with the same filter paper wedge. ⑤ ⑥

6. Touch to the negative stain droplet and remove excess stain with the filter paper wedge. Allow the thin layer of adsorbed sample plus negative stain to dry at room temperature.

7. Position the grid, sometimes with the help of a needle point, onto a filter paper in a Petri dish. Record the sample and negative stain details alongside each grid.

8. Position all grids prepared during the specimen preparation session within a storage box, for transport to the EM room, with appropriate documentation of the date, sample and negative stain information.

Notes

This procedure takes approximately 5 min per specimen grid.

① Several other negative staining salts can be used, alone or in combination with 0.1% trehalose, e.g. uranyl formate, sodium phosphotungstate and silicotungstate, methylamine tungstate. Methylamine vanadate (Nanovan®) is particularly useful for low-density negative staining of nanogold-labelled samples.

② Carbon-coated EM grids are prepared by evaporating carbon *in vacuo* onto mica or onto grids positioned on a floating plastic film, such as formvar, see *Protocol 2.1*. Carbon alone is best, but some breakage of this support film may be encountered during the negative staining procedure.

③ In practice it is best to restrict the number of specimens prepared in any one session to ∼10.

④ Equipment for glow-discharge treatment of carbon support films is usually available in EM laboratories. Glow-discharge improves the hydrophilicity of the carbon and assists sample and stain spreading.

⑤ Depending upon the total salt and other solute concentration in the sample, fewer or a greater number of water washes may be required. Excessive washing may break the carbon film.

⑥ For negative staining on holey carbon films (see ref. 6), a higher sample concentration is desirable (i.e. 1 mg/ml). The washings are best done using 0.1% w/v) trehalose and negative stain should also contain 0.1% trehalose. The final removal of negative stain should be performed more rigorously, in order to produce a very thin film of sample + stain spanning the holes.

Immunonegative staining

Reagents

Antibodies

Antibody–colloidal gold conjugate

Protein A-/protein G–colloidal gold conjugate

Phosphate buffered saline (PBS) or Tris-buffered saline (TBS)

1% Bovine serum albumin (BSA) in PBS (blocker)

1% species ① or fetal calf serum in PBS (blocker)

2.5% v/v Glutaraldehyde in PBS (diluted from 25% glutaraldehyde stock)

Negative stain solutions (as for *Protocol 2.2*)

Equipment

Nickel or gold EM grids with formvar-carbon or carbon support films *plus* adsorbed biological material (e.g. cells, organelles, proteins, see *Protocol 2.2* or cleaved cell/organelles, see *Protocol 2.4*)

Fine curved forceps with rubber closing ring (or reverse action forceps)

Parafilm

Pipette and tips

Filter paper wedges

Petri dishes

Procedure [10, 11]

1. Wash sample grids with 20 μl droplets of PBS on a parafilm surface, to remove unbound material. ②

2. Incubate at RT on 20 μl droplet of 1% species or fetal calf serum in PBS for ~10–30 min. ③

3. Droplet-wash with 1% BSA in PBS.

4. Incubate by floating on a 20 μl droplet of appropriately diluted primary antibody (use 1% BSA in PBS) for 15–120 min. ④

5. Droplet wash with 1% BSA in PBS.

6. Incubate on 20 μl droplet of appropriately diluted secondary antibody–colloidal gold conjugate. ⑤

7. Droplet wash with 1% BSA in PBS.

8. Fix briefly in 2.5% glutaraldehyde in PBS.

9. Droplet wash with distilled water.

10. Droplet negative stain (as in *Protocol 2.2*). ⑥

Notes

This procedure takes approximately 2–3 h, depending upon incubation times.

① Non-immune serum from the same species as the antibody is a good 'blocker'.

② Carbon films alone will be found to be rather fragile and should be handled very gently if used; carbon-formvar films are more robust. Cellular material can also be incubated in antibody and antibody–gold conjugates whilst attached to the Alcian blue-coated mica, i.e. before carbon-coating, for external labelling (see *Protocol 2.4*)

③ This step is important to block non-specific interactions. During grid incubation, the droplets on parafilm should be covered with an inverted Petri dish to reduce evaporation.

④ If the primary antibody is already gold conjugated, go straight to step 7.

⑤ Protein A- or protein G–gold conjugate can also be used, or avidin/streptavidin–gold if the sample material, primary or bridging antibody is biotinylated.

⑥ It is essential to perform all possible control incubations in parallel, to rule out non-specific labelling.

The negative staining-carbon film technique: cell and organelle cleavage

Reagents

0.01% (w/v) Alcian blue in distilled water

Cell or organelle suspension

25% (v/v) Glutaraldehyde

20% (v/v) Glycerol in 0.155 M ammonium acetate

Negative stain solutions (see *Protocol 2.2*)

PBS and 0.1% (v/v) glutaraldehyde in PBS

Small pieces of freshly cleaved mica (e.g. 1 × 2 cm)

Equipment

EM grids (400 mesh)

Filter paper (e.g. Whatman No. 1)

Fine curved forceps with rubber closing ring (or use reverse-action forceps)

Parafilm™

Petri dishes

Pipettes and tips

Vacuum coating unit

Procedure [13]

1. Flood the clean surface of freshly cleaved mica with 0.01% Alcian blue.

2. After approx. 1 min wash thoroughly in distilled water and air-dry. ❶

3. Apply 20 µl cell suspension and spread over mica surface.

4. Leave horizontal for approx. 2–5 min.

5. Wash thoroughly with PBS to remove unbound cells. ①

6. Wash thoroughly with 20% glycerol-0.155 M ammonium acetate to remove non-volatile salts.

7. Drain off excess fluid onto a filter paper.

8. Dry *in vacuo* ② in coating unit and coat with a thin layer of carbon.

9. Remove mica from vacuum apparatus.

10. Float off the carbon layer + adsorbed membranes onto distilled water in a Petri dish. This physically cleaves the cells. ③

11. Transfer pieces of carbon onto individual bare EM grid held in fine forceps, from beneath the floating carbon.

12. Negatively stain the material attached to the *lower* surface of the carbon with a 20 µl droplet on parafilm™ (see *Protocol 2.2*).

13. Remove excess negative stain with a filter paper wedge and air-dry.

14. Repeat until all the floating carbon has been transferred to grids and negatively stained.

Notes

This procedure takes approximately 2 h.

① Attachment of cells to mica can be monitored by light microscopy. Brief fixation of the attached cells may be performed using 0.1% glutaraldehyde in PBS. Avoid excessive fixation, as this will interfere with the cell cleavage. External labelling can be applied to mica-bound cells prior to cleavage.

② Usually at least 1 h at 10^{-5} Torr will be required to remove all the glycerol.

③ This cell cleavage stage of the procedure can also be performed in 4 mm diam. micro-wells in Teflon™ block, using small pieces of mica (e.g. 3 mm squares with the corners cut off). Also, droplet immunolabelling can be performed at this stage, either directly or after brief glutaraldehyde fixation as in. ①

Pause point

❶ Alcian blue-treated mica can be stored in dust-free conditions until required.

Preparation of unstained and negatively stained vitrified specimens

Reagents

Aqueous suspension of biological material (~1.0 mg/ml)

Liquid nitrogen

Liquid ethane

16% (w/v) ammonium molybdate (to ~ pH 7.0 with NaOH)

Equipment

Holey carbon support films ①

Plunge-freezing apparatus (available commercially or workshop-made)

Fume extraction hood

Polystyrene container for liquid nitrogen

Small metal container for liquid ethane

Fine straight forceps, with rubber closing ring (or reverse action forceps)

Filter paper (e.g. Whatman No. 4)

Procedure ②

1. Adjust in advance the fall of the plunge freezing apparatus within a fume extraction hood so that a specimen grid held by fine straight forceps will enter the small metal container and not be damaged.

2. Fill the small container, surrounded by liquid nitrogen in the polystyrene container, with liquid ethane from a commercial cylinder of cryogen. Position vertically beneath the plunge-freezing mechanism.

3. Apply a ~4 μl droplet of sample (ideally in water or a low ionic strength buffer) to a holey carbon film held in straight forceps by the closing ring. For cryo-negative staining [8], the grid should be floated on a 100 μl droplet of 16% ammonium molybdate solution for ~60 s.

4. Position and clamp the forceps, with holey carbon support film, in the holder of the plunge freezer.

5. Blot away the excess fluid by touching a filter paper onto the sample droplet for 1 or 2 s. ③

6. Remove the filter paper and instantly release the plunge mechanism.

7. Release the forceps and transfer the grid from the liquid ethane to liquid nitrogen. ④

8. Store the thin vitrified specimen under liquid nitrogen in an appropriate container or transfer direct to a liquid nitrogen-cooled cryoholder. ⑤

Notes

These procedures take approximately 30 min.

① Holey carbon support films are readily prepared by established procedures available in most EM laboratories [3] (see *Protocol 2.1*); they are also available commercially. Quantofoil® micromachined holey carbon grids can also be used.

② These procedures, developed initially by Marc Adrian and his colleagues [7, 8] usually require a protein/lipid/ nucleic acid concentration somewhat in excess of 1 mg/ml. The study of specimens prepared by this technique requires a suitable cryoelectron micro- scope with cryoholder and cryotrans- fer system. Specimens have to be stud- ied in the electron microscope under strict low electron dose conditions. The necessary financial investment and time involved are considerable; supplementary skill in image process- ing is usually required.

③ A very thin layer of fluid is left across the holes of the holey carbon support film. Satisfactory blotting time can only be determined by trial and error.

④ As much liquid ethane as possible should be removed at this stage, since any remaining on the grid sur- face solidifies on entering the liquid nitrogen.

⑤ Further detail on cryotransfer and cryoelectron microscopy can best be learnt by visiting a laboratory rou- tinely performing this procedure. (See also ref [3])

PROTOCOL 2.6

Metal shadowing of biological specimens

Reagents

Ammonium acetate

Ammonium carbonate

Biological sample (protein/nucleic acid/ lipid/polysaccharide, ~0.05 mg/ml) ①

Glycerol (redistilled)

Uranyl acetate

Equipment ②

Carbon rod, carbon fibre or electron beam carbon source

Carbon rod sharpener

Curved forceps

Dialysis tubing

EM grids

Emery paper

Mica

Petri dishes

Pipettes and tips

Platinum/platinum-palladium/tungsten wire

Rotary specimen holder

Small nebulizer

Vacuum coating apparatus

Procedure

1. Prepare a pointed carbon rod with a molten droplet (~2 mm diameter) of platinum at the end. ③

2. Position the platinum rod against an angled carbon rod within the carbon source. Both carbon rods should be smoothed with emery paper for close electrical contact.

3. Prepare suitably spread sample material ① on a freshly cleaved clean mica surface, by spraying or evenly spreading with a pipette tip on its side.

4. Air-dry the sample. ④

5. Position the mica + sample within vacuum coating apparatus, at an angle of *c*. 45–60° to the horizontal, within a stationary or rotary holder.

6. Evacuate chamber to 10^{-5} Torr or better.

7. Preheat carbon rods gently to outgas with a shutter between the source and specimen. Remove shutter and rapidly evaporate the platinum carbon until a desired thickness has been achieved. ⑤

8. Remove mica and float the platinum–carbon replica onto water in a Petri dish. ⑥

9. Prepare specimens by bringing EM grids individually from beneath the floating replica. Wipe gently on a filter paper to remove any overhanging replica and air-dry.

Notes

This procedure will take between 1 and 20 h, depending upon the removal of any glycerol and volatile buffer from the sample *in vacuo*.

① The sample should be dialysed against bidistilled water or 0.155 M ammonium acetate/bicarbonate at the desired pH, to remove non-volatile salts. It may be necessary to fix some samples with glutaraldehyde prior to water dialysis to reduce particle flattening, particularly if air-drying is to be employed. Uranyl acetate stabilization may also be useful to prevent structural collapse.

② Many technical variants of the metal shadowing procedure are available to suit different biological samples. The protocol presented represents a simplified example that will produce reproducible results, but should be modified to suit the sample material and the vacuum coating equipment that is available.

③ The globule of platinum at the end of a carbon rod is prepared from a small ring of platinum wire, by gentle heating, in vacuo, until it melts.

④ If glycerol (e.g. 20–50%) and/or ammonium acetate/bicarbonate are present, drying should be performed *in vacuo* for an extended period of time (several hours or overnight).

⑤ A thickness monitor is not necessary at this stage, but may be useful if a further vertically evaporated layer of carbon is added to stabilize the replica.

⑥ Alternatively, the mica can be cut into small (~3 mm) squares and the replica floated off onto water in 4 mm diam. micro-wells in a Teflon™ block or in a small Petri dish, before transfer to EM grids. With care, on-grid immunolabelling can also be performed at this stage.

A routine schedule for tissue processing and resin embedding

Reagents

Acetone

0.075 M cacodylate buffer (pH 7.4)

Embedding resin

25% (v/v) Glutaraldehyde

3% (v/v) Glutaraldehyde in 0.075 M sodium cacodylate buffer (pH 7.4)

Graded methanol–distilled water solutions (e.g. 30–90%)

Methanol

1% (w/v) Osmium tetroxide (osmic acid) in 0.075 M cacodylate buffer (pH 7.4)

Propylene oxide (1,2-epoxy propane)

Uranyl acetate

Equipment

Forceps

Glass vials

Scalpel

Tissues

Procedure

1. Place 1–2 mm pieces of tissue in glass vials and fix in 10–20 ml glutaraldehyde in cacodylate buffer (pH 7.4), usually overnight at 4 °C, or a somewhat shorter time at room temperature.

2. Wash tissue pieces in 10 ml cacodylate buffer, with at least four changes and finally at 4 °C overnight. ❶

3. Post-fix tissue in 1% osmium tetroxide in cacodylate buffer at 4 °C for 1–2 h.

4. Wash tissue in 10–20 ml distilled water for 20 min and repeat twice. ❷

5. Dehydrate tissue using graded methanol solutions: 10 min in 30% methanol, ① then 10 min in each of 60, 70, 80 and 90% methanol solutions and finally 100% methanol.

6. Wash with propylene oxide, 2 × 10 min.

7. Infiltrate tissue pieces with resin at room temperature, for at least 30 min.

8. Repeat infiltration with fresh resin overnight, at room temperature.

9. Embed tissue pieces in fresh resin and polymerize at 60 °C for 24–48 h, depending upon the resin used. ②

10. Thin section and post-stain sections (see *Protocol 2.9*).

Notes

This procedure will take approximately 3 days.

① Block staining can be included at this stage using 2% uranyl acetate in 30% methanol, for 30 min, followed by washing for 10 min in 30% methanol before continuing the dehydration sequence.

② The resin used will depend upon the nature of the tissue and experiments. For routine embedding, the epoxy resins (Araldite™, Epon™, Spurr's™) are all suitable, and should be prepared according to the manufacturer's instructions. For post-embedding immunolabelling, with tissue prepared at a reduced level of fixation and dehydration to maintain tissue antigenicity, the glycol methacrylate 'London' resins (LR White™ and LR Gold™), Unicryl™ or cross-linked acrylate-methacrylate Lowicryl™ resins should be used (see *Protocol 2.9*).

Pause points

❶ Glutaraldehyde-fixed tissue can be stored for several days after washing.

❷ Tissue treated with osmium tetroxide can be stored in water for several days before dehydration.

Agarose encapsulation for cell and organelle suspensions

Reagents

Distilled water

3% (v/v) Glutaraldehyde in 0.075 M cacodylate buffer (pH 7.4)

High purity, low gelling temperature agarose (e.g. SeaPlaque®, Sigma, Oxoid)

Phosphate buffered saline (PBS)

Equipment

Microcentrifuge and tubes

Pipettes and plastic tips

Single-sided razor blade/scalpel

Procedure

1. Fix specimen material in suspension using 3% glutaraldehyde. ① ②

2. Prepare a 1.5–2% solution of agarose in distilled water, by heating until agarose has just melted. Allow to cool to ~40 °C, whilst remaining a solution.

3. Pellet the specimen material, by centrifugation if necessary, and remove most of the fixative solution from above the pellet.

4. With a plastic pipette add a small amount of agarose to the pellet, mix gently and centrifuge at 40 °C to form a loose pellet of the cells or organelles.

5. Allow the agarose to harden at room temperature or 4 °C. ❶

6. Remove the agarose gel from the tube and cut off the bottom region containing the pelleted sample into ~2 mm cubes using a razor or scalpel blade.

7. Process the cubes of agarose as for a tissue sample (see *Protocol 2.7*).

Notes

This procedure will take approximately 2 h.

① The agarose procedure can also be used for cell monolayers in culture flasks or Petri dishes, with recovery of the monolayer for flat-embedding.

② Unfixed cell samples can be processed, if the agarose is prepared in PBS or Medium 199. Glutaraldehyde fixation can then be applied to the cubes of agarose.

Pause point

❶ Agarose gel can be stored for several days as long as drying is avoided.

Routine staining of thin sections for electron microscopy

Reagents

Distilled water

Lead citrate–sodium citrate solution, according to Reynolds [14] ①

NaOH Pellets and 1 N solution

Uranyl acetate (saturated aqueous solution; filter or centrifuge before use)

Equipment

Beakers

Dental wax/Parafilm™

Filter funnel and filter paper

Fine forceps

Glass Petri dishes

Microcentrifuge and tubes

Pipette and tips

Plastic wash bottle containing distilled water

Procedure

1. To stain sections with uranyl acetate ②: put 20 µl droplets of uranyl acetate solution onto paraffin wax or Parafilm™ in a Petri dish.

2. Float the grids with section side down ③ on the stain droplets in a covered Petri dish. The usual staining time ranges from 30 min to 2 h at room temperature.

3. Wash the grids thoroughly individually with flowing distilled water from a pipette, while holding them in forceps, or rinse in three consecutive 100 ml beakers of distilled water.

4. Stain sections with lead citrate: place some NaOH pellets (to minimize atmospheric CO_2) in the Petri dish alongside 40 µl droplets of freshly filtered/centrifuged stain solution.

5. Float the grids, section side down on the stain droplets, cover firmly with Petri dish lid and stain for 5 min (Araldite™, Spurr™, Methacrylate™), 15–30 min for Epon™.

6. Wash grids thoroughly with distilled water, as step 3, or with 0.02 M NaOH ④ followed by distilled water.

Notes

This procedure will take approximately 1 h.

① Dissolve 1.33 g lead citrate in 30 ml distilled water and add 1.76 g sodium citrate. Mix the resulting suspension for 1 min and allow to stand for 30 min. Add 8.0 ml 1 N NaOH and make up to 50 ml with distilled water and mix. Store the solution in a well-sealed dark bottle at 4 °C and centrifuge/filter before use [14].

② Omit steps 1–3 if the specimen material has been block stained with uranyl acetate.

③ Sections may be mounted on formvar-coated EM grids, for greater stability.

④ Use NaOH washing if lead citrate staining produces precipitate on the sections.

Post-embedding indirect immunolabelling of thin sections

Reagents

1% (w/v) Bovine serum albumin (BSA) ① in PBS or TBS

5% (v/v) Fetal calf serum (blocker) in PBS or TBS

Phosphate buffered saline (PBS) or Tris buffered saline (TBS)

Primary IgG antibody (often monoclonal)

Protein A/protein G colloidal gold conjugate

Secondary antibody, antibody–ferritin or antibody–colloidal gold conjugate ②

Sodium *m*-periodate

Tween 20

Equipment

Parafilm™ or spotting plates

Petri dishes

Pipette and tips

Thin sectioned tissue/cells, usually in methacrylate resin on nickel or gold EM grids

Procedure[15]

1. Etch the plastic of the sections by floating the grids on droplets ③ of freshly prepared saturated sodium *m*-periodate to expose protein antigens; 5 min is usually sufficient for LR White and LR Gold.

2. Wash grids three to five times with distilled water or PBS.

3. Block non-specific interactions by incubating sections with 1% BSA or 5% fetal calf serum in PBS for 15 min.

4. Rinse five times with PBS.

5. Incubate sections for 1–2 h ③ at room temperature (RT) on 20 μl droplets of the primary antibody diluted with 1% BSA in PBS, on parafilm in a covered Petri dish.

6. Rinse five times with PBS.

7. Incubate grid for 30 min at RT with appropriately diluted secondary antibody colloidal gold conjugate and go to step 11 below, *or*

8. If an unlabelled secondary antibody is being used as a bridging antibody, incubate grids on the secondary antibody diluted with 1% BSA in PBS for 30 min at room temperature. ④

9. Rinse five times with PBS.

10. Incubate grids for 30 min at RT on droplets colloidal gold conjugated with protein A, protein G or immunoglobulin, or if the secondary antibody is biotinylated, streptavidin-gold. ⑤

11. Rinse again five times with PBS and then a further five times with deionized distilled water.

12. Study grids briefly in EM to assess the level of labelling. ⑥

13. Post-stain the sections briefly with uranyl acetate and lead citrate, if additional tissue density is required, before detailed EM study.

Notes

① Tween 20 can also be added to the PBS, to 0.1% (v/v).
This procedure takes several hours, depending upon incubation times.

② Colloidal gold conjugates are usually purchased commercially. See also ref. 10.

③ 20 μl droplets of reagent and PBS on parafilm will be found to be satisfactory.

④ Incubation times can be varied to suit antibody avidity, in order to obtain adequate labelling.

⑤ Biotin–streptavidin binding affinity is strong and can lead to excellent labelling.

⑥ All necessary controls, i.e. without primary/secondary antibodies, should be incorporated alongside grids passed through the complete protocol.

PROTOCOL 2.11

Imaging the nuclear matrix and cytoskeleton by embedment-free electron microscopy

Jeffrey A. Nickerson and **Jean Underwood**

Introduction

Conventional embedded section electron microscopy is well suited for the visualization in cells and tissues of membrane-bounded organelles, whose sectioning generates characteristic membrane profiles. Among other uses, embedded sections also allow simple selective staining for cell components using electron dense elements, such as EDTA-regressive or Terbium staining for RNA [16, 17] and allow the imaging of chromatin packaging densities, for example, distinguishing euchromatin and heterochromatin. Embedded sections, however, are not an ideal technique for viewing filamentous networks such as the nuclear matrix or cytoskeleton. After sectioning and staining, filaments are usually seen in cross-section. Planes of section where a filament is near the surface of the section and parallel to that surface show the filament as a filament but they are rare. It can take imagination to understand the three-dimensional complexity of filamentous cell structures given only embedded section images. It has been suggested that this feature of conventional electron microscopy has led casual observers to undervalue the importance of these structures in cell organization [18].

Embedment-free electron microscopy permits the visualization of structural networks in all their three-dimensional complexity. In whole mount electron microscopy, cells are grown on grids and visualized without any sectioning [19, 20]. In resinless section electron microscopy, cells are embedded and sectioned but the embedding material is removed before visualization [21]. Neither technique requires staining with electron dense compounds; proteins are sufficiently electron dense to form images of high contrast to vacuum. Conventional sections require staining because the electron density of cell molecules and the embedding material is similar. Embedment-free-techniques can, however, be combined with immunogold antibody staining to localize specific proteins within structures [22].

Reagents

Auroprobe EM grade gold-conjugated second antibodies, obtained from Amersham Biosciences Corp. (Piscataway, NJ)

Delafield's Hematoxylin, available as a solution from Exaxol Chemical Corp. (Clearwater, FL)

Diethylene Glycol Distearate (DGD), available from Polysciences (Warrington, PA) and Electron Microscopy Sciences (Fort Washington, PA)

All other electron microscopy chemicals and supplies can be purchased from either Ted Pella (Redding, CA) or from Electron Microscopy Sciences

Solutions

1. *TBS-1: (10 mM Tris HCl, pH 7.7, 150 mM NaCl, 3 mM KCl, 1.5 mM MgCl$_2$, 0.05% (v/v) Tween 20, 0.1% (w/v) bovine serum albumin, 0.2% (w/v) glycine)*

   To make 1 l of buffer, use 10 ml Tris-HCl (from a 1 M Tris HCl stock solution); 8.766 g NaCl; 0.224 g KCl; 0.352 g MgCl$_2$·8H$_2$O; 1 g bovine serum albumin; 2 g glycine; and 500 µl of Tween 20. Filter with a 0.22µ bottle top filter in a sterile hood and freeze in aliquots at −20°C.

2. *TBS-2: (20 mM Tris HCl, pH 8.2, 140 mM NaCl, 0.1% (w/v) bovine serum albumin)*

   To make 1 l of buffer, use 20 ml Tris HCl (from a 1 M Tris HCl stock solution); 8.176 g NaCl; and 1 g bovine serum albumin. Filter with a 0.22µ bottle top filter in a sterile hood and freeze in aliquots at −20°C.

3. *Normal goat serum: (10% and other dilutions)*

   Make 10% normal goat serum with (v/v) heat inactivated normal goat serum in TBS-1 to the desired volume. Filter just before use with a 0.22µ filter; then use this filtered 10% preparation to make all further dilutions needed.

4. *Cacodylate buffer: (The stock buffer is 0.2 M sodium cacodylate, pH 7.2–7.4)*

   The following method makes 1 l of 0.2 M cacodylate buffer.

   Solution A:

sodium cacodylate	42.8 g
(Na(CH$_3$)$_2$AsO$_2$.3H$_2$O)	
distilled water	100.0 ml

   Solution B (0.2 M HCl):

conc. HCl (36–38%)	10 ml
distilled water	603 ml

   The stock solution of the desired pH can be obtained by adding Solution B as shown below to 20 ml of Solution A and diluting to a total volume of 200 ml.

Solution B (ml)	pH of buffer
23.2	7.0
17.2	7.2
11.2	7.4

   The stock 0.2 M sodium cacodylate buffer is stable for a few months and should be kept at 4°C. The washing buffer is 0.1 M sodium cacodylate and is prepared by mixing together 1 : 1 v/v 0.2 M sodium cacodylate and distilled water as needed.

5. *Glutaraldehyde fixative: (2.5% solution)*

   The 2.5% glutaraldehyde fixative is freshly prepared prior to use in 0.1 M cacodylate buffer and can be stored for only several hours at 4°C. Only EM grade glutaraldehyde should be used. Glutaraldehyde is packaged in 1 ml ampules of 8 or 25% aqueous solutions.

6. *Osmium fixative: (1–2% solution; optional fixative)*

   A solution of 1–2% osmium tetroxide in 0.1 M cacodylate buffer, pH 7.2–7.4, is the optional second fixative. Osmium tetroxide can be purchased from most electron microscopy suppliers as a stock solution or as a crystal in sealed ampules from which a stock solution is made with distilled water. The osmium stock solution is stable for 1–2 months at 4°C. The osmium fixative is freshly prepared by mixing the osmium stock solution with 0.2 M cacodylate buffer and distilled water to the desired concentration before use.

Equipment and supplies

Carbon evaporator

Colloidal gold-conjugated second antibodies, 5 and 10 nm

Critical point dryer

Diamond knife or glass knives

Gold and copper EM grids, 200 or 300 mesh

Oven (50–60 °C)

Thermanox coverslips

Transmission electron microscope

Ultramicrotome

Procedure

Method 1 Whole mounts of cells for electron microscopy

This technique allows the imaging of the three-dimensional structure of the nuclear matrix and cytoskeletal without sectioning and is most appropriate for very thin cells. For nuclear matrix extraction procedures, see *Protocol 6.5, Uncovering the Nuclear Matrix in Cultured Cells*. For visualizing the internal nuclear matrix by this technique, cells should be both thin and have a nuclear lamina of relatively low density, so that internal nuclear components are not obscured. SAOS-2 cells HeLa cells and fibroblasts have yielded good results.

Formvar (0.25% w/v in ethylene dichloride) support films are applied on gold 200 or 300 mesh grids, lightly carbon coated, and then sterilized in culture dishes under ultraviolet light in a tissue culture hood for a minimum of 2 h. Cells are cultured in subconfluent monolayers on the films on these grids, carbon-coated side up. Cells are extracted, fixed and processed through all steps *in situ* while still attached to the formvar film. Grids are either moved between the different solutions, or solutions are gently removed and exchanged by pipette, without ever allowing the surface of the grids to dry. This is particularly important following nuclear matrix or cytoskeletal extractions because the resulting structures are constructed of fragile filaments.

1. This method can be adapted to examine cell structures extracted in many ways. For this protocol, however, we will discuss cells that have been extracted to reveal the nuclear matrix by *either one* of the following methods:

    (a) *Cross-link stabilized nuclear matrix method.* ① This preparation has already been cross-linked with formaldehyde.

    (b) *Classical nuclear matrix method.* ① If this preparation will be used for immunolocalization of individual proteins, then it will have been fixed in a way affording preservation of both ultrastructure and antigenicity.

2a. Cells can be examined for morphology by directly proceeding on to step 11, 'Post-fixation'. ①

    or

2b. Selected proteins in these preparations can be localized by electron microscopy using specific primary antibodies and gold-conjugated second antibodies according to the following method:

Immunogold localization of proteins

This is a sandwich procedure using a primary antibody and a colloidal gold conjugated second antibody. Colloidal gold conjugates are available in a variety of sizes. In general, smaller beads yield a higher density of labelling. We have achieved good results with 5 and 10 nm beads. Good preservation of ultrastructure requires fixation before primary antibody staining. The cross-link stabilized nuclear matrix ① has already been extensively cross-linked, and the classical nuclear matrix ① has already been fixed in 4% formaldehyde in cytoskeletal buffer for 50 min at 4°C. The formaldehyde fixative must be freshly prepared, from a stock solution of 16% formaldehyde (EM-grade). Appropriate controls include samples with no primary antibody in the first incubation. For double label experiments using gold beads of two different diameters, useful controls for cross-reactivity of the second antibodies include two samples, each with only one primary antibody, and then incubated with both conjugated second antibodies.

3. Wash in TBS-1 at room temperature, twice in 5 min. Glycine in the TBS-1 will quench free aldehyde groups.

4. *Blocking* In order to block non-specific staining, the samples are incubated at room temperature in 10% normal goat serum in TBS-1 (freshly filtered with a 0.22µ filter) for 1 h.

5. *First antibody* Incubations with the first antibody are done in a moist chamber for 1–3 h at room temperature, or overnight at 4°C. The first antibody is diluted to the desired concentration in TBS-1 containing 1% normal goat serum.

6. The samples are rinsed in TBS-1, three times, 5–10 min each time.

7. Block with 5% normal goat serum in TBS-1, at room temperature for 30 min.

8. *Second antibody* Without rinsing, the samples are incubated in the appropriate gold-conjugated second antibody diluted 1:3 to 1:10 in TBS-2. This incubation is usually performed in a moist chamber for 1–3 h at room

temperature or at 37°C. This gold concentration works well for Auroprobe conjugated second antibodies from Amersham. Other gold-bead reagents need to be titrated to determine ideal concentrations.

9. Rinse the samples with TBS-1, four times, 10 min each time.

10. Rinse in 0.1 M cacodylate buffer twice for 3 min.

Post-fixation

11. Incubate in 2.5% glutaraldehyde fixative at 4°C for 1 h. If antibody staining is not performed and the preparation is to be used only for visualization of ultrastructure, then the formaldehyde fixation of the classical nuclear matrix required for antibody staining should be omitted and the samples immediately fixed in glutaraldehyde.

12. Wash the fixed nuclei in 0.1 M cacodylate buffer at 4°C twice in 5 min. *Steps 13 and 14 are optional.*

13. Incubate in 1–2% osmium fixative at 4°C for 30 min.

14. Wash out the osmium with 0.1 M cacodylate buffer at room temperature twice in 5 min.
    The fixed nuclear matrices can be stored overnight, at 4°C in 0.1 M cacodylate buffer.

Dehydration

This is performed at room temperature.

15. Briefly transfer the samples from 0.1 M cacodylate buffer through increasing ethanol concentrations (30–95%).

16. Then dehydrate the samples ending with three changes of 100% ethanol for 10 min each.

The dehydrated matrices can be stored overnight, at 4 °C. They can then be processed after bringing them to room temperature.

Final sample preparation

Samples are dehydrated by critical point drying from 100% ethanol to air. This must be done to prevent surface tension generated by air-drying from damaging the delicate architecture of filament networks.

17. *Critical point drying* Transfer the grids to 100% ethanol in the critical point dryer and process according to the apparatus instructions. Store the grids desiccated until they are (optionally) carbon coated.

18. *Carbon coating* (optional but preferable) When dry, the grids can be viewed in the electron microscope, but the samples are quite delicate and can be vulnerable to beam damage. The specimens can be stabilized by a light coating with carbon in a standard carbon coating apparatus.

Method 2 Resinless sections

This method was designed to observe the structure of the nuclear matrix in cells which have been extracted by the cross-link stabilized nuclear matrix method ① or the *classical nuclear matrix* method ① to reveal the underlying network of branched filaments. It is easily adapted for other, for example cytoskeletal, preparations. Individual proteins can be localized in the nuclear matrix or cytoskeleton using specific antibodies and gold-conjugated second antibodies.

Resinless sections can best be described as a method for infiltrating the delicate structural network of the extracted nucleus with a wax (DGD) to provide support for the fragile structures while they are being sectioned. Thin sections are cut and the

wax is removed, leaving a well-preserved structure without any remaining embedding material.

This method combines the advantages of the whole mount for imaging three-dimensional networks with the ability to section. Sectioning is important to image internal nuclear structures that in a whole mount may be obscured by the nuclear lamina. As with whole mount preparations, resinless sections are particularly suitable for the collection of stereo paired images.

Cells are grown in culture on Thermanox coverslips in subconfluent monolayers. They are extracted to reveal the nuclear matrix structure, fixed and processed through all steps *in situ* while still attached to the coverslip (see *Protocol 6.5, Uncovering the Nuclear Matrix in Cultured Cells*). Coverslips are not separated from the cells until after they are embedded with the wax (DGD).

Cultured cells grown in suspension are harvested, extracted and fixed as a small pellet. After the fixation the pellet is placed into a BEEM capsule for further processing. The height of the pellet in the BEEM capsule should be less than 2 mm; if larger, then divide the sample or allow more time for individual steps. ①

1. Coverslips or cell pellets can be processed through all of the same steps (with or without the antibody staining) which have been described above in Method 1. 'Whole mounts of cells for electron microscopy', step 1 through post-fixation, step 14.

2. *Block staining* (optional but preferred) The following staining procedure is not necessary for the ultrastructural visualization of the sample but is very helpful for locating cells in the DGD block prior to sectioning. One of two methods can be selected to stain the cells; both are done at room temperature.

2a. Staining with Delafield's Hematoxylin. Before beginning cell dehydration, Method 1, step 15, incubate in hematoxylin for 20 min. Continue the dehydration as in Method 1, steps 15 and 16.

or

2b. Staining with Eosin Y. Begin cell dehydration through graded ethanol concentrations, Method 1, step 15, and at the 70% ethanol stage, transfer the samples to a freshly prepared, saturated solution of Eosin Y in 70% ethanol for 20 min. Briefly wash the samples in 70% ethanol and then continue the dehydration as in Method 1, steps 15 and 16.

Transitional solvent

Ethanol and DGD are not miscible so a transition fluid, n-butanol, is used.

3. Transfer the samples to a 1 : 1 v/v mixture of 100% ethanol : n-butanol for 5 min.

4. Transfer cells to 100% n-butanol, two changes, 5 min each. Put the samples in a 56–60 °C oven.

DGD infiltration

The infiltration with DGD is performed at 56–60 °C. DGD is melted to this temperature in an oven and samples can be handled in the oven or on the lab bench under an infrared lamp.

5. Prepare a mixture of 1 : 1 v/v DGD : n-butanol and pour it over the samples (use pre-warmed transitional fluid and melted DGD). Leave the samples in the oven, without a cover, for 30 min to allow the transitional solvent to evaporate.

6. Replace the mixture with two changes of pure melted DGD for 30–60 min each change to ensure proper infiltration.

7a. For cells on Thermanox coverslips, remove the coverslips from the oven and add a large drop of DGD over the top of the cells. At room temperature, let them cool briefly to solidify. Then, immediately peel the Thermanox coverslips from the solid DGD. Cell structures should come off the coverslip, adhering to the surface of the DGD. Cut squares (about 16 mm^2) from the area that contains cells and mount them on DGD stubs with a drop of melted DGD. Trim the blocks for *en face* EM sectioning.

or

7b. For samples in BEEM capsules (pelleted cells from suspensions), cut away the capsule plastic and the trim the remaining block for EM sectioning.

8. The sectioning is done with glass or diamond knives with troughs filled with distilled water on an ultramicrotome. The sections are collected on carbon coated formvar films supported on copper 200–300 mesh grids. The section thickness, approximately 150 nm, is estimated by the continuous interference colour which is essentially the same as in epon sections.

DGD removal

9. Immerse the grids in n-butanol and incubate them at room temperature. The n-butanol effects a slow extraction and samples are best left for a few hours or overnight.

10. Transfer the grids with sections to a 1 : 1 v/v mixture of 100% ethanol with the dewaxing solvent n-butanol for 5 minutes

11. Transfer grids to three changes of 100% ethanol, 10 min each.

Final sample preparation

This is the same as in Method 1. Follow the steps for critical point

Drying, Method 1, step 17, and for Carbon Coating, Method 1, step 18.

Note

① Follow the procedures for the cross-link stabilized nuclear matrix or the classical nuclear matrix methods described in *Protocol 6.5*, 'Uncovering the nuclear matrix in cultured cells'.

Acknowledgement

We thank Sheldon Penman, Gabriela Krockmalnic and David Capco, who developed these techniques and passed them along to us.

References

1. Harris, J. R. (ed.) (1991) *Electron Microscopy in Biology: A Practical Approach*. IRL Press at Oxford University Press, Oxford.
2. Sommerville, J. and U. Scheer (eds) (1987) *Electron Microscopy in Molecular Biology: A Practical Approach*, IRL Press, Oxford.
3. Harris, J. R. (1997) *Negative Staining and Cryoelectron Microscopy: The Thin Film Techniques*. RMS Handbook No. 35, BIOS Scientific Publishers, Oxford.
4. Hajibagheri, M. A. N. (ed.) (1999) *Electron Microscopy Methods and Protocols*. Humana Press, Totowa, New Jersey, USA.
5. Harris, J. R. and Rickwood, D. (eds) (1999) *Multimedia Methods in Cell Biology, Version 1:0*. Chapman & Hall/CRCnetBASE.
6. Harris, J. R. and Scheffler, D. (2001) Routine preparation of air-dried negatively stained and unstained specimens on holey carbon support films: a review of applications. *Micron*, **33**, 461–480.
7. Adrian, M., Dubochet, J., Lepault, J. and McDowall, A. W. (1984) Cryo-electron microscopy of viruses. *Nature*, **308**, 32–36.
8. Adrian, M., Dubochet, J., Fuller, S. D. and Harris, J. R. (1998). Cryo-negative staining. *Micron*, **29**, 145–160.
9. Heuser, J. E. (1983) Procedure for freeze-drying molecules adsorbed to mica flakes. *J. Mol. Biol.*, **169**, 155–195.
10. Hyatt, A. D. (1991) Immunogold labelling techniques. In: *Electron microscopy in Biology: A Practical Approach* (J. R. Harris, ed.), pp. 59–81. IRL Press at Oxford University Press, Oxford.
11. Petry, F. and Harris, J. R. (1999) Ultrastructure, fractionation and biochemical analysis of *Cryptosporidium parvum* sporozoites. *Int. J. Parasitol.*, **29**, 1249–1260.
12. Harris, J. R. and Horne, R. W. (1994) Negative staining: a brief assessment of current technical benefits, limitations and future possibilities. *Micron*, **25**, 5–13.
13. Harris, J. R. (1992) Negative staining-carbon film technique: new cellular and molecular applications. *J. Elect. Microsc. Tech.*, **18**, 269–276.
14. Reynolds, E. S. (1963) The use of lead citrate at high pH as an electron opaque stain in electron microscopy. *J. Cell Biol.*, **67**, 208–210.
15. Oliver, C. (1994) In: *Methods in Molecular Biology* Vol. 34: *Immunological Methods and Protocols* (L. C. Javois, ed.), pp. 291–328. Humana Press, Inc.
16. Bernhard, W. (1969) A new staining procedure for electron microscopical cytology. *J. Ultrastruct. Res.*, **27**, 250–265.
17. Biggiogera, M. and Fakan, S. (1998) Fine structural specific visualization of RNA on ultrathin sections. *J. Histochem. Cytochem.*, **46**, 389–395.
18. Penman, S. (1995) Rethinking cell structure. *Proc. Natl Acad. Sci. USA*, **92**, 5251–5257.
19. Capco, D. G. and Penman, S. (1983) Mitotic architecture of the cell: the filament networks of the nucleus and cytoplasm. *J. Cell Biol.*, **96**, 896–906.
20. Capco, D. G., Wan, K. M. and Penman, S. (1982) The nuclear matrix: three-dimensional architecture and protein composition. *Cell*, **29**, 847–858.
21. Capco, D. G., Krockmalnic, G. and Penman, S. (1984) A new method of preparing embedment-free sections for transmission electron microscopy: applications to the cytoskeletal framework and other three-dimensional networks. *J. Cell Biol.*, **98**, 1878–1885.
22. Nickerson, J. A., Krockmalnic, G., He, D. C. and Penman, S. (1990) Immunolocalization in three dimensions: immunogold staining of cytoskeletal and nuclear matrix proteins in resinless electron microscopy sections. *Proc. Nat. Acad. Sci. USA*, **87**, 2259–2263.

3

Cell Culture

Anne Wilson and John Graham

Protocol 3.1	Tissue disaggregation by mechanical mincing or chopping	54
Protocol 3.2	Tissue disaggregation by warm trypsinization	56
Protocol 3.3	Cold trypsinization	58
Protocol 3.4	Disaggregation using collagenase or dispase	60
Protocol 3.5	Recovery of cells from effusions	63
Protocol 3.6	Removal of red blood cells by snap lysis	64
Protocol 3.7	Removal of red blood cells (rbc) and dead cells using isopycnic centrifugation	65
Protocol 3.8	Quantitation of cell counts and viability	67
Protocol 3.9	Recovery of cells from monolayer cultures	71
Protocol 3.10	Freezing cells	74
Protocol 3.11	Thawing cells	76
Protocol 3.12	Purification of human PBMCs on a density barrier	80
Protocol 3.13	Purification of human PBMCs using a mixer technique	82
Protocol 3.14	Purification of human PBMCs using a barrier flotation technique	83

Cells: isolation and analysis – Anne Wilson

Introduction

The main subcellular components separated from isolated cells include: (i) compact organelles such as mitochondria, nuclei and lysosomes, (ii) membrane systems including the cell membrane, Golgi complex and endoplasmic reticulum, (iii) cytoskeletal elements and (iv) proteins and small cytosolic particles. The four main types of tissue comprise: epithelial, connective, muscle and nervous. All tissues contain specialized cell types which are organized into functional units. The integrity of the tissue is maintained by

Cell Biology Protocols. Edited by J. Robin Harris, John Graham, David Rickwood
© 2006 John Wiley & Sons, Ltd

structural and chemical adhesion mechanisms, both between adjacent cells and to the extracellular matrix or basal lamina. Cells may be joined by a variety of specialized junctional complexes involving different cytoskeletal filaments and different families of adhesion molecules, including cadherins and adherins which are often dependent upon divalent cations (Mg^{++} and Ca^{++}). Dissociation of tissue requires disruption of adhesion with minimal damage to the cells. Since the process of separation may itself compromise cellular integrity and function, awareness of the effect of cell isolation on the viability and function of the intact cells is necessary for optimizing separation conditions for specific cell components.

In turn, the validity of the results obtained from experimental studies on subcellular components is dependent upon the quality of the isolated cells from which those components were derived. Methods used for isolating single cells should maintain viability and metabolic activity of the population; a quantitative baseline to which results can be related and inter-experimental results compared is essential. Determination of cell number, cell viability and protein concentration provides this baseline.

Solid tissues, biological fluids including blood, pleural and abdominal effusions and continuous cell lines are all potential sources for isolated cells. Cells recovered from fresh tissues are likely to be heterogeneous and methods for identifying the isolated cell types are required. Confirmation of cell type derived from cell lines is also advisable.

Heterogeneity of structure between different tissues precludes the use of generic protocols for tissue disaggregation, though features common to tissue-specific protocols include a mechanical means of disaggregating the tissue into small fragments and enzymatic digestion of the supporting matrix for release of individual cells. Variations in enzyme type, enzyme concentrations, exposure time and temperature allow optimization of conditions for different tissues.

A variety of methods described elsewhere [1], for different normal animal tissues are summarized in Table 3.1. Although primarily for setting up cell cultures, they are still relevant in that cell culture requires viable cells with preserved metabolic and reproductive function. Explant cultures are used when the required cell population is represented by only a small number of cells and expansion of numbers is needed.

Rather than offering different protocols for specific tissues, protocols are included that have been used for the disaggregation of human ovarian malignancies. These tumours are sufficiently variable in their characteristics as to require a range of techniques and they provide a basic starting point for mechanical disaggregation and enzyme digestion. The protocols can be followed for different tissue types and used as a baseline for establishing optimal conditions for disaggregation.

Mechanical disaggregation of tissue

Introduction

Mincing or chopping tissue in liquid medium releases cells from the supporting matrix; the isolated cell population often includes large numbers of red blood cells and the viability of nucleated cells may be low as a result of mechanical trauma. However, most disaggregation procedures begin with a mechanical step to reduce the tissue to small fragments, providing a larger surface area for exposure to the digesting enzymes and facilitating penetration of enzymes into the tissue. Soft tissues can be minced finely

Table 3.1 A summary of methods used for disaggregating different tissues [1]

Tissue	Description of disaggregation procedure
Epidermis	Cold trypsinization to separate dermis and epidermis followed by pipetting and scraping of separate layers to release keratinocytes
Breast	Recovery of cells from early lactation and post-weaning milk
Cervix	As epidermis *or* initiation of explant cultures from chopped punch biopsies and harvesting of epithelial cells migrating from explants
Liver	Perfusion of blood vessels with a low concentration of collagenase
Pancreas	Digestion with a mix of trypsin and collagenase and filtration to separate acini from single cells
Kidney	Digestion of finely chopped kidney with collagenase. Differential sedimentation to separate heavier tubules and glomeruli from stromal cells
Bronchus	Growth of epithelial monolayers from bronchial explants using specialized media and modified growth surface
Prostate	Digestion of minced tissue with collagenase
Muscle	Digestion of minced tissue with collagenase
Cartilage	Consecutive digestion of chopped fragments with hyaluronidase, then trypsin, then collagenase
Bone	Digestion of chopped trabecular bone with collagenase and trypsin
Endothelium	Addition of collagenase to lumen of isolated, intact blood vessel with its free ends ligated to retain solution whilst digestion takes place
Brain	Trypsinization of minced tissue

with scissors and harder tissues chopped with opposing scalpel blades. Further release of cells from minced tissue is achieved by vigorous pipetting of the smaller fragments in the suspending fluid, creating shear forces strong enough to separate cells from the fragments.

Tissue disaggregation by mechanical mincing or chopping

Introduction

The biohazards associated with the tissue must be ascertained and appropriate equipment used for handling it [2–5]. In the following protocols, a Class II cabinet is used, protecting both user and tissue from contamination. The practice of aseptic technique is recommended because it ensures that stored media retain sterility and the risk of spurious results from the presence of micro-organisms is minimized. If cells are to be cultured, it is a necessity.

Equipment

All equipment coming into direct contact with tissue must be sterile.

Dissecting scissors *(Note: Curved scissors are easier to use in a Petri dish than straight scissors because they can be laid flat along the surface of the dish and thus provide a more extensive cutting surface.)*

Forceps

Disposable scalpels

Glass Petri dishes *(Note: Dissecting instruments damage plastic dishes and debris may traumatize the cells.)*

Pastettes (short-form and long-form)

Universal containers (conical-bottomed) with labels showing volume gradations

Rack for universal containers

Bench centrifuge

Refrigerator

Class II safety cabinet

Reagents

Media coming into direct contact with cells must be sterile.

Hanks balanced salt solution without calcium and magnesium (HBSS) *(Note: HBSS is purchased either as a single-strength solution or as a 10× strength solution. It is more economical to make up single-strength solution from 10× strength using tissue culture grade sterile water and adding 5 ml of sterile 7.5% sodium bicarbonate per litre [final concentration of 0.37 g of sodium bicarbonate per litre].)*

Storage medium *(Note: HBSS has a low nutrient content and, if cells are to be stored for several hours, their viability may be impaired. A variety of culture media are available which can be supplemented with serum to give a semi-defined formulation. More frequently used media include MEM, TC199 and RPMI 1640 [6]. If a fully defined formulation, is required, serum-free media can be used [7].)*

Procedure

- Place the tissue into the glass Petri dish and moisten its surface with a small volume of HBSS.

- Dissect away fat, necrotic tissue and connective tissue using a scalpel or scissors and forceps and discard it. (Necrotic tissue is creamy-yellow with a soft, friable texture; it may contain foci of pus.)

- Add a further small volume of HBSS to the tissue and cut it into small fragments. *(Note 1: It is easier to cut or mince the tissue if the volume of liquid is kept small because the fragments then stay on the surface of the dish and cannot float away from the dissecting instruments. Note 2: Mince soft tissue with scissors and chop harder tissue with two opposed scalpel blades.)*

- Tilt the dish to an angle of $\sim 30°$ from the horizontal and allow the medium and released cells to wash down into the angle of the dish. Use a sterile pastette to aspirate the cell suspension and transfer it to a sterile universal container.

- Add a further small volume of HBSS to the dish with a pastette and wash it over the fragments and surface to release and recover more cells.

- Repeat the last three steps two more times, aiming to reduce the tissue to fragments ~ 2 mm across. Pool the released cells from the three mincing/chopping steps.

- Centrifuge the cell suspension at 1000 rpm for 5 min. *(Note 1: $g = 1.12R$ $(rpm/1000)^2$, where $R = radius$ of rotor arm from centre of rotor to centre of centrifuge bucket. Note 2: If the final cell suspension is of low viability it may be enhanced by keeping the recovered cells at $4°C$ and also centrifuging at this temperature. Note 3: Released cells from a mechanically produced suspension may be of low viability because damaged cells separate more easily from the surrounding matrix and the physical forces involved damage cells. However, some tissues are disaggregated effectively with mechanical methods.)*

- Decant the supernatant from the cell pellet and resuspend the cells in 10 ml of storage medium. Store in the refrigerator at $4°C$. *(Note: The volume of medium added should be increased to 20 ml if the cell pellet is ≥ 5 ml.)*

- Go to either *Protocol 3.8* for cell count and viability determination or to *Protocols 3.2–3.4* for enzyme disaggregation.

Tissue disaggregation by warm trypsinization

Introduction

Trypsin has been used for disaggregating tissue for many years [8]. The enzyme damages cells if exposure is prolonged and care must be taken to minimize this when devising tissue-specific protocols. Trypsinization can be carried out at 37 °C (warm trypsinization) or at 4 °C (cold trypsinization) [9]. DNA released from damaged cells becomes viscous and its presence impairs the recovery of free cells. DNAse is therefore included at low concentrations to digest DNA and reduce viscosity. Serum contains trypsin inhibitors and these must be washed from the tissue before digestion is started. Conversely, the action of trypsin is neutralized by the addition of serum. Under serum-free conditions, soybean trypsin inhibitor can be used instead.

Equipment

All equipment which comes into contact with cells must be sterile.

Wide-necked glass conical flask with a cap made of a double layer of foil extending ~4 cm down the neck of the flask

Spatula or teaspoon (Note: A teaspoon is convenient for handling larger volumes of tissue.)

Magnetic stirrer bar (Note: Soak in 70% ethanol/30% water and rinse in sterile HBSS to sterilize before use.)

Hotplate/magnetic stirrer (with the heat setting adjusted to maintain a temperature of ~37 °C or a shaking water bath set at ~37 °C. (Note 1: Care needs to be taken with a hot plate because of the risk of overheating. To avoid using the heater element the hotplate can be put into an incubator set at 37 °C and disaggregation carried out using only the magnetic stirrer. In the absence of a mechanical means of agitating the enzyme/tissue mixture the flask can be incubated at ~37 °C and shaken gently by hand at 5 min intervals. Note 2: A water shaker bath is the most efficient method for maintaining temperature and minimizing mechanical damage during disaggregation.)

Pastettes (long-form and short-form)

Universal containers (conical-bottomed) with a label showing volume gradations

Rack for universal containers

Bijoux

Bench centrifuge

Refrigerator

Freezer (−20 °C) for reagent storage

Reagents

All reagents must be sterile.

Trypsin – 2.5% (Note: Thaw out 100 ml of stock solution (2.5%) and store as 10 ml aliquots in sterile universal containers at −20 °C.)

DNAse – 0.4% (Note: Make up stock solution of DNAse Type I in HBSS at

0.4% and filter-sterilize through a sterile 0.22μ filter membrane (eg. Millipore). Store as 1 ml aliquots in sterile bijoux containers at −20 °C.)

Hanks balanced salt solution without calcium and magnesium (HBSS)

HBSS or storage medium (see *Protocol 3.1*) containing 10% serum (e.g. fetal calf or calf) *(Note: Soybean trypsin inhibitor can be used instead of serum if serum-free conditions are being used.)*

70% methanol/30% water

Procedure

- Thaw out a 10 ml aliquot of 2.5% trypsin and add to 90 ml of HBSS to give a working concentration of 0.25% trypsin. Mix well by swirling the bottle gently. (*Note: Do not invert the bottle to mix because it increases the risk of contamination around the neck due to leakage.*)

- Thaw out a 1 ml aliquot of DNAse and add to the 100 ml solution of 0.25% trypsin in HBSS to give a working concentration of 0.004%.

- Rinse the minced fragments from mechanical disaggregation in 3 × 10 ml washes of HBSS to remove trypsin inhibitors in the serum and discard the washings. (*Note: Viable cells may be recovered from mechanical disaggregation (see Protocol 3.1) but this must be confirmed by cell counts and viability determinations.*)

- Transfer the tissue mince from the Petri dish to the conical flask containing a sterile magnetic stirrer bar. (*Note: Use the sterile spatula or teaspoon to do this. Use of a wide-necked conical flask makes it easier to drop the fragments to the bottom of the flask without losing them around the neck of the flask.*)

- Add ~10 volumes of 0.25% trypsin/ 0.004% DNAse (pre-warmed to ~37 °C) to ~1 volume of tissue. (*Note: 100 ml of solution should be sufficient for three to four trypsinizations of a piece of tissue which was originally about 2 cm³ in volume.*)

- Re-cap the flask with the aluminium foil lid and place it on the stirrer. Set the stir speed to keep the fragments in suspension; there should be a small vortex with no frothing.

- Carry out trypsinization for 20–30 min.

- Remove the flask from the stirrer and allow the larger undigested fragments to settle. Remove the foil cap and carefully decant the enzyme solution containing released cells into a universal container. This is most easily done by tilting the flask to allow the fragments to settle into the angle of the flask and using either a long-form pastette or a 10 ml pipette with a bulb to aspirate the cell suspension from the fragments.

- Add an equal volume of serum-containing medium to the universal to neutralize the effects of the trypsin. (*Note: 50 ml centrifuge tubes can be substituted for universals.*)

- Centrifuge the universal or centrifuge tube containing released cells at 1000 rpm for 5 min and resuspend the cells in medium. Label as T1 and store at 4 °C.

- Add fresh, pre-warmed 0.25% trypsin/ 0.004% DNAse to the flask and repeat the last two steps twice more (T2 and T3). (*Note: Trypsinization is complete when only fluffy white fragments of connective tissue remain.*)

- Go to *Protocol 3.8* for cell counting and viability. Determine the viability of the individual harvests and pool those with high cell yield and viability.

Cold trypsinization

Equipment

All equipment that comes into direct contact with cells must be sterile.

Wide-necked glass conical flask with a cap made of a double layer of foil extending ~4 cm down the neck of the flask

Spatula or teaspoon (*Note: A teaspoon is convenient for handling larger volumes of tissue.*)

Magnetic stirrer bar (*Note: Soak in 70% ethanol/30% water and rinse in sterile HBSS to sterilize before use.*)

Hotplate/magnetic stirrer, with the heat setting adjusted to maintain a temperature of ~37 °C or a shaking water bath set at ~37 °C. (*Note 1: Care needs to be taken with a hot plate because of the risk of overheating. To avoid using the heater element the hotplate can be put into an incubator set at 37 °C and disaggregation carried out using only the magnetic stirrer. In the absence of a mechanical means of agitating the enzyme/tissue solution the flask can be incubated at ~37 °C and shaken gently by hand at 5 min intervals. Note 2: A water shaker bath is the most efficient method for maintaining temperature and minimizing mechanical damage during disaggregation.*)

Pastettes (long-form and short-form)

Universal containers (conical-bottomed) with a label showing volume gradations

Rack for universal containers

Bijoux

Bench centrifuge

Refrigerator

Freezer (−20 °C) for reagent storage

Reagents

All reagents must be sterile.

Trypsin – 2.5% (*Note: Thaw out 100 ml of stock solution (2.5%) and store as 10 ml aliquots in sterile universal containers at −20 °C.*)

DNAse – 0.4% (*Note: Make up stock solution of DNAse Type I in HBSS at 0.4% and filter-sterilize through a sterile 0.22μ filter membrane (e.g. Millipore). Store as 1 ml aliquots in sterile bijoux containers at −20 °C.*)

Hanks balanced salt solution without calcium and magnesium (HBSS)

HBSS or storage medium (see *Protocol 3.1*) containing 10% serum (e.g. fetal calf or calf) (*Note: Soybean trypsin inhibitor can be used to maintain serum-free conditions.*)

Procedure

• Follow the steps for mechanical disaggregation (*Protocol 3.1*) to reduce the tissue to fragments.

• Prepare a 100 ml working solution of 0.25% trypsin in HBSS (10 ml of 2.5%

stock + 90 ml HBSS) and add 1 ml of 0.4% DNAse to give a final concentration of 0.004%.

- Rinse the tissue fragments in 3 × 10 ml of HBSS to remove trypsin inhibitors and discard the washings.

- Transfer the tissue mince to a sterile universal and add ~5 volumes of 0.25% trypsin/0.004% DNAse. Store overnight at 4 °C; store the remainder of the prepared working solution of 0.25% trypsin/0.004% DNAse at 4 °C also.

- The following day warm the unused working enzyme solution to ~37 °C and transfer the tissue and enzyme solution in the universal to a wide-necked conical flask. Add the warmed trypsin/DNAse (~10 volumes per 1 volume of tissue) and carry out the warm trypsinization procedure as described in *Protocol 3.2*. (*Note 1: The overnight soaking of tissue in trypsin minimizes proteolytic damage and speeds up the release of cells from the connective tissue matrix. One trypsinization step may be enough to release all the cells. Note 2: Trypsinization is complete when only fluffy white fragments of connective tissue remain.*)

- Go to *Protocol 3.8* for cell counts and viability determination.

Disaggregation using collagenase or dispase

Collagenase and dispase (neutral protease) are less harmful to cells than trypsin and digestion with these enzymes can be carried out in the presence of serum. Contaminating fibroblasts are more likely in a collagenase digest because it is more effective against the connective tissue matrix in which the fibroblasts are found; this may be irrelevant if cells are not to be cultured, since there will be no opportunity for fibroblast over-growth. Collagen exists in several molecular forms, the main ones being Type I–Type V. Type I, found in skin, tendon, bone, cartilage, ligaments and internal organs, constitutes ∼90% of the total collagen of the body. Type IV is found predominantly in basal lamina [10]. Collagenase Type Ia and collagenase Type IV are mainly used for tissue disaggregation, though crude collagenase is sometimes recommended because it contains small amounts of non-specific proteases which contribute to matrix breakdown. Enzyme activity is given as units per gram of enzyme and this is given on the specification sheet for the batch. The enzyme is usually diluted to give a defined number of units per ml. Examples of concentrations used for different tissues are shown in Table 3.2.

Equipment

All equipment which comes into direct contact with cells must be sterile.

Tissue culture flasks (25 or 75 cm^2)

Spatula or teaspoon for transferring tissue to flask

Pastettes (long-form and short-form)

Universal containers (conical-bottomed) with a label showing volume gradations

Rack for universals

Bench centrifuge

Refrigerator

Freezer

Sterile 0.45μ and 0.22μ filter assemblies

20 ml plastic syringe

Reagents

All reagents must be sterile.

Hanks balanced salt solution without calcium and magnesium (HBSS)

Collagenase (crude, Type I or IV) *(Note 1: Dissolve the contents of the entire bottle in HBSS to give a concentration 10× the working concentration required. Filter-sterilize it through 0.45μ and 0.22μ filters using the plastic syringe. Dispense into aliquots suitable for the amount of tissue to be disaggregated and store at −20 °C. Note 2: Do not use serum in the diluting medium because it is difficult to filter-sterilize when the protein content is high. Note 3: Follow the same procedure to prepare a stock solution of dispase.)*

Incubation medium (see *Protocol 3.1* for notes on storage medium) containing 10% serum (e.g. fetal calf or calf)

Procedure

• Transfer the fragments from mechanical disaggregation into a culture flask using

Table 3.2 Type and concentration of collagenase and dispase used for the disaggregation of different tissues

Enzyme	Concentration	Tissue	Reference
Collagenase	400–600 U/ml	Bladder cancer	11
Collagenase Type Ia	200 U/ml	Brain tumours	12
Collagenase Type Ia	300 U/ml	Colon tumours	13
Dispase	1 U/ml		
Collagenase Type I	1 mg/ml	Normal and malignant gastric epithelium	14
Collagenase Type IV	0.1%	Lung tumours	15
Collagenase	100 U/ml	Myoepithelium	16
Dispase	5 U/ml		
Crude collagenase	200 U/ml	General	17
Collagenase Type I	675 U/ml	Prostate	18

either a sterile spatula or teaspoon. Use a size of flask appropriate to the volume of tissue (aim to add ~10 volumes of enzyme solution per 1 volume of tissue fragments). (*Note: Culture flasks provide a sterile non-toxic environment for disaggregation; use a 25 cm² flask for 10 ml and a 75 cm² flask for 25 ml incubation volumes.*)

- Make a ten-fold dilution of the thawed stock solution of collagenase (or dispase) in the incubation medium. Add ~10 volumes of the enzyme medium to 1 volume of tissue in the culture flask.

- Incubate the flask horizontally in a 37 °C incubator overnight. (*Note: Different media have different buffering capacities and the metabolic activity and the incubation period need to be taken into consideration when deciding about the medium to use. Phenol red is used as a pH indicator and the buffering system is based on bicarbonate concentration. Hanks = 0.35 g/l: Earle's = 2.2 g/l; Dulbecco's = 3.75 g/l. HBSS is suitable when metabolic activity and CO_2 production is low and a low buffering capacity is sufficient. Earle's is used when tissue/cells are more metabolically active and a higher buffering capacity is needed; most culture media are based on Earle's salts.*)

- The following day use a wide-bore pastette to aspirate the fragments several times and release loosened cells. Check the flask under an inverted phase-contrast microscope for cell release; continue digestion if cell yield is low and/or cells can still be seen within the tissue fragments. (*Note: If the medium becomes viscous, add DNAse at 0.004% as in Protocols 3.2 and 3.3*)

- When digestion has been completed, transfer the cell suspension to a universal container or 50 ml centrifuge tube (depending on volume) and centrifuge at 1000 rpm for 5 min.

- Discard the supernatant and resuspend the cell pellet in culture medium.

- Go to *Protocol 3.8* for cell counting and viability determination.

Isolation of cells from body fluids – Anne Wilson

Introduction

Body fluids are a source of cells for a number of normal and pathological conditions.

Cells can be recovered from breast milk, urine, blood, and effusions resulting from a variety of pathologies. Abnormal accumulations of fluid in the pleural and abdominal cavities may occur in cancer and such fluids contain both cancer cells and inflammatory cells. Volumes vary from 20 ml up to 5 l, though a large volume is not necessarily indicative of a high cell yield.

Protocol 3.5

Recovery of cells from effusions

Equipment

All equipment coming into direct contact with cells must be sterile.

50 ml centrifuge tubes

Pastettes (long-form and short-form)

Universal containers (conical-bottomed) with labels showing volume gradations

Racks for universal containers and centrifuge tubes

Bench centrifuge

Bottles for waste liquid disposal

Reagents

All reagents must be sterile.

Hanks balanced salt solution without calcium and magnesium (HBSS)

Storage medium (see *Protocol 3.1*)

Procedure

- Dispense the fluid into the centrifuge tubes and centrifuge at 1500 rpm for 20 min.

- Pour off the supernatant into a waste bottle for disposal according to local protocols for the disposal of biohazardous waste.

- Resuspend the cell pellets in 5 ml of HBSS or storage medium and transfer into universal containers, pooling the pellets from several tubes.

- Go to *Protocol 3.8* for determination of cell number and viability.

Removal of red blood cells – Anne Wilson

Introduction

Red blood cells released during tissue disaggregation may be removed either by snap lysis, use of a lysis buffer or by isopycnic centrifugation. The latter method has the advantage of removing red blood cells and dead cells at the same time, thus increasing the viability of the cell population [19].

Removal of red blood cells by snap lysis

Equipment

All equipment coming into direct contact with cells must be sterile.

Universal containers (conical-bottomed) with a label showing volume gradations

Pastettes (long-form and short-form)

Bench centrifuge

Reagents

All reagents must be sterile.

Culture-grade water

Double strength HBSS (Dispense 20 ml of $10\times$ HBSS into a bottle and add 80 ml of culture-grade water. Mix gently and add 1 ml of 7.5% sodium bicarbonate to buffer to pH 7.4. Store at $4\,^{\circ}$C.)

Storage medium (see *Protocol 3.1*) containing 10% serum (e.g. fetal calf or calf)

Procedure

- Centrifuge the cell suspension at 1000 rpm for 5 min and discard the supernatant.

- Add 10 ml of culture-grade water to the cell pellet and pipette it gently up and down several times to resuspend the cells. *(Note: Cells vary in their ability to withstand hypotonicity. Keep the exposure time down to <15 s.)*

- Immediately add an equal volume of buffered $2\times$ HBSS to restore tonicity and mix the cell suspension gently with a pastette.

- Centrifuge again at 1000 rpm for 5 min and check the cell pellet for red blood cells. Repeat the procedure if they are still present in significant numbers. Otherwise resuspend the cleaned cell pellet in storage medium.

- Go to *Protocol 3.8* for cell counts and viability determination.

Removal of red blood cells (rbc) and dead cells using isopycnic centrifugation

Introduction

The removal of red blood cells by snap lysis has already been described but rbc may also be removed using isopycnic centrifugation. This method has the added advantage that dead cells are removed with the rbc and the percentage viability of the cell population is thus increased [19].

Equipment

Universal containers (conical-bottomed) with labels showing volume gradations

Pastettes (long-form and short-form)

10 ml graduated pipettes and bulb or pipetting device

Bench centrifuge

Reagents

Lymphoprep or similar

Culture medium

Procedure

- Resuspend the cells at a high cell concentration $\sim 5 \times 10^6$/ml-10^7/ml. Add 10 ml of Lymphoprep to a universal container and carefully layer over $\sim$5 ml of cell suspension using either a pastette or a 10 ml pipette. Hold the universal at an angle and slowly trickle the cell suspension down the side of the universal rather than directly onto the surface of the Lymphoprep. The aim is to get a clean boundary between the two liquids to achieve optimal separation.

- Centrifuge the tubes at 1000 rpm for 20 min with the brake off. *(Note: A gentle slowing of the centrifuge maintains a clean boundary between the Lymphoprep and the culture medium.)*

- Carefully remove the containers from the centrifuge and observe the interface between the two liquids. There will be a creamy yellow band of cells visible. Red blood cells and dead cells form a pellet at the bottom of the container.

- Use a pastette to remove the viable cell layer at the interface and transfer to a universal container.

- Add fresh culture medium to the harvested cells and mix well. Centrifuge at 1000 rpm for 5 min and repeat twice to wash the Lymphoprep from the cells.

- Resuspend the cells in a known volume of medium and carry out cell counting and viability determinations (see *Protocol 3.8*).

Cell counting and cell viability – Anne Wilson

The number of cells in a cell suspension is usually determined microscopically, though a Coulter counter can be used [20]. Viability of the cells is determined using a dye exclusion method. These vary in sophistication but are usually based on the principle that dead cells take up vital dyes whereas living cells with intact membranes exclude them. The ratio of stained and unstained cells thus gives a reasonable estimate of the cell viability. However, cells may be metabolically impaired and dying yet still exclude the dye, thus overestimating viability [21, 22].

Other problems may result from the presence of large clumps of cells, especially from the disaggregation of epithelial tissue. These may be too big to go under the coverslip of the counting chamber, resulting in underestimation of cell number. For cloning experiments in which single cells are required, their exclusion may also give inaccurate information on the purity of a single cell suspension. Clumps can be broken down by passing the cell suspension through needles of decreasing gauge several times, but this tends to reduce viability. Another way around the problem is to lyse the cells, stain the released nucleii with crystal violet and carry out a nuclear count instead of a whole cell count.

The most commonly used vital dye is trypan blue, though others such as erythromycin and nigrosin [23] may be used. A range of commercial kits designed for cytotoxicity testing are available, using a variety of end points for assessing cytotoxicity and viability [24], some based on different fluorescent dyes [25, 26].

Quantitation of cell counts and viability

Equipment

Inverted phase-contrast microscope with ×10 objective

Universal containers (conical bottomed) with labels showing volume gradations

Improved Neubauer haemocytometer chamber

Coverslips for chamber (*Note: The coverslips are thicker than those used for standard histology, but still break easily if pressure is not distributed evenly when applying the coverslip. It is a good idea to carry a stock of spares.*)

Bijoux or equivalent plastic container for small volumes, e.g. test tube or centrifuge tube

Pastettes (long-form and short-form)

Variable volume micropipette (100–250 μl)

Sterile tips for micropipettes

Sterile graduated pipettes, 1–10 ml, with bulb or automatic pipetting device

Reagents

Trypan blue – 1% (w/v)

HBSS or storage medium (see *Protocol 3.1*)

Procedure

• Prepare the chamber for counting. Lightly moisten the area either side of the central grid and slide the coverslip on horizontally from the edge using firm pressure with both thumbs. The coverslip is correctly positioned when Newton's rings (rainbow colours) can be seen on either side adjacent to the central marked area. *(Note 1: The depth of the improved Neubauer chamber is 0.1 mm when the coverslip is correctly positioned; the calculation for determining the cell number is based upon a volume of $0.1 \times 1 \times 1$ mm^3. The calculated cell number will be wrong if the coverslip is incorrectly positioned. Note 2: If a different type of chamber is used, check the product information for calculation of cell numbers.)*

• Centrifuge the cell suspension at 1000 rpm for 5 min and discard the supernatant. Resuspend the cells in a known volume of medium.

• Mix the suspension by aspirating it up and down several times with a pastette to make sure that the cells are evenly dispersed.

• Take 250 μl of the mixed cell suspension up with a micropipette and transfer it to a bijou. Use a clean pipette tip to transfer the same volume of 1% trypan blue solution to the cell suspension. Mix it well by flicking the bijou several times with the thumb and forefinger. Immediately take up a drop of the cell suspension into the tip of a pastette and gently place it at the very edge of the coverslip adjacent to the counting area.

Capillary action will draw the cell suspension under the coverslip, filling the area between the central and side gullies. Repeat this on the other side of the coverslip. *(Note: If the coverslip is not positioned correctly both chambers and the gully either side of the chambers will flood with cell suspension and accurate cell counting will not be possible.)*

- Place the chamber on the microscope platform and move it around until you have located the central grid of 25 small squares bounded by three parallel lines. Count the total number of cells (stained and unstained) in these 25 squares. (They are each subdivided into 16 smaller squares to facilitate counting.) In order to avoid counting the same cell twice, count the cells on the top line, the left-hand line and the centre of each small square bounded by triple lines. Use a tally counter if the numbers are large. *(Note 1: The area counted depends upon the cell density. If the cell number reaches ~100 in the central square this is sufficient for accuracy. If the cell density is much lower, count the total number of cells in the four corners and take an average. For further accuracy it is advisable to repeat the total cell count and take an average of the two readings. Note 2: If you are using phase contrast on the inverted microscope, live and dead cells can be distinguished without staining. Live cells are brightly refractile with a distinct outline; dead cells are flat and grey with an uneven edge.)*

- When you have counted the total number of cells, repeat the counts in the same areas, counting only the blue-stained (dead) cells. Calculate the average number of dead cells and subtract from the total number of cells to obtain the number of live cells. *(Note: Accuracy can be confirmed by repeating the count on unstained cells only.)*

- The viability of the cell population is calculated using the formula:

$$\text{Viability} = \frac{\text{number of live cells}}{\text{number of live} + \text{dead cells}} \times 100\%$$

- The concentration of cells in the cell suspension is calculated as follows:

Average number of cells in one large square (25 small squares) $\times 10^4$ = no. of cells/ml

(Note: This is either the number of cells in the central large square or the mean of the counts in the four large corner squares.)

- The total number of cells in the cell suspension = no. of cells/ml × total volume.

- The total number of live cells in the suspension = total number of cells × % viability *(Note: If you are calculating the number of cells from the trypan blue stained population, remember to allow for the twofold dilution factor introduced by diluting the cell suspension in the trypan blue solution.)*

- Example: (see Figure 3.1)

Volume of cell suspension = 20 ml

Total cell count of live cells in 4 corner squares = 160

Total cell count of live and dead cells in 4 corner squares = 200

Mean cell count of live cells in one large square = 40

Mean cell count of live and dead cells in one large square = 50

Viability of cell population
$$= 40/50 \times 100\%$$
$$= \underline{\mathbf{80\%}}$$

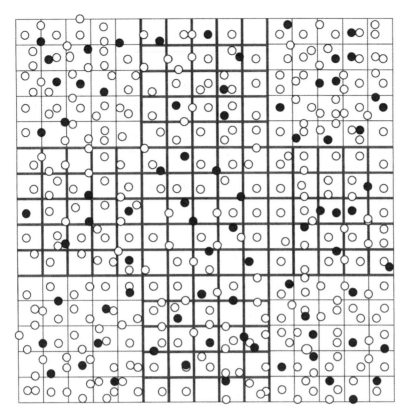

Figure 3.1 A haemocytometer chamber (see Protocol 9, worked example). Live cells are clear and dead cells black

Concentration of viable cells

$$= \frac{50 \times 10^4 \times 80}{100} \text{cells/ml}$$

$$= 40 \times 10^4 \text{ viable cells/ml}$$

$$= \mathbf{4 \times 10^5 \text{ viable cells/ml}}$$

Multiply this ×2 to allow for dilution factor = **8 × 10⁵ viable cells/ml**

Total number of cells in original cell suspension $= 8 \times 10^5 \times 20$

$$= \mathbf{16 \times 10^6 \text{cells}}$$

To adjust the cell number to 10^6/ml, centrifuge the cell suspension at 1000 rpm for 5 min. Pour off the supernatant and resuspend the cells in 16 ml of medium. Confirm the cell concentration by recounting the number of cells.

To adjust the cell number to 1.5×10^5 /ml:

Original cell count $= 8 \times 10^5$/ml, dilution factor $= 8/1.5 = 5.3$

Therefore dilute 1 ml of 8×10^5/ml to 5.3 ml.

Use of cell cultures – Anne Wilson

Introduction

A wide range of cell lines representing diverse cell types and species are available commercially from national collections in Europe and America (ECACC, ATCC, DSmZ). In theory, cells grown in culture provide a continuous, replicate supply of material of a homogeneous characterized cell population. However, precautions need to be taken to ensure that the latter is true in practice. Cross-contamination of cell lines can occur during culture and the line may not therefore be as described [27, 28]. Particular attention needs to be paid to this possibility if lines are exchanged between laboratories rather than obtained from commercial sources. Isoenzymes, karyotypes and DNA fingerprinting are all appropriate tools for speciating cell lines [29–31]. A wide range of antibodies are also available for the immunocyto-chemical identification of phenotypic characteristics such as cytoskeletal proteins, surface markers and biosynthetic products.

Cell lines in long-term culture can show drifts in both phenotype and genotype. When using cell lines for a series of related experiments it is advisable to use cells from similar passage numbers. Generally these are within 10 passages and once the stock has passed this number fresh working stock should be thawed out for use. Guidelines for handling cancer cell lines have been published by UKCCR [32].

Continuous cultures of cell lines may become contaminated with mycoplasma. The presence of these micro-organisms affects a number of cell properties. Surface changes may alter the antigenicity of cells and their presence results in competition for a variety of metabolites needed for DNA synthesis, protein synthesis and other essential metabolic pathways. Signs of contamination include a decrease in growth rate, increased acidity of medium, deterioration of cells and the appearance of a grainy background. There are several different species found in culture and assay for the detection of mycoplasma should be a regular routine during maintenance of cell lines in long-term culture. A number of methods are available for detecting mycoplasma in culture, the simplest being the use of Hoechst 33258, a fluorescent stain which binds to DNA and reveals the presence of extracellular and extranuclear DNA [33]. A variety of commercially available kits can also be obtained [34].

Recovery of cells from monolayer cultures

Equipment

All equipment coming into direct contact with cells must be sterile.

Class II cabinet

Incubator set at 37 °C

Tissue culture flasks

Universal containers (conical-bottomed) with labels showing volume gradations

Pastettes (long-form and short-form)

Inverted phase contrast microscope with ×10 objective

Water bath set at 37 °C

Waste bottle for spent culture medium

Permanent black marker pen

Reagents

All reagents must be sterile.

Hanks balanced salt solution without calcium and magnesium (HBSS)

Culture medium appropriate to cell type

Trypsin/versene solution (1–2 ml/25 cm^2) *(Note 1: If this has been purchased as a 10× strength solution, thaw out stock, dilute to 1× strength in HBSS and dispense 5 ml aliquots into sterile bijoux. Store at −20 °C.)*

Procedure

- Switch on the Class II cabinet 10 min before starting and swab down its surfaces with 70% methanol/30% water. Any equipment to be used in the cabinet should be placed towards the back so that air flow is not obstructed and sterility is maintained.

- Warm the reagents to be used to 37 °C.

- Remove the flasks for harvesting from the incubator and check them for health under the microscope.

- Decant the spent medium into a sterile bottle for waste medium and rinse the monolayer gently with 3 × 5 ml washes of HBSS, pouring the wash medium from the flask into the waste bottle after each rinse. If the medium dribbles down the neck of the flask use a tissue soaked in 70% alcohol/30% water to wipe it off.

- Add 1–2 ml of trypsin-versene to the flask, re-cap it and tilt the flask horizontally in several directions to ensure the entire cell monolayer is bathed in enzyme solution.

- Place the flask horizontally into the incubator for 2–5 min.

- Remove it and check visually for cell detachment. If the cells have become rounded and are detaching from the cell surface this will be visible to the naked eye. If the cells are still attached return the flask to the incubator for several more minutes and check again for detaching cells. *(Note: Cell lines vary in the speed with which they detach; times may vary from 1 to 20 min. Harvesting procedure is usually specified in details accompanying purchased cell lines.)*

- Once the cells have begun to detach add ~5–10 ml of culture medium containing serum to the flask. Re-cap it and tap the side of the flask against the palm of your hand several times; the shearing forces cause further detachment of the cells. Use the microscope to check for complete detachment of cells.

- Use a pastette or pipette to dispense the detached cells into a universal. Rinse the surface of the flask with another 5 ml of medium and transfer the remainder of the detached cells into the universal container. Check the flask under the microscope to make sure that all the cells have detached and been recovered.

- Centrifuge the cells at 1000 rpm for 5 min and resuspend them in a known volume of medium for cell counts and viability determination (see *Protocol 3.8*).

Notes

① Familiarize yourself with the normal appearance of the cells in the different stages of culture so that you notice abnormalities. These may be due to a number of factors including: pH too acid or too alkaline, nutrient deficiency due to infrequent media changes, cells seeded at too low a density in the starting culture, cells left in stationary phase for too long, contamination by micro-organisms (bacteria, yeast, fungi, mycoplasma), cross-contamination with another cell line, or natural drift in phenotype and/or genotype.

② Cells should be harvested at the same time point in the life of the culture, ideally when they are reaching the end of the exponential growth phase and before the culture has become stationary, though this may be

dependent upon the state of the cells required for subfractionation.

③ Take the following precautions to avoid cross-contamination between cell lines: (a) Handle one cell line at a time. (b) Use separate media, solutions, pipettes and waste medium disposal bottles for each cell line. (c) **Never** put a pipette that has been in contact with cells back into a medium or reagent bottle. (d) Clean the cabinet before handling the next cell line. (e) Handle fast-growing cell lines (e.g. HeLa) that grow from a small inoculum of cells last.)

Cryopreservation – Anne Wilson

Stocks of cells can be kept for long periods by freezing them down and storing them in liquid nitrogen.

Critical factors in the successful freezing of cells include the rate of freezing, the type and concentration of cryoprotectant and the cell concentration [35–39]. Although optimal results are obtained using a programmable cell freezer it is possible to control the rate of freezing in other ways. These include the use of a thick-walled polystyrene box stored in a −80 °C freezer or a special device designed to fit into the neck of a liquid nitrogen cell storage bank, thus holding the cells in vapour phase and allowing them to cool down slowly. These methods do not give such a high viability on recovery of cells from frozen stock as does the programmable freezer. Avoidance of ice crystal formation within the cell is critical and this is minimized in several ways, including incorporation of a cryoprotectant such as dimethylsulfoxide (DMSO) or glycerol, an increased serum concentration, a high cell number and a controlled slow rate of freezing. For most mammalian cells, the optimum rate is −1 °C/min. DMSO at 10% is the most commonly used cryoprotectant and the protocol outlined here describes its use. However, it may be toxic to some cells. (*Note: DMSO is a polar solvent and can penetrate the human skin, together with any substances dissolved in it. Take care when handling it and wear protective gloves.*)

The method of recovery of cells is also critical and should aim to avoid osmotic shock. Frozen cells need to be thawed quickly to $37\,^{\circ}C$ and then diluted very slowly in culture medium.

Protocol 3.10

Freezing cells

Equipment

All equipment coming into direct contact with cells must be sterile.

Plastic freezing vials (2 ml)

Pastettes (long-form and short-form)

Universal containers (conical-bottomed) with labels showing volume gradations

Black permanent marker pen

Waste bottle for medium

Improved Neubauer haemocytometer chamber with coverslip

Bench centrifuge

Inverted phase contrast microscope with ×10 objective

Programmable cell freezer or

Thick-walled polystyrene box or

Device for neck of liquid nitrogen cell bank or storage vessel

Dewar flask with vented lid suitable for the storage of liquid nitrogen

Liquid nitrogen cell bank

Storage vessel for liquid nitrogen

Trays or straws for storing ampoules in cell bank

Face protection

Insulated protective gloves

Long forceps

Micropipette (1–5 ml)

Tips for micropipette

Reagents

All reagents must be sterile.

Hanks balanced salt solution without calcium and magnesium (HBSS)

Dimethylsulfoxide (DMSO)

Culture medium containing 10–20% serum

1% Trypan blue

Liquid nitrogen in Dewar flask

Procedure

- Determine the viability and cell count of the population for freezing using *Protocol 3.8*.

- Calculate the volume of diluent needed to resuspend the cells at 5×10^6/ml.

- Prepare this volume of a 10% solution of DMSO in culture medium and mix well.

- Label the appropriate number of freezing vials clearly with the date, pass number and cell type using the black marker pen.

- Centrifuge the cell suspension at 1000 rpm for 5 min, discard the supernatant and resuspend the cells in the required volume of 10% DMSO. Mix well.

- Dispense the cell suspension as 1.8 ml aliquots into the freezer tubes using a micropipette and cap them securely.

- For programmable freezing the vials can be transferred directly to the freezer and the freezing cycle started (see manufacturer's instructions).

- Transfer the vials to the long-term storage facility when freezing is complete.

- For freezing with a device for the neck of the cell bank or storage vessel, try the following as a starting point: Keep the filled freezing vials at 4 °C for 60 min, transfer to the freezing device and keep in vapour phase overnight, then transfer them to the long-term storage facility.

- For freezing with a polystyrene box, try the following as a starting point: Keep the filled freezing vials at 4 °C for 60 min, transfer to the box and hold the box at −20 °C for 60 min, transfer the box to −80 °C overnight, then transfer them to the long-term storage facility. (Note: Frozen cells can be stored at −80 °C for several months.)

Thawing cells

Equipment

All equipment coming into direct contact with cells must be sterile.

Water bath set at 37 °C

Protective face mask

Insulated protective gloves

Long forceps

Universal containers (conical-bottomed) with labels showing volume gradations

Watch with second hand

Pastettes (long-form and short-form)

Bench centrifuge

Dewar flask for storage of liquid nitrogen

Reagents

All reagents must be sterile.

Culture medium warmed to 37 °C

Liquid nitrogen in Dewar flask

70% methanol/30% water

Procedure

- Half-fill the Dewar flask with liquid nitrogen.

- Remove vials from the nitrogen cell bank and transfer to the Dewar flask containing liquid nitrogen for transport to the lab. (*Note: There is a risk of vials exploding when removed from the nitrogen bank. Protective face mask and gloves must be worn until the vial has been completely thawed.*)

- Remove one freezing vial from the flask using forceps.

- Holding the vial at arm's length, plunge it into the water bath and agitate the vial in the water to thaw it quickly and bring it up to 37 °C.

- Immerse the thawed vial in 70% methanol/30% water to sterilize it and wipe dry with a tissue. (*Note: Make sure you have previously noted the details written on the side of the vial in case they get washed off.*)

- Remove the lid of the vial in the Class II cabinet and transfer the thawed contents of the vial into a universal container, using a pastette.

- Take up some warmed culture medium into the pastette and add one drop ($\sim$100 μl) to the cell suspension. Mix it gently and add another drop $\sim$15 s later. Continue until the volume has been doubled. Then add $\sim$500 μl every 15 s with gentle mixing until the volume has doubled again. Complete the dilution by adding 1000 μl every 15 s up to $\sim$16 ml. (*Note: The rate of addition of medium is gradually increased as the volume increases because the risk of osmotic shock decreases.*)

- When dilution is complete, centrifuge the cell suspension at 1000 rpm for 5 min and resuspend in 15–20 ml medium. Mix gently and centrifuge again at 1000 rpm for 5 min. Repeat this step to wash off residual DMSO.

- Resuspend the cells in a known volume of medium and determine cell number and viability using *Protocol 3.8.*

Isolation of human peripheral blood mononuclear cells – John Graham

Introduction

Using a density barrier

The density of most human peripheral blood mononuclear cells (PBMCs) is <1.078 g/ml. In order to separate these cells from the denser erythrocytes and polymorphonuclear leukocytes (PMNs), Boyum [40, 41] therefore used an isoosmotic density barrier solution of metrizoate and a polysaccharide (Ficoll®) of density 1.077–1.078 g/ml. The polysaccharide is included to aggregate the erythrocytes and so make them easier to sediment. Blood (diluted 1 : 1 with saline) is layered over the barrier and the mononuclear cells are harvested from the interface after centrifugation.

More recently metrizoate has been replaced by the chemically similar diatrizoate (see Figure 3.2) and there are several commercially available media (e.g. Lymphoprep™, Ficoll-Hypaque™ and Histopaque™) that have the same density

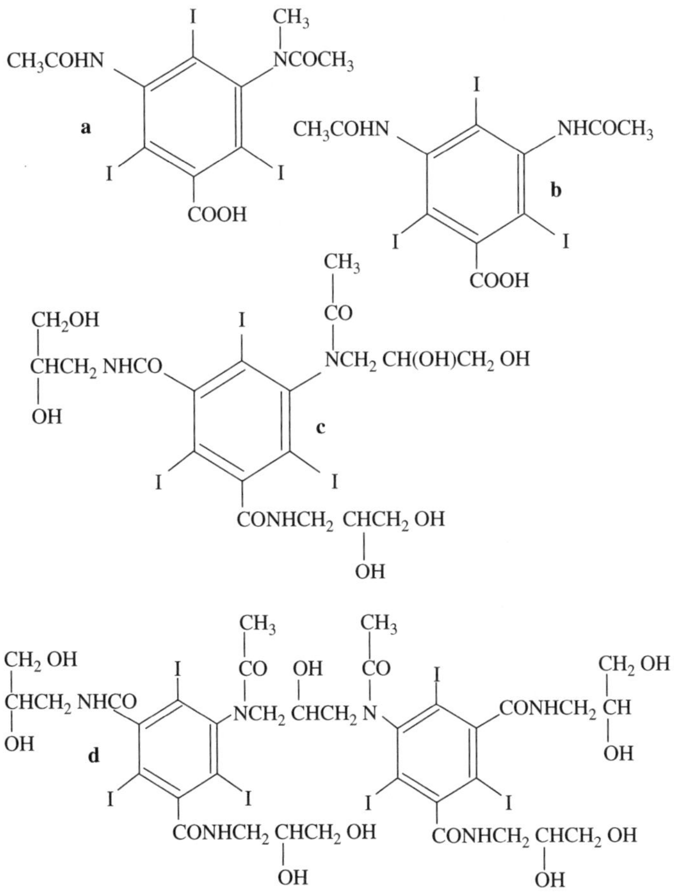

Figure 3.2 Molecular structure of iodinates density gradient media: (a) metrizoate, (b) diatrizoate, (c) Nycodenz®, (d) iodixanol

(1.077 g/ml) and contain 9.6% (w/v) diatrizoate and 5.6% polysucrose (or Ficoll®). The only difference between the media is the endotoxin level – this is lower (<1 EU/ml) in Lymphoprep™ than in other comparable commercial media.

There is, however, evidence that the polysaccharide may interact with the surface of lymphocytes [42]. Moreover, the presence of an impermeant ion (diatrizoate) in the medium may also affect the Gibbs-Donnan equilibrium of ions across the membrane. An alternative medium was therefore developed which was non-ionic and contained no polysaccharide. This isoosmotic medium comprises 14.1% (w/v) Nycodenz®, 0.44% NaCl and 5 mM Tricine-NaOH, pH 7.0 [43, 44]. It is available commercially as NycoPrep™ 1.077; it has the same density as Lymphoprep™ and separation of the PBMCs is achieved in exactly the same manner, although in the absence of a polysaccharide a slightly longer centrifugation time is required to achieve satisfactory pelleting of the erythrocytes.

The density barrier can alternatively be produced by dilution of one of the commercial general-purpose media, NycoPrep™ Universal (60%, w/v Nycodenz®) or OptiPrep™ (60% w/v iodixanol). See Figure 3.2 for the molecular structure of Nycodenz® and iodixanol.

Using a mixer technique

An alternative strategy, devised by Ford and Rickwood [45], considerably simplifies the procedure. A 19% (w/v) Nycodenz® solution ($\rho = 1.100$ g/ml) was added to an equal volume of whole blood in order to raise the density of the plasma in whole blood to 1.077 g/ml. During centrifugation at $1500g$ for 30 min at $20\,°C$ the erythrocytes and polymorphonuclear leukocytes (PMNs) will sediment while the PBMCs float to the top and they

are recovered from the meniscus and the medium below it. The ease of operation makes the mixer–flotation technique particularly attractive when handling large numbers of potentially pathogenic samples.

Recoveries and purity of PBMCs isolated by flotation are almost identical to those obtained with Lymphoprep™. Kaden et al. [46] compared Lymphoprep™ with a mixer based on Nycodenz®; these workers found that the PBMC harvests were essentially identical by both techniques.

A commercial solution (NycoPrep™ Mixer) is no longer produced and the high-density solution for mixing with the blood must be produced by dilution of either NycoPrep™ Universal or OptiPrep™. A convenient alternative is simply to add a smaller volume of OptiPrep™ (1.32 g/ml) itself – the lower osmolality of OptiPrep™ compared to NycoPrep™ Universal allows the use of this strategy and it is the one that is described below.

An interesting observation is that if the blood has to be stored before fractionation, then, as long as the density of the blood is raised by addition of the dense medium immediately after drawing, the loss of recovery and purity of the PBMCs that is observed with density barrier techniques is much less marked. This is probably related to the fact that once the density of the plasma has been raised, the PBMCs do not settle out upon standing [47].

Using a barrier flotation technique

In a third alternative (a variation of the mixer technique), the blood is adjusted to a density considerably higher than that of the PBMCs ($\rho = 1.090$ g/ml) and layered beneath a $\rho = 1.078$ g/ml density barrier. As with the mixer, the PBMCs float to the surface, but this is the only system in which the cells do not band adjacent to the plasma-containing sample layer. In both of the other techniques the PBMCs are heavily contaminated by platelets (thrombocytes). In this method the great majority of platelets remain in the plasma layer [48]. The low-density barrier thus acts as a 'buffer-zone' between the PBMCs and all the other blood components – erythrocytes, PMNs, platelets and plasma proteins.

Purification of human PBMCs on a density barrier

Reagents

Anticoagulant: 100 mM EDTA or 3.8% citrate ①

Hepes-buffered saline (HBS): 0.85% (w/v) NaCl, 10 mM Hepes-NaOH, pH 7.4 ②

Lymphocyte separation medium (e.g. Lymphoprep™ or NycoPrep™ 1.077) or OptiPrep™ diluted with HBS (5 vol. + 17 vol. respectively)

Equipment

Low-speed bench centrifuge (preferably temperature controlled) with swinging-bucket rotor to take 15 ml clear screw-capped polystyrene centrifuge tubes

Plastic Pasteur pipettes

Syringe (5–10 ml) and metal filling cannula (optional) ③

Procedure

1. Collect venous blood using either EDTA (2 mM final concentration) or citrate (0.38% final concentration) as anticoagulant.

2. Dilute blood with an equal volume of HBS by repeated gentle inversions. ④

3. Using a plastic Pasteur pipette, layer 6 ml of this diluted blood over 3 ml of the chosen medium in the centrifuge tube. ⑤

4. Centrifuge at 700g for 15 min (Lymphoprep™) or 700g for 20 min (NycoPrep™ 1.077 or iodixanol solution) at 20 °C. ⑥ ⑦

5. Allow the rotor to decelerate using a slow deceleration program or without the brake. ⑧

6. Collect the cells at the interface using a Pasteur pipette.

7. Dilute with two volumes of HBS and harvest the cells at 200–500g for 15 min.

Notes

① Heparin is also suitable but the most widely used anticoagulant is EDTA or citrate.

② Tricine-buffered saline or any suitable balanced salt solution is permissible.

③ A syringe and metal cannula can be used to underlay the blood with the medium (step 3) – the preferred method particularly for larger volumes.

④ High yields (>95%) of PBMCs are only obtained if the whole blood is diluted with saline. With undiluted blood yields are reduced to <85% because the interface between the sample and the medium is less stable and there is a tendency for the blood to 'stream' through the medium, carrying erythrocytes and mononuclear cells into the pellet.

⑤ For larger volumes use 20 ml of diluted blood over 10 ml of medium in a 50 ml tube.

⑥ The longer time is recommended for density barriers not containing polysaccharide.

⑦ If the separation is executed at 4 °C add another 5 min centrifugation time.

⑧ Slow smooth deceleration is recommended. Rapid changes in the rpm create vortices in the liquid and 'swirling' of the pellet and banded cells.

Purification of human PBMCs using a mixer technique

Reagents

Anticoagulant: 100 mM EDTA or 3.8% citrate ①

Hepes-buffered saline (HBS): 0.85% (w/v) NaCl, 10 mM Hepes-NaOH, pH 7.4 ②

OptiPrep™

Equipment

Low-speed bench centrifuge (preferably temperature controlled) with swinging-bucket rotor to take 15 ml clear screw-capped polystyrene centrifuge tubes

Plastic Pasteur pipettes

Procedure

1. Collect venous blood using either EDTA (2 mM final concentration) or citrate (0.38% final concentration) as anticoagulant.

2. Mix whole blood gently but thoroughly (by repeated inversion) with OptiPrep™ (10 vol. + 1.25 vol. respectively) in a suitable capped centrifuge tube (15 ml tubes for 5–10 ml samples). ③

3. Layer approx 0.5 ml of HBS on top and centrifuge at $1500g_{av}$ for 30 min at 20 °C. ④

4. Allow the rotor to decelerate using a slow deceleration program or without the brake. ⑤

5. Collect the PBMCs from the meniscus downwards to about 1 cm from the cell pellet. ⑥

6. Dilute the collected material with two volumes of buffered-saline and pellet the cells at $250–500g$ for 5–10 min.

Notes

① Heparin is also suitable but the most widely used anticoagulant is EDTA or citrate.

② Tricine-buffered saline or any suitable balanced salt solution is permissible.

③ The actual increase in density of the plasma will depend on the haematocrit of the blood, if contamination of the PBMCs by PMNs is unacceptable, the amount of OptiPrep™ added should be reduced. If the yield of PBMCs is very low, the amount of OptiPrep™ added should be increased.

④ The layer of HBS on top of the blood is not critical to the separation, but it facilitates the harvesting of the PBMCs from the meniscus.

⑤ Slow smooth deceleration is recommended. Rapid changes in the rpm create vortices in the liquid and 'swirling' of the pellet and banded cells.

⑥ Approximately 90% of the PBMCs are in the top third of the plasma.

Purification of human PBMCs using a barrier flotation technique

Reagents

Anticoagulant: 100 mM EDTA or 3.8% citrate ①

Hepes-buffered saline (HBS): 0.85% (w/v) NaCl, 10 mM Hepes-NaOH, pH 7.4 ②

OptiPrep™

Equipment

Low-speed bench centrifuge (preferably temperature controlled) with swinging-bucket rotor to take 15 ml clear screw-capped polystyrene centrifuge tubes

Plastic Pasteur pipettes

Syringe (5–10 ml) and metal filling cannula (optional) ③

Procedure

1. Collect venous blood using either EDTA (2 mM final concentration) or citrate (0.38% final concentration) as anticoagulant.

2. Make a working solution of 40% (w/v) iodixanol: dilute 4 ml of OptiPrep™ with 2 ml of HBS.

3. Adjust the plasma of whole blood to approx $\rho = 1.095$ g/ml by adding 2.7 ml of the working solution to 10 ml of whole undiluted blood and mix by repeated gentle inversion.

4. Prepare the $\rho = 1.078$ g/ml density barrier solution by diluting 5 ml of working solution with 9.6 ml of HBS.

5. Using a syringe and metal cannula, underlayer 5 ml of the density barrier with 5 ml of blood in a 15 ml centrifuge tube.

6. Layer approx 0.5 ml of HBS on top and centrifuge at 700 g_{av} for 30 min at 20 °C. ④ ⑤

7. Allow the rotor to decelerate using a slow deceleration program or without the brake. ⑥

8. The PBMCs band on the top of the 1.078 g/ml barrier. Remove the band with a pipette.

Notes

① Heparin is also suitable but the most widely used anticoagulant is EDTA or citrate.

② Tricine-buffered saline or any suitable balanced salt solution is permissible.

③ A syringe and metal cannula is the preferred method for underlayering the medium with the blood (step 5) but a Pasteur pipette may be used for overlayering the blood with the medium.

④ The layer of HBS on top of the blood is not critical to the separation, but it facilitates the harvesting of the PBMCs from the meniscus.

(5) The separation may be executed at 4 °C.

(6) Slow smooth deceleration is recommended. Rapid changes in the rpm create vortices in the liquid and 'swirling' of the pellet and banded cells.

References

1. Freshney, R. I. (1987) *Culture of Animal Cells. A Manual of Basic Technique*, 2nd edn, pp. 257–288. Wiley-Liss.
2. James, J. J., Johnstone, L. and Milman, A. (1997) *Laboratory Manager*. Croner Publications Ltd.
3. Health Services Advisory Committee (1991) *Safe Working and the Prevention of Infection in Clinical Laboratories*. HMSO, London.
4. Health Services Advisory Committee (1991) *Safety in Health Services Laboratories. Safe Working and the Prevention of Infection in Clinical Laboratories. Model Rules*. HMSO, London.
5. Howie Report (1978) *Code of Practice for Prevention of Infection in Clinical Laboratories and Post Mortem Rooms*. HMSO, London.
6. Freshney, R. I. (1987) *Culture of Animal Cells. A Manual of Basic Technique*, 2nd edn, pp. 57–84. Wiley-Liss.
7. Karmiol, S. (2000) Development of serum-free medium. In: *Animal Cell Culture. A Practical Approach* (John R. W. Masters, ed.), pp. 105–121. Oxford University Press.
8. Waymouth, C. (1974) To disaggregate or not to disaggregate. Injury and cell disaggregation, transient or permanent? *In vitro*, **10**, 97–111.
9. Freshney, R. I. (1987). *Culture of Animal Cells. A Manual of Basic Technique*, 2nd edn, pp. 115–118. Wiley-Liss.
10. Alberts, B., Bray, D., Lewis, J., Raff, M., Roberts, K. and Watson, J. D. (1983) In: *Molecular Biology of the Cell*, pp. 673–715. Garland Publishing Inc.
11. X Fu, V., Schwarze S. R., Reznikoff, C. A. and Jarrard, D. F. (2004) In: *Culture of Human Tumor Cells* (Roswitha Pfagner and R. Ian Freshney, eds), pp. 101–123. Wiley-Liss.
12. Darling, J. L. (2004) In: *Culture of Human Tumor Cells* (Roswitha Pfagner and R. Ian Freshney, eds), pp. 349–372. Wiley-Liss.
13. Whitehead, R. H. (2004) In: *Culture of Human Tumor Cells* (Roswitha Pfagner and R. Ian Freshney, eds), pp. 67–80. Wiley-Liss.
14. Park, J-G., Ku, J-L., Kim, H-S., Park, S-Y. and Rutten, M. J. (2004) In: *Culture of Human Tumor Cells* (Roswitha Pfagner and R. Ian Freshney, eds), pp. 23–66. Wiley-Liss.
15. Wu, R. (2004) In: *Culture of Human Tumor Cells* (Roswitha Pfagner and R. Ian Freshney, eds), pp. 1–21. Wiley-Liss.
16. Barsky, S. H. and Alpaugh, M. L. (2004) In: *Culture of Human Tumor Cells* (Roswitha Pfagner and R. Ian Freshney, eds), pp. 221–260. Wiley-Liss.
17. Freshney, R. I. (1987) *Culture of Animal Cells. A Manual of Basic Technique*, 2nd edn, pp. 122–124. Wiley-Liss.
18. Freshney, R. I. (1987) *Culture of Animal Cells. A Manual of Basic Technique*, 2nd edn, pp. 268–270. Wiley-Liss.
19. Vries, J. E., Bentham, M. and Rumke, P. (1973) Separation of viable from non-viable tumor cells by flotation on a Ficoll-triosil mixture. *Transplantation*, **15**, 409–410.
20. Freshney, R. I. (1987) *Culture of Animal Cells. A Manual of Basic Technique*, 2nd edn. pp. 268–270. Wiley-Liss.
21. Roper, P. R. and Drewinko, B. (1976) *Cancer Res.*, **36**, 2182.
22. Weisenthal, L. M., Lalude, A. O. T. and Miller, J. B. (1983) In vitro chemosensitivity of human bladder cancer. *Cancer*, **51**, 1490–1496.
23. Kaltenbach, J. P., Kaltenbach, M. H. and Lyons, W. B. (1958) Nigrosin as a dye for differentiating live and dead ascites cells. *Exp. Cell Res.*, **15**, 112–117.
24. Wilson, A. P. (2000) In: *Animal Cell Culture*, 3rd edn, (John R. W. Masters, ed.), pp. 185–186. Oxford University Press.
25. Rotman, B. and Papermaster, B. W. (1966) Membrane properties of living mammalian cells as studied by enzymatic hydrolysis of fluorogenic esters. *Proc. Natl Acad. Sci. USA*, **55**, 134–141.

26. Patel, D. (1999) Viable cell counting using ethidium bromide and acridine orange. In: *Multimedia Methods in Cell Biology* (J. R. Harris and D. Rickwood, eds), p. 91. Chapman & Hall, CRCnetBase.

27. Gartler, S. M. (1967) Genetic markers as tracers in cell culture. In: *Second Biennial Review Conference on Cell, Tissue and Organ Culture.* NCI Monographs, pp. 167–195.

28. Nelson-Rees, W. and Flandermeyer, R. R. (1977) Inter- and intraspecies contamination of human breast tumour cell lines HBC and Br Ca5 and other cell cultures. *Science,* **195,** 1343–1344.

29. Hay, R. J. (1992) Cell line preservation and characterisation. In: *Animal Cell Culture* (R. Ian Freshney, ed.), IRL Press, Oxford.

30. Freshney, R. I. (1987) *Animal Cell Culture. A Manual of Basic Techniques,* 2nd edn, pp. 169–185. Wiley-Liss.

31. Macy, M. (1978) Identification of cell line species by isoenzyme analysis. *Man. Am. Tissue Culture Assoc.,* **4,** 833–836.

32. UKCCR (1999) *Guidelines for the Use of Cell Lines in Cancer Research.* Published by UKCCR, PO Box 123, Lincoln's Inn Fields, London WC2A 3PX.

33. Chen, T. R. (1977) In situ detection of mycoplasma contamination in cell cultures by fluorescent Hoechst 33258 staining. *Exp. Cell Res.,* **104,** 255.

34. Wilson, A. P., Garner, C. M. and Hubbold, L. (1999) Hoechst staining for the detection of mycoplasma contamination. In: *Multimedia Methods in Cell Biology* (J. R. Harris and D. Rickwood, eds), p. 182. Chapman & Hall, CRCnetBase.

35. Ashwood-Smith, M. J. and Farrant, J. (1980) *Low Temperature Preservation in Medicine and Biology.* Pitman Medical.

36. Foreman, J. and Pegg, D. E. (1979) Cell preservation in a programmed cooling machine: the effects of variations in supercooling. *Cryobiology,* **16,** 315–321.

37. Green, A. E., Athreya, B., Lehr, H. B. and Coriell, L. L. (1967) Viability of cell cultures following extended preservation in liquid nitrogen. *Proc. Soc. Exp. Biol. Med.,* **124,** 1302–1307.

38. Harris, L. W. and Griffiths, J. B. (1977) Relative effects of cooling and warming rates on mammalian cells during the freeze–thaw cycle. *Cryobiology,* **14,** 662–669.

39. Leibo, S. P. and Mazur, P. (1971) The role of cooling rates in low-temperature preservation. *Cryobiology,* **8,** 447–452.

40. Boyum, A. (1964) *Nature,* **204,** 793–794.

41. Boyum, A. (1968) *Scan. J. Lab. Clin. Invest Suppl.,* **97,** 51–76.

42. Feucht, H. E., Hadan, M. R., Frank, F. and Reithmuller, G. (1980) *J. Immunol Methods,* **38,** 43–51.

43. Ford, T. C. and Rickwood, D. (1982) *Anal. Biochem.* **123,** 293–298.

44. Boyum, A. (1983) In: *Iodinated Density Gradient Media – a Practical Approach* (Rickwood, D., ed.), Oxford University Press, pp. 147–171.

45. Ford, T. C. and Rickwood, D. (1990) *J. Immunol. Methods* **134,** 237–241.

46. Kaden, J., Schönemann, C., Leverenz, S. and Koch, B. (1994) *Allergologie Jahrgang,* **17,** 429–433.

47. Ford, T. C. and Rickwood, D. (1992) *Clin. Chim. Acta,* **206,** 249–252.

48. Ahmed, Y., Walton, L. J. and Graham, J. M. (2004) *12th International Congress of Immunology,* Montreal, Abstr. No. 1758.

4

Isolation and Functional Analysis of Organelles

John Graham

Protocol 4.1	Isolation of nuclei from mammalian liver in an iodixanol gradient (with notes on cultured cells)	98
Protocol 4.2	Isolation of metaphase chromosomes	100
Protocol 4.3	Isolation of the nuclear envelope	102
Protocol 4.4	Nuclear pore complex isolation	104
Protocol 4.5	Preparation of nuclear matrix	106
Protocol 4.6	Preparation of nucleoli	107
Protocol 4.7	Isolation of a heavy mitochondrial fraction from rat liver by differential centrifugation	108
Protocol 4.8	Preparation of a light mitochondrial fraction from tissues and cultured cells	110
Protocol 4.9	Purification of yeast mitochondria in a discontinuous Nycodenz® gradient	112
Protocol 4.10	Purification of mitochondria from mammalian liver or cultured cells in a median-loaded discontinuous Nycodenz® gradient	114
Protocol 4.11	Succinate–INT reductase assay	116
Protocol 4.12	Isolation of lysosomes in a discontinuous Nycodenz® gradient	117
Protocol 4.13	β-Galactosidase (spectrophotometric assay)	119
Protocol 4.14	β-Galactosidase (fluorometric assay)	120
Protocol 4.15	Isolation of mammalian peroxisomes in an iodixanol gradient	121
Protocol 4.16	Catalase assay	123
Protocol 4.17	Analysis of major organelles in a preformed iodixanol gradient	124
Protocol 4.18	Separation of smooth and rough ER in preformed sucrose gradients	127

Cell Biology Protocols.   Edited by J. Robin Harris, John Graham, David Rickwood
© 2006 John Wiley & Sons, Ltd

Protocol 4.19 Separation of smooth and rough ER in a self-generated
                   iodixanol gradient   129
Protocol 4.20 NADPH-cytochrome c reductase assay   131
Protocol 4.21 Glucose-6-phosphatase assay   132
Protocol 4.22 RNA analysis   133
Protocol 4.23 Isolation of Golgi membranes from liver   134
Protocol 4.24 Assay of UDP-galactose galactosyl
                   transferase   136
Protocol 4.25 Purification of human erythrocyte 'ghosts'   137
Protocol 4.26 Isolation of plasma membrane sheets from rat liver   139
Protocol 4.27 Assay for 5′-nucleotidase   141
Protocol 4.28 Assay for alkaline phosphodiesterase   143
Protocol 4.29 Assay for ouabain-sensitive Na^+/K^+-ATPase   144
Protocol 4.30 Isolation of chloroplasts from green leaves or pea
                   seedlings   145
Protocol 4.31 Measurement of chloroplast chlorophyll   147
Protocol 4.32 Assessment of chloroplast integrity   148

Introduction

This chapter is concerned primarily with the purification of the major subcellular membranes and organelles and organelle components from mammalian tissues and cultured cells and from plant tissues; it therefore features nuclei, mitochondria, lysosomes, peroxisomes, Golgi, smooth and rough endoplasmic reticulum (ER), plasma membrane and chloroplasts. The protocols are primarily concerned with the isolation of these organelles on a preparative basis, while Chapter 5 is concerned with analytical gradients for studies on membrane trafficking and cell signalling.

With few exceptions, subcellular organelles cannot be purified satisfactorily from a tissue or cell homogenate by differential centrifugation alone (see below); instead, the pelleted material from differential centrifugation is further purified in a density gradient, usually on the basis of buoyant density but occasionally on the basis of sedimentation velocity.

Homogenization

For most purposes, an isoosmotic medium containing 0.25 M sucrose, 1 mM EDTA, 10 mM Tris-HCl, pH 7.4 may be used for any soft tissue or cultured animal cell. There are a number of minor variations such as adding EGTA instead of, or in addition to, EDTA and substituting Tris by an alternative organic buffer (Hepes or Tricine). Such modifications do not materially affect the result of the fractionation procedure; they are normally introduced to be more compatible with some subsequent add-on procedure or analysis. Others, summarized in Table 4.1, are more critical to the isolation of a particular organelle or use of a particular tissue or cell type. The presence of divalent cations is detrimental to the functioning of mitochondria hence EDTA is regularly included in media for the isolation of these organelles (and also lysosomes and peroxisomes). Mannitol-containing media (sometimes also supplemented with K^+) are also suited to the preservation of oxidative phosphorylation in mitochondria. On the other hand, media omitting any chelating agent but containing K^+ and Mg^{2+} ions are essential for the preservation of nuclear structure. Omission of EDTA is also common in the isolation of rough ER, while inclusion of 0.1% ethanol promotes good preservation of peroxisome structure and function. High K^+ media are also often used if the homogenate tends to be rather gelatinous (e.g. skeletal muscle). For a more detailed commentary on organelle-specific media, see ref. 2.

The tissue or cell type often dictates the homogenization hardware and the homogenization medium (Table 4.1). Most soft tissues can be disrupted using a Potter-Elvehjem (Teflon-glass) homogenizer driven by a high-torque electric motor (at 500–700 rpm). For hard tissue such as skeletal muscle to be homogenized by this device, it is often first softened using protease treatment. This approach is widely used for isolation of organelles such as mitochondria [4, 5], but for the isolation of other membranes such as plasma membrane from muscle, a Polytron™ (rotating blades) homogenizer is more commonly used [12, 13].

Cultured cells grown as a monolayer can often be disrupted using a tight-fitting Dounce homogenizer (all glass) in an isoosmotic medium (see Table 4.1), but a ball-bearing homogenizer (cell-cracker) or several passages through a 25-gauge syringe needle are frequently used, often more effective, alternatives for any type of cultured cell. The ball-bearing homogenizer is recognized as not only effective in disrupting these cells, but at the same time delicate organelles such as lysosomes seem well preserved [14]. For more information on these devices, see ref. 2.

Table 4.1 Homogenization media

Homogenization medium	Application	Refs
0.2 M mannitol, 50 mM sucrose, 1 mM EDTA, 10 mM, Hepes-NaOH, pH 7.4	Liver mitochondria	1, 2
100 mM KCl, 20 mM MOPS, 5 mM $MgSO_4$, 1 mM ATP, 5 mM EDTA, 0.2% bovine serum albumin	Muscle mitochondria	3
0.1 M sucrose, 10 mM EDTA, 46 mM KCl, 100 mM Tris-HCl, pH 7.4	Muscle mitochondria	4
0.21 M mannitol, 70 mM sucrose, 10 mM EDTA, 10 mM Tris-HCl, pH 7.4, containing 2.4 mg trypsin/100 ml	Muscle mitochondria	5
0.25 M sucrose, 1 mM EDTA, 0.1% ethanol, 10 mM Tris-HCl, pH 7.5	Peroxisomes	6
0.25 M sucrose, 25 mM KCl, 5 mM $MgCl_2$, 20 mM Tris-HCl, pH 7.4	Nuclei	7
0.32 M sucrose, 1 mM EDTA, 10 mM Hepes-NaOH, pH 7.4	Brain mitochondria, synaptosomes	2
0.25 M sucrose, 1 mM EDTA, 10 mM triethanolamine-acetic acid, pH 7.6	Monolayer cells	8
15 mM KCl, 1.5 mM Mg(OAc), 1 mM DTT, 10 mM Hepes-KOH, pH 7.5, adjusted to 75 mM KCl, 5 mM Mg(OAc), 1 mM DTT, 45 mM Hepes-KOH, pH 7.5 after homogenization	Cultured cells	9, 10
1.5 mM KCl, 5 mM $MgCl_2$, 5 mM Tris-HCl, pH 7.4, adjusted to 1 mM EGTA after swelling but before homogenization	Lymphocytes	11

Some suspension culture cells, particularly those containing a very high nucleus : cytoplasm volume ratio, such as lymphocytes or Jurkat cells, are often impossible to disrupt by any means, unless they are first swollen in a frankly hypoosmotic medium (see Table 4.1). This means that the medium may not be ideally suited to organelle isolation. The method that employs a hypoosmotic magnesium acetate solution (Table 4.1) incorporates a strategy for minimizing possible deleterious effects of the medium. After the cells have been swollen in a large volume of the buffer, they are pelleted and most of the liquid removed prior to homogenization [9, 10]; in this way the isoosmotic cytosol is minimally diluted, and so affords some protection to the released organelles. Further protection is provided by immediate addition of a hyperosmotic medium to return the homogenate to an isoosmotic condition. This approach might be used with any method using a hypoosmotic medium. Frequently, however, the less satisfactory option of pelleting the organelle prior to suspension in a suitable isoosmotic medium is chosen.

Differential centrifugation

Differential centrifugation [15] is used to prepare crude fractions of subcellular particles from homogenates by imposing a series of centrifugations at sequentially increasing g-forces and/or centrifugation times, the supernatant from each step being passed to the next round of centrifugation. It is often a necessary first step subsequent to a purification of one or more organelles on a density gradient. Routinely it involves four centrifugations and it separates particles mainly by size (sedimentation velocity):

1. $500-1000g$ for $5-10$ min (nuclear pellet)
2. $3000g$ for 10 min (heavy mitochondrial pellet)
3. $12\,000-15\,000g$ for $10-20$ min (light mitochondrial pellet)
4. $100\,000g$ for $45-60$ min (microsomal pellet)

There are two problems with differential centrifugation: poor recoveries and contamination of pellets by smaller particles. In step 1 for example, the nuclei at the top of the sample are exposed, initially at least, to the lowest g-force and have the furthest to travel, and they will therefore not become part of the stable pellet during the centrifugation. On the other hand, smaller particles (e.g. mitochondria), initially close to the bottom of the tube, which have the shortest distance to travel and are exposed to the highest g-forces, will become part of the pellet. These problems are most severe in swinging-bucket rotors because of their long sedimentation path lengths and consequent large differences between g_{min} and g_{max}. Fixed-angle rotors, of the same tube volume, have much shorter sedimentation path lengths and are the recommended ones for differential centrifugation.

The contamination of a pellet by smaller particles can be reduced by 'washing'. This involves resuspending the pellet in the homogenization buffer before repeating the centrifugation and combining the two supernatants for the next round of centrifugation. Sometimes the washing is repeated. The procedure of washing has its own attendant problems: unless the resuspension of the pellet is carried out very gently (preferably in a loose-fitting Dounce homogenizer, using very few slow strokes of the pestle), the organelles in the pellet themselves can be further disrupted. It is also time consuming.

Nevertheless there are a few cases where differential centrifugation is the only purification procedure; the preparation of the so-called 'heavy mitochondrial fraction' from rat liver or beef heart is one of them (see *Protocol 4.7*). In this case, the ability of differential centrifugation to produce a mitochondrial fraction of acceptable purity, rapidly and under isoosmotic conditions in a 'mitochondrial-friendly medium', is essential to the retention of coupled oxidative phosphorylation. The use of a density gradient might provide additional purity, but the additional time and the stress imposed on the organelles by the increased hydrostatic pressure associated with higher g-forces needed for density gradient centrifugation, can cause a serious loss of functional integrity.

Differential centrifugation is also important simply because it does separate particles mainly by size. Buoyant density gradient purification of peroxisomes or lysosomes for example is almost invariably carried out on a light mitochondrial fraction (see *Protocol 4.8*) to eliminate smaller particles that may have similar densities. For the same reason intact stacks of Golgi membranes are similarly purified from a $5000g$ fraction of a liver homogenate (see *Protocol 4.23*). On the other hand, unless they are first removed, large rapidly sedimenting particles in homogenates may disturb shallow gradients designed to fractionate smaller low-density particles.

Equipment

With the exception of the microsomal pellet, all differential centrifugation can be executed in a refrigerated high-speed centrifuge. To minimize the contamination of pellets by smaller particles (see above) fixed-angle rotors are preferred to swinging-bucket rotors (of the same tube volume) because of their shorter sedimentation path length and the smaller difference in g-force between the meniscus and the bottom of the tube [15, 16]. Clearly the smaller the angle of the rotor (to the vertical) the more efficient the separation, but the other effect of a smaller angle – the tendency of the pellet to be less compact and well-defined is a disadvantage. In practice, rotors with angles of approx 30° represent a happy medium. The only exception to this preference for a fixed-angle rotor is the formation of the nuclear pellet (step 1); swinging-bucket rotors tend to be used for this step simply because it has routinely been carried out in low-speed centrifuges, which are predominantly fitted with such rotors. Moreover, the difference in sedimentation rate between the nuclei and the next largest organelles (mitochondria) is so large, that rotor type is less important.

Density gradient centrifugation

It is now widely recognized, particularly for some of the larger organelles (nuclei, mitochondria, lysosomes, peroxisomes), which can be pelleted at relatively low g-forces ($1000-15\,000g$), that the use of viscous, hyperosmotic sucrose gradients at g-forces of more than $100\,000g$ in long sedimentation path length swinging-bucket rotors, is both inconvenient and potentially deleterious to the organelles. Consequently, there is a pronounced trend to the use of low-viscosity gradient media such as Nycodenz® or iodixanol [17], whose gradients have a much lower osmolality than those of sucrose (or even be isoosmotic). Although the traditional swinging-bucket rotor is still widely used for density gradient centrifugation, the short sedimentation path length of vertical or near-vertical rotors not only reduces the centrifugation time, it also minimizes considerably the hydrostatic pressure imposed on the particles in the gradients. The new protocols in this chapter reflect these trends and concerns. The means by which discontinuous and continuous gradients are prepared and how they are harvested is too large a subject to be considered in this text; ref. 18 contains detailed accounts of these procedures.

Some protocols take the advantage of the ability of some of the modern density gradient media to form self-generated gradients. The colloidal silica particles in Percoll® can form such gradients at g-forces above 20 000g, while other gradient media such as iodixanol, which is a true solute, requires g-forces in excess of 180 000g. Because some of the colloidal silica particles will pellet during the centrifugation, vertical rotors (normally the most efficient rotors for such gradients) are not recommended for self-generated Percoll® gradients – fixed-angle rotors are more commonly used.

A disadvantage of Percoll® gradients is that it is normally a requirement to remove the colloidal silica particles prior to any analysis; the particles are light scattering and can interfere with any spectrophotometric measurements; they also cause irregular banding of proteins in SDS-PAGE. Moreover, removal of the silica colloid by sedimentation can also lead to the loss of organelles into the gel pellet that is formed [19]. With the exception of spectrophotometric assays in the UV, if the organelle is at a sufficiently high concentration to permit direct analysis on the gradient harvest, then it is usually unnecessary to remove solutes such as sucrose, Nycodenz® or iodixanol. If solute removal is required then the organelles can simply be sedimented from the gradient harvest after dilution with two volumes of buffer, without loss of material [20].

Nuclei and nuclear components (see *Protocols 4.1–4.6*)

Nuclei

Detergent-free methods for the purification of nuclei have routinely involved the pelleting of the organelles from a crude nuclear pellet suspension, through a dense (approx 2 M) sucrose barrier. The nuclei lose water to the grossly hyperosmotic medium and attain a density >1.32 g/ml. Pelleting the nuclei is thus the only option since this is the limiting density of sucrose solutions. The high viscosity of the solutions also requires a g-force, normally of 100 000g or higher, for at least 1 h. It is possible that DNA–protein interactions become destabilized by the loss of water and the high hydrostatic pressure at the bottom of the tube. On the other hand, in a discontinuous gradient of iodixanol (centrifuged at 10 000g for 20 min), which is close to being isoosmotic, the nuclei exhibit a much lower density (approx 1.23 g/ml) and can therefore band at an interface. The method, developed for mammalian liver [20–23] has been used for other tissues such as brain [24–26], a wide variety of cultured cells including CHO cells [27], HeLa cells [28], HFK cells [29], human fibroblasts [30–32], human carcinoma cells [33] and non-mammalian sources such as mussel [34] and wheatgerm [35].

Nuclear components

Isolation of the nuclear envelope poses a problem; should the relatively intact nuclear envelope 'ghost' or the totally fragmented nuclear envelope be considered the desired product? In the former situation, both inner and outer nuclear membranes and their interconnecting pore complexes are likely to be present, together with the nuclear lamina and perhaps some residual heterochromatin. In the second situation, fragmentation of the nuclear envelope, for instance by high salt extraction or heparin treatment, may separate the two nuclear membranes and produce damage to the nuclear pore complexes, but contamination with residual heterochromatin is likely to be less. For nuclear envelope isolation, it is usual to decondense the chromatin at low ionic strength or slightly raised pH in the absence of divalent cations prior to deoxyribonuclease (DNase) digestion, whereas for the production of the nuclear matrix, the DNase digestion is performed on condensed chromatin within which nucleoli are retained. The nuclear matrix elements are apparently destroyed under the low ionic strength decondensation and DNase digestion conditions and are released from the nuclear envelope. Nevertheless, these two apparently contradictory approaches enable different subnuclear fractions to be produced. It is generally assumed that the outer nuclear membrane bears considerable similarity to rough ER. Studies on the isolated nucleolus have been overshadowed by other investigations in recent years, but the isolation of this subnuclear organelle is undoubtedly required before detailed macromolecular information can be obtained to complement the vast amount of data from *in situ* antibody labelling studies on the nucleolus.

Mitochondria (see *Protocols 4.7, 4.9* and *4.10*)

Common sources of mitochondria for respiratory studies are rat liver and beef heart. In these studies, yield of organelle is not the primary consideration, which is, instead, that the method should be as rapid and gentle as possible. Preparation of a 'heavy fraction' of mitochondria using a favourable medium and a g-force of only $3000g$ conform to these requirements (*Protocol 4.7*). A more complete recovery of the mitochondria from a 'heavy + light mitochondrial' fraction can be obtained by density gradient centrifugation. Because of the high osmolality of sucrose gradients, mitochondria, lysosomes and peroxisomes have density ranges that overlap considerably. Isoosmotic gradients of both Percoll® [36], Nycodenz® [37] and iodixanol [20] provide far superior resolution. The density of mitochondria in Nycodenz® is slightly higher than in iodixanol, thus improving their resolution from the lighter lysosomes. As yeast mitochondria are now widely used in many functional and structural studies, *Protocol 4.9* describes the widely used purification of these mitochondria in Nycodenz® gradients [38]. Often resolution has been improved by flotation from a bottom-loaded discontinuous gradient and such a method has been introduced for yeast mitochondria [39]. It has been observed with mammalian mitochondria, however, that by placing the crude mitochondrial fraction at the bottom of the gradient, the high hydrostatic pressure can cause some partial disruption of the mitochondrial structure [40], hence the development of the median loading strategy described in *Protocol 4.10* [41, 42]. See also *Protocol 4.17*, which describes gradients for purifying all the major organelles from a light mitochondrial fraction.

Enzyme markers

Any easily measurable component of the tricarboxylic acid cycle can be used as a functional marker, the most frequently used being succinate dehydrogenase with either the native electron acceptor (cytochrome *c*) or an artificial one, such as *p*-iodonitrotetrazolium violet (INT). The latter is certainly the easier to use; reduction of cytochrome *c* requires the use of cyanide to inhibit reoxidation by oxygen.

Lysosomes

Sucrose gradients only provide optimal separation of rat liver mitochondria and lysosomes if the density of the latter is perturbed by loading with Triton WR1339 [43]. In both Percoll® [44] and iodinated density gradients [20, 37] the density of lysosomes is sufficiently distinct that their isolation is relatively easy. Note, however, that in Percoll®, lysosomes are denser than mitochondria, while in iodinated density gradient media, they are lighter. A simple flotation through a discontinuous gradient of metrizamide, originally described by Wattiaux and Wattiaux-De Coninck [39], has more recently been adapted to Nycodenz® by Olsson *et al.* [45] as metrizamide is no longer available commercially. Although Percoll® is popular for the purification of lysosomes, the higher relative enrichment of lysosomes enzyme markers in the Nycodenz® prepared organelles makes this the method of choice (see *Protocol 4.12*). Because, in the discontinuous gradient, the lysosomes band at the top interface, the harvesting of the organelles is very easy. See also *Protocol 4.17*, which describes gradients for purifying all the major organelles from a light mitochondrial fraction.

Enzyme markers

Acid phosphatase [46] is the classical marker for lysosomes. Although very easy to assay, the presence of this enzyme in the cytosol makes this marker less than satisfactory. Use of one of the lysosomal enzymes associated with the breakdown of glycosylated molecules such as β-galactosidase or β-*N*-acetylglucosaminidase has become more popular. For most of these enzymes, the *p*-nitrophenol derivative is commercially available, making spectrophotometric assay by measurement of the released nitrophenol very easy (*Protocol 4.13*). An alternative fluorometric assay, which is rather more sensitive, is also given (*Protocol 4.14*).

Peroxisomes

Except in Percoll®, peroxisomes tend to be the densest of the organelles in the light mitochondrial fraction in any density gradient medium since they do not have an enclosed osmotic space as the limiting membrane is permeable to small molecules. Thus in sucrose, Nycodenz® and iodixanol peroxisomes attain a density that is essentially that of the membrane itself, but it is only in the iodinated density gradient media that they are easily resolved from both mitochondria and endoplasmic reticulum, which are less dense in Nycodenz® and iodixanol (particularly in the latter), than in sucrose [20, 37]. In Percoll®, on the other hand, although peroxisomes are distinctively less dense than the other organelles of the light mitochondrial fraction, they co-band with the endoplasmic reticulum [47]. Iodixanol gradients are now therefore the gradients of choice for the isolation of mammalian peroxisomes (*Protocol 4.15*). However, because Nycodenz®

has been available for a much longer time than iodixanol, a huge literature on the use of these gradients for the purification of both mammalian and yeast peroxisomes exists. Some of these methods are summarized in the extensive Notes section in *Protocol 4.15*. See also *Protocol 4.17*, which describes gradients for purifying all the major organelles from a mammalian light mitochondrial fraction.

Enzyme markers

Peroxisomes contain many enzymes associated with lipid metabolism but the most commonly used marker is catalase. Catalase oxidizes H_2O_2 to oxygen and water; it is routinely assayed by back titration of the residual H_2O_2, which forms a stable yellow product with titanium oxysulfate (see *Protocol 4.16*). This method is easier to execute than the alternative one using potassium permanganate.

Rough and smooth endoplasmic reticulum (ER)

ER is often analysed along with other membrane compartments such as Golgi, endosomes and the *trans*-Golgi network (TGN) during studies into membrane trafficking and cell signalling (see Chapter 5). *Protocols 4.18* and *4.19* in this chapter, by contrast, are concerned primarily with the bulk (preparative) isolation of these membranes.

The microsome fraction (see 'Differential centrifugation' above) will contain, in addition to the membrane vesicles derived from the smooth and rough ER, contaminating organelles that failed to sediment at $15\,000g$, together with core structures from peroxisomes, which have lysed during the previous manipulations. The smooth vesicle fraction will also contain vesicles derived from other sources such as the Golgi and TGN. Rough ER and smooth ER can be effectively separated in a continuous sucrose gradient [48] as described in *Protocol 4.18*. To resolve the rough ER from the peroxisomal cores (and other contaminating organelles), the density of the RER vesicles is reduced by stripping off the ribosomes with pyrophosphate and they are purified on a second continuous sucrose density gradient [48]. As both the SER and RER band broadly, the method may offer some scope for subfractionation of both membranes. *Protocol 4.19* offers an alternative method for separating these membranes in a self-generated iodixanol gradient.

Enzyme and chemical markers

Commonly used markers for the endoplasmic reticulum were worked out for rat liver, notably NADPH-cytochrome *c* reductase (*Protocol 4.20*) and glucose-6-phosphatase (*Protocol 4.21*). Kidney is the only other tissue in which glucose-6-phosphatase has a clear functional role, yet there are a number of instances where it has been used as a marker for the endoplasmic reticulum from cultured cells, even though there is no clear reason why it should be present in such cells. Glucose-6-phosphatase is determined by the measurement of the released P_i using one of many of the standard protocols using molybdate [49]. NADPH cytochrome c reductase is determined by measuring the production of reduced cytochrome c [50]. Inclusion of rotenone in the assay avoids any interference from mitochondrial electron transport. It is normal to use a recording spectrophotometer to monitor the rate of increase of A_{550} over a short period of about 5 min.

Another enzyme which is associated with the early processing of the oligosaccharide chains of glycoproteins in the endoplasmic reticulum is ß-D glucosidase which is sometimes used as an endoplasmic reticulum marker, particularly in cultured cells [51]. A simple chemical assay for RNA (rough ER) is given in *Protocol 4.22*.

Golgi membranes

Golgi membranes are often analysed along with other membrane compartments such as ER, ERGIC (the ER-Golgi intermediate compartment) and endosomes and during studies into membrane trafficking and cell signalling (see Chapter 5). *Protocol 4.23.* in this chapter, by contrast, is concerned primarily with the bulk (preparative) isolation of these membranes from rat liver; it uses a medium containing dextran to promote the retention of Golgi stacks during homogenization. The intact Golgi stacks from mammalian liver can be sedimented at 5000g and subsequently purified by banding at a 1.2 M sucrose barrier [52]. Subsequently the Golgi can be "unstacked" by hydrolysing the dextran with a mixture of amylases.

Enzyme marker

Galactosyl transferase is the most widely used marker for the Golgi membranes but it is strictly only a marker for the *trans*-Golgi where it is involved in the terminal galactosylation of oligosaccharide chains of glycoproteins. All methods use UDP-galactose radiolabelled in the sugar moiety as the donor; the methods vary in the type of acceptor. One method uses N-acetyl-glucosamine as an artificial acceptor and the radiolabelled product, N-acetyllactosamine, must be separated from the UDP-galactose by passing through a small Dowex ion exchange resin [53] - this is a simple technique but very tedious with large numbers of samples. The method described in *Protocol 24* uses ovalbumin as the acceptor; the radiolabelled product from the incubation is simply TCA-precipitated onto a filter disc, washed, dried and counted [54].

Plasma membrane (see *Protocols 4.25* and *4.26*)

As with ER and Golgi, the plasma membrane is often analysed along with other membrane compartments such as endosomes and the *trans*-Golgi network (TGN) during studies into membrane trafficking and cell signalling (see Chapter 5). *Protocols 4.25 and 4.26* in this chapter, by contrast, are concerned primarily with the bulk (preparative) isolation of these membranes. There are two principal sources for bulk isolation of mammalian plasma membrane that have been used for studies of membrane composition, function and architecture: the human erythrocyte and rat liver. The human erythrocyte membrane in particular is easy to isolate by hypotonic lysis [55] and the lack of internal organelles means that density gradient fractionation is not required (*Protocol 4.25*). Bacterial limiting membranes also pose relatively few problems [56]. Structures such as desmosomes, tight junctions or an extensive cytoskeleton, effectively stabilize certain plasma membrane domains from polarized tissues against fragmentation during homogenization. Thus, the contiguous membrane of rat liver or the brush border of intestinal epithelial cells may also be isolated relatively easily, in the form of sheets, which are rapidly sedimenting and easily separated from most other organelles and membrane vesicles. *Protocol 4.26* describes the isolation of the contiguous membrane from rat liver [57].

Plasma membrane analysis using enzyme markers (Protocols 4.27–4.29)

Routine functional analysis of plasma membrane relies on enzyme assays, principally for Na^+/K^+-ATPase, $5'$-nucleotidase, alkaline phosphatase or alkaline phosphodiesterase. There is considerable variation in levels of these enzymes in plasma membranes from different sources and in the case of cells from polarized tissues; they are often domain-specific. Thus, the Na^+/K^+-ATPase is readily measured in the human erythrocyte membrane, much less easily measured in the plasma membrane from cultured cells and considerably enriched in the basolateral domains (compared to apical domains) from polarized cells. Alkaline phosphatase, alkaline phosphodiesterase and $5'$-nucleotidase are ubiquitous, highly enriched in apical domains and easier to measure than the Na^+/K^+-ATPase.

Chloroplasts (see *Protocol 4.30*)

There are two approaches to the isolation of chloroplasts: (1) by mechanical disruption of green leaves or (2) more gentle disruption of protoplasts prepared from the plant tissue [58]. Due to the fragility of chloroplasts, the former is restricted to soft green leaves, such as those of spinach. Tissues rich in oxalate, phenolics or starch cannot be used. The advantage of the mechanical shear method (*Protocol 4.30*) is that large amounts of material can be processed quickly, but for the best preparations, it is necessary to remove partially disrupted chloroplasts using a Percoll® gradient. The protoplast method can be used with a wide range of tissues, but it is lengthy, expensive and inconvenient for large amounts of chloroplasts.

Chloroplast functional assays

To determine the integrity of the chloroplast preparation it is normal to perform two assays: measurement of the chlorophyll content (*Protocol 4.31*) and the rate of oxygen evolution in the presence of ferricyanide and NH_4Cl (to uncouple electron flow) before and after hypoosmotic disruption of the chloroplasts (*Protocol 4.32*).

Isolation of nuclei from mammalian liver in an iodixanol gradient (with notes on cultured cells)

Reagents

OptiPrep™ (60%, w/v, iodixanol, $\rho = 1.32$ g/ml)

OptiPrep™ Diluent (OD): 150 mM KCl, 30 mM MgCl$_2$, 120 mM Tricine-KOH, pH 7.8

Homogenization medium (HM): 0.25 M sucrose, 25 mM KCl, 5 mM MgCl$_2$, 20 mM Tricine-KOH, pH 7.8. ①

Add protease inhibitors to the OptiPrep™ diluent and HM as required.

Equipment

High-torque overhead electric motor (thyristor-controlled)

Potter-Elvehjem homogenizer, 20–40 ml, clearance ~0.07 mm ②

High-speed centrifuge with swinging-bucket rotor with approx. 15 or 50 ml tubes

Syringe with metal cannula (internal diameter approx. 0.8 mm) ③

Nylon gauze

Procedure [20]

Carry out all operations at 0–4 °C.

1. Prepare a 50% (w/v) iodixanol working solution: dilute 5 vol. of OptiPrep™ with 1 vol. of OD.

2. Gradient solutions: Prepare two gradient solutions of 30 and 35% (w/v) iodixanol by diluting 50% (w/v) iodixanol working solution with HM (6 vol. + 4 vol. and 7 vol. + 3 vol. respectively). Keep on ice.

3. Excise the liver and chop with scissors to produce a fine mince.

4. Transfer approx. half of the liver to the Potter-Elvehjem homogenizer in 20 ml of HM and disrupt the tissue using approx. eight strokes of the pestle rotating at approx. 500 rpm. ④

5. Repeat with the other half of the liver mince.

6. Filter through the nylon gauze.

7. Adjust the homogenate to 25% (w/v) iodixanol by mixing with and equal volume of the 50% iodixanol working solution. ⑤

8. Transfer 10–15 ml of the homogenate to a 50 ml tube and underlayer with 10 ml each of the 30 and 35% iodixanol gradient solutions.

9. Centrifuge at $10\,000g_{av}$ for 20 min. ⑥

10. The nuclei band at the 30/35% iodixanol interface, using a syringe and metal cannula aspirate the nuclei.

11. Dilute the suspension with two volumes of HM and pellet the nuclei at $2000g$ for 10 min.

Notes

This procedure will take about 1 h.

① This homogenization medium has probably a wide application to mammalian tissues and to a few cultured cells, e.g. X57 clonal striatal cells [59], but it is probably unsuitable for many cultured cells. Using an isoosmotic sucrose-containing medium, such cells can often only be disrupted in the presence of EDTA, which is not ideally suited to the recovery of intact nuclei. In such cases, the use of a divalent cation-containing hypoosmotic medium may be preferred. Some examples are: 10 mM $MgCl_2$, 1 mM KCl, 5 mM Hepes-NaOH, pH 7.4 for CHO cells [27], 1.5 mM $MgCl_2$, 10 mM KCl, 10 mM Hepes-NaOH, pH 7.6 for human fibroblasts and endothelial cells [30] and 2.5 mM $MgCl_2$, 1.5 mM KCl, 5 mM Tris-HCl, pH 7.4 for lymphocytes [11].

② If cultured cells are being used rather than a soft tissue such as liver, a Dounce (tight-fitting, Wheaton type A) homogenizer, ball-bearing homogenizer or syringe fitted with a 23 or 25 gauge needle should be substituted. The optimal conditions for the homogenization of cultured cells vary with the type of cell; different media and homogenization devices will have to be tested for their suitability. In the authors' experience, nitrogen cavitation, although very effective for most cell types, tends to render the nuclei rather fragile. The aim is to obtain at least 90% cell rupture while maintaining nuclear integrity (as judged by phase contrast microscopy).

③ Flat-tipped metal filling cannulas can be obtained from most surgical equipment supply companies.

④ Optimal homogenization conditions need to be worked out for any particular cell type. The number of passes of the pestle of a Dounce homogenizer should not exceed 20 and either 4–5 passes through the ball-bearing homogenizer or 12 passes through a syringe needle should be sufficient.

⑤ In the case of cultured cells which have been homogenized in a hypoosmotic medium, it is common to pellet the nuclei at $800–2000g$ for 5–10 min before resuspending in the liver homogenization medium. Where possible it may acceptable instead to adjust the composition of cultured cell homogenate to that of the liver homogenization medium.

⑥ It is rarely necessary to adjust the centrifugation parameters. The efficacy of the separation depends more on the rapid rate of sedimentation rather than on their buoyant density and smaller nuclei may require an increase in centrifugation time to 30 or 40 min. Only for mussel nuclei [34] were the g-force and time increased significantly ($100\,000g$ for 1 h).

Isolation of metaphase chromosomes [60]

Reagents

Colcemid

Hexylene glycol buffer: 1 M hexylene glycol, 0.5 mM CaCl$_2$, 0.1 mM Pipes-NaOH, pH 6.8

Gradient solutions: 28, 58 and 60% (w/v) Nycodenz® in hexylene glycol buffer ①

Dimethyldichlorosilane (1% in carbon tetrachloride)

Equipment

Low and high-speed centrifuge

High-speed centrifuge with swinging-bucket rotor for 15 ml tubes

Corex tubes (15 ml) for swinging-bucket rotor

Syringe with 23-gauge needle

Phase contrast microscope

Two-chamber gradient maker or Gradient Master™ ②

Procedure

1. Treat the Corex tubes with dimethyldichlorosilane for 5 min; then dry and bake at 60 °C for 24 h.

2. Arrest the cultured cells in metaphase by the addition of 0.06 μg/ml of colcemid to the culture medium and incubate for 3 h.

3. Selectively detach the metaphase cells by gentle shaking and cool the detached cells for 20 min at 4 °C prior to pelleting them by centrifugation at 300g for 5 min.

4. Wash the cells by suspending them in the hexylene glycol buffer and pellet them as in step 3.

5. Resuspend the cells in the same buffer (cell concentration <2 mg/ml) and incubate them for 10 min at 37 °C before lysing the cells by passing the cell suspension several times through a 23-gauge needle. Monitor cell lysis by phase contrast microscopy. ③

6. Pellet the chromosomes by centrifugation at 2000g for 10 min.

7. Make a 10 ml gradient in the treated Corex tubes from equal volumes of 28 and 58% Nycodenz® and underlayer with the chromosome pellet suspended in 60% Nycodenz®.

8. Centrifuge at 16 300g for 10 min [60]. The chromosomes band at approx. 1.23 g/ml, approx two-thirds way down the gradient.

9. The metaphase chromosomes can be stained by one of the banding methods described in the literature [61].

Notes

The procedure will takes about 5 h, including the incubation with colcemid.

① Wray [60] used metrizamide, which is no longer commercially available.

Nycodenz®, which is a very similar iodinated density gradient solute, has been substituted.

② If neither of these devices is available, make a discontinuous gradient from 28, 38, 48 and 58% Nycodenz® and allow it to diffuse.

③ Other homogenization techniques are permissible, for example nitrogen cavitation at 250 psi (17 bar).

Isolation of the nuclear envelope

J. Robin Harris

Reagents

Deoxyribonuclease I

KCl (1 M in 20 mM Tris-HCl, pH 7.4)

MgCl$_2$ (1 M stock solution)

PMSF (1 M stock solution in dried ethanol)

Purified nuclei (see *Protocol 4.1*)

Buffer: 2 mM Tris-HCl, pH 7.4

Envelope suspension buffer (ESB): 0.25 sucrose, 2 mM Tris-HCl, pH 7.4

Gradient solutions: 1.0 M, 1.5 M, 1.8 M, 2.0 M sucrose in 2 mM Tris-HCl, pH 7.4

All solutions (except the MgCl$_2$) should contain 0.1 mM PMSF

Equipment

Abbé refractometer

High-speed centrifuge with fixed-angle rotor and tubes

Ultracentrifuge with fixed-angle and swinging-bucket rotors (and tubes)

Phase contrast microscope

UV spectrophotometer (silica cuvettes)

Procedure [62–65]

Except where indicated carry out all operations at 0–4 °C.

1. Swell/decondense nuclear chromatin by incubation of purified nuclei in buffer for 10 min, followed by centrifugation at $\sim$10 000g for 10 min. Resuspend the pellet in buffer and repeat twice. ①

2. Resuspend the gelatinous pellet of swollen nuclei in buffer containing deoxyribonuclease I (10 μg/ml). Incubate at room temperature for 30 min with mixing. Halfway through the incubation adjust the suspension to $\sim$0.1 mM MgCl$_2$. The gelatinous nature of the chromatin should be rapidly lost as DNA cleavage occurs. Monitor the production of nuclear 'ghosts' by phase contrast microscopy. ②

3. Centrifuge at $\sim$60 000g for 30 min in a fixed-angle or swinging-bucket rotor.

4. Resuspend the pellet of crude nuclear envelopes in buffer and centrifuge again. Repeat this step several times until the release of DNA, monitored by the absorbance of the supernatant at 280 nm, has reached a low plateau. ③

5. Optionally, wash the nuclear envelope with the KCl solution to assist the removal of ionically bound proteins, histones in particular, followed by centrifugation at $\sim$60 000g for 30 min. ④

6. Resuspend the washed nuclear envelope in ESB and overlay on a step gradient made from the four gradient solutions, in a swinging-bucket rotor. Centrifuge for several hours (or overnight) to band the nuclear envelope isopycnically. ⑤ ⑥

7. The purified envelopes should band predominantly at the 1.5 M/1.8 M sucrose interface. Carefully remove the nuclear envelope layer from the gradient using a Pasteur pipette.

8. Monitor the integrity of the nuclear envelope by phase contrast microscopy and by negative staining and/or thin section electron microscopy using the agarose embedment procedure (*Protocol 2.8*). ⑦

9. Perform the required biochemical analysis on the purified preparation, for example SDS-PAGE and enzymology; endogenous peroxidase is a reasonable marker enzyme for the nuclear envelope from rat liver, Mg-ATPase and many ER enzymes are also present [66].

Notes

This procedure will take approximately 1 working day.

① Removal of divalent cations causes chromatin and nucleolar decondensation. This is necessary to enable a reasonably complete removal of chromatin from the nuclear envelope to be achieved.

② A trace amount of Mg^{2+} is required for DNase I activity; addition of $MgCl_2$ ensures a more complete digestion of DNA.

③ The partially purified nuclear envelope obtained at this stage may be sufficient for many requirements.

④ High-salt washing usually causes considerable disruption of nuclear envelopes/nuclear 'ghosts'. This may be acceptable for many investigations.

⑤ Use the refractometer to produce accurate sucrose molarities.

⑥ If a considerable quantity of heterochromatin remains bound to the nuclear envelope, the membrane may appear at the 1.8 M/2.0 M sucrose interface.

⑦ Only electron microscopy can provide evidence for the presence of inner and outer nuclear membranes, absence of residual heterochromatin and nuclear matrix and nuclear pore integrity.

PROTOCOL 4.4

Nuclear pore complex isolation

J. Robin Harris

Reagents

Heparin

PMSF (1 M stock solution in ethanol)

Purified nuclear envelope (*Protocol 4.3*)

Extraction buffer (EB): 1% Triton X-100 in 10 mM Tris-HCl, pH 7.4, 0.1 mM PMSF

Tris solutions: 2 mM Tris-HCl, pH 7.4, 10 mM Tris-HCl, pH 7.4

Gradient solutions: 0.25 M sucrose and 1.5 M sucrose in 10 mM Tris-HCl, pH 7.4

Adjust all solutions to 0.1 mM PMSF.

Equipment

Gradient collector ①

High-speed centrifuge with fixed-angle rotor and tubes

Two-chamber gradient maker or Gradient Master™

Ultracentrifuge with swinging-bucket rotor and tubes

Ultrasonicator (probe-type, with timer)

Procedure [67]

1. Extract the nuclear envelope for 30 min at 4 °C with EB. ②

2. Centrifuge at 20 000g for 30 min to pellet loosely the nuclear pore complex-lamina [68]. Resuspend the pellet gently in 2 mM Tris-HCl (pH 7.4) and repeat the centrifugation.

3. Resuspend the pellet in a small volume (i.e. 1 ml) of 10 mM Tris-HCl (pH 7.4) and ultrasonicate, with the sample in an ice-bath. Repeated short treatments (e.g. 10 s) are desirable, followed by cooling periods, because of sample heating. ③

4. Monitor the disruption of the nuclear envelope/nuclear pore complex release by preparing negatively stained specimens, e.g. using 2% ammonium molybdate (pH 7.0) or 5% ammonium molybdate-0.1% trehalose (pH 7.0) for EM study (see *Protocol 2.2*) with immediate assessment and continuation of ultrasonication, if required.

5. Layer the sample onto a 0.25–1.5 M sucrose-10 mM Tris-HCl (pH 7.4) linear density gradient and centrifuge at ~60 000g for 1–2 h at 4 °C. ④

6. Collect fractions from the sucrose gradient. ⑤

7. Monitor the content of the fractions by EM of negatively stained specimens to locate the position on the sucrose gradient to which the nuclear pore complexes have sedimented, with assessment of their integrity.

8. Perform SDS-PAGE of the gradient fractions to monitor the protein separation achieved. ⑥

Notes

This procedure will take approximately 4 h.

① The Beckman Fraction Recovery System is a commonly used device.

② The preparation of nuclear pore complexes can be performed with intact nuclear envelope, but some membranes will remain attached to the pore complexes.

③ Ultrasonication can damage proteins, but may leave some lamins attached to the nuclear pore complexes. Treatment of yeast nuclear pore complex lamina with heparin has also been found to successfully separate the pore complex [69], but heparin has been found to cause structural damage to nuclear pore complexes in other systems.

④ The separation is by sedimentation velocity but the centrifugation time is not critical. Ideally, the nuclear pore complexes should sediment approx. two-thirds of the way down the centrifuge tube. If the centrifugation is performed using a total nuclear envelope ultrasonicate, membrane fragments derived from the inner and outer nuclear membranes can be recovered from the gradient. This may present the possibility of assessing the two nuclear membranes individually, by isopycnic centrifugation.

⑤ The Beckman Fraction Recovery System can be used to collect fractions dense end first by tube puncture or low density end first by upward displacement with a dense medium or (if this device is not available) by very careful pipetting directly from the top of the tube.

⑥ Supernatants and pellets can be monitored for DNA, RNA and protein (SDS-PAGE) at all stages of the procedure, if desired.

PROTOCOL 4.5

Preparation of nuclear matrix [68–70]

Reagents

Ammonium sulphate (1 M and 0.2 M in 10 mM Tris-HCl, pH 7.4, 0.2 mM $MgCl_2$)

Magnesium chloride (1 M stock solution)

PMSF (1 M stock solution in dried ethanol)

Purified nuclei (see *Protocol 4.1*) in 0.25 M sucrose, 10 mM Tris-HCl, pH 7.4, 5 mM $MgCl_2$

DNase I

Tris-HCl (1 M stock solution, pH 7.4)

Suspension buffer (SB): 5 mM $MgCl_2$, 10 mM Tris-HCl, pH 7.4 (containing 0.1 mM PMSF)

Triton X-100

Equipment

Low-speed centrifuge with fixed-angle rotor and tubes

Procedure

Bring all solutions to, and carry out all operations at, 0–4 °C.

1. Add Triton X-100 to the purified nuclear suspension to a concentration of 1%. Incubate for 30 min with mixing by pipette (avoid foaming).

2. Centrifuge at 1000 g for 10 min and discard the supernatant. Resuspend the extracted nuclear pellet in a small volume (e.g. 5 ml) of SB.

3. Digest the nuclei with DNase I (30 U/mg DNA) for 30 min. ①

4. Increase the salt concentration by slowly adding the 1 M ammonium sulphate solution to a final ionic strength of 0.6 (0.2 M ammonium sulphate); make up to 40 ml with the 0.2 M ammonium sulphate solution and incubate for 30 min. ②

5. Centrifuge at 1000g for 15 min to pellet the nuclear matrices. Discard the supernatant.

6. Resuspend the pellet in SB. Monitor the product by phase contrast microscopy and thin section electron microscopy. ③

Notes

The procedure will take about 3 h.

① Note that the nuclear chromatin has not been decondensed.

② Alternatively, NaCl can be used for the high salt extraction.

③ Nucleoli should be visible within the less dense matrix, together with intact nuclear pore complex lamina.

PROTOCOL 4.6

Preparation of nucleoli [65, 71]

Reagents

Suspension buffer (SB): 0.34 M sucrose, 0.05 mM MgCl$_2$, 10 mM Tris-HCl, pH 7.5

Density barrier: 0.88 M sucrose, 0.05 mM MgCl$_2$, 10 mM Tris-HCl, pH 7.5

Include protease inhibitors in these solutions as required

Equipment

Refrigerated low-speed or high-speed centrifuge with swinging-bucket rotor

Ultrasonicator

Light microscope (phase contrast or normal optics)

Procedure

Carry out all operations at 0–4 °C.

1. Suspend the purified nuclei (see *Protocol 4.1*) in SB.

2. Sonicate the suspension for 10–15 s, followed by a rest period of 30–40 s.

3. Monitor the nuclear breakage under the microscope; use phase contrast or normal optics with methylene blue (0.1%) stained material. Disruption of all the nuclei usually requires 40–60 s of total sonication time.

4. In a centrifuge tube, underlayer the sonicate with half the volume of the density barrier solution and centrifuge at 2000 g for 20 min and harvest the pellet of nucleoli. ① ②

Notes

The procedure should take approx. 90 min.

① An alternative to the gradient is to pellet the nucleoli at 10 000g for 15 min; then resuspend the pellet in 5 mM MgCl$_2$ 10 mM Tris-HCl, pH 7.4. Repeat this washing until the release of DNA and non-nucleolar matrix proteins reaches a minimal level.

② Monitor the product by RNA and DNA analysis, SDS-PAGE and thin section electron microscopy after the addition of low gelling temperature agarose.

PROTOCOL 4.7

Isolation of a heavy mitochondrial fraction from rat liver by differential centrifugation

Reagents

Mannitol buffer: 0.2 M mannitol, 50 mM sucrose, 10 mM KCl, 1 mM EDTA, 10 mM Hepes-NaOH, pH 7.4 ①

Equipment

Glass rod

Dounce homogenizer, 20–30 ml (loose-fitting, Wheaton type B)

High-torque overhead motor (thyristor controlled)

High-speed centrifuge with fixed-angle rotor (to take 40–50 ml tubes)

Low-speed refrigerated centrifuge (40–50 ml tubes) ②

Potter-Elvehjem homogenizer, 20–40 ml, clearance ~0.07 mm

Syringe with metal cannula (internal diameter approx. 0.8 mm)

Procedure

The volumes given are for a liver of approx. 10 g. After step 2 carry out all operations at 0–4 °C.

1. Starve a male rat overnight to deplete the glycogen. ③

2. Sacrifice the animal by cervical dislocation and quickly excise the liver and weigh it. ④

3. Transfer the liver to a beaker on ice and mince it finely with scissors (the pieces of liver should be no more than ~3 mm³ in size). Suspend the mince in the mannitol buffer (approx. 4 ml per g of liver).

4. Decant half the liquid and mince into the Potter-Elvehjem homogenizer, and homogenize using five to six up-and-down strokes of the pestle (rotating at ~700 rpm). Repeat with the other half of the sample. ⑤

5. Divide the homogenate between two centrifuge tubes and centrifuge at 1000g for 10 min. ⑥

6. Decant the supernatant and keep on ice. Resuspend the pellet in ~30 ml of buffer using two to three strokes of the pestle of the Potter-Elvehjem homogenizer and repeat step 5. ⑦

7. Decant and combine the supernatants and recentrifuge at 1000g for 10 min.

8. Remove the supernatants using a metal cannula attached to a 20 ml syringe. ⑧

9. Centrifuge the supernatants at 4000g for 15 min.

10. Remove the 4000g supernatant using the metal cannula and syringe, keeping the tip of the cannula as close to the meniscus as possible in order to remove as much of the floating lipid as possible. ⑨

11. Also carefully aspirate the loosely packed pinkish material on top of the more compact brown pellet and wipe the inside of the tube with a tissue to remove any adhering lipid.

12. Add ~10 ml of buffer to each tube and crudely resuspend the pellet using a glass rod.

13. Complete the resuspension using two to three gentle strokes of the pestle of the Dounce homogenizer.

14. Make up to the original volume with buffer and decant into new tubes.

15. Repeat steps 9–14 three times. ⑩

16. Finally, resuspend the purified heavy mitochondria in a suitable medium compatible with the subsequent analysis.

Notes

This procedure will take approx. 1.5 h.

① For other tissues, such as beef heart, a sucrose/succinate/EDTA/Tris buffer is common – see ref. 72.

② Steps 5 and 7 could be executed using a high-speed centrifuge rather than a low-speed centrifuge.

③ Starving the animal must be carried out in accordance with national animal care requirements.

④ For respiratory studies, it is obviously important to avoid any anaesthetic techniques in sacrificing the animal. Decapitation is a useful alternative, which also serves to exsanguinate the liver, but whatever technique is used, it must be carried out by a trained, licensed individual.

⑤ Do not force the pestle through any compacted tissue pieces at the bottom of the homogenizer; lower the glass vessel to resuspend the tissue in the vortex from the revolving pestle.

⑥ It is advisable not to fill the tubes of any rotor, a swinging bucket rotor in particular, completely full, because of the problems associated with use of these rotors (see 'Differential centrifugation' above).

⑦ Washing the nuclear pellet will recover any trapped mitochondria.

⑧ Removal of this supernatant with a syringe, rather than by decantation, reduces contamination by nuclei from the pellet.

⑨ The presence of free fatty acids can cause uncoupling, so it is important to remove as much as possible of the lipid.

⑩ The washing procedure removes contaminating lysosomes.

Preparation of a light mitochondrial fraction from tissues and cultured cells

Reagents

Homogenization medium (HM): 0.25 M sucrose, 1 mM EDTA, 10 mM Hepes-NaOH, pH 7.4 ①

Add protease inhibitors to the above medium as required

Phosphate buffered saline (*cells only*)

Equipment

Low-speed refrigerated centrifuge with swinging-bucket rotor for 30–40 ml tubes ②

High-speed centrifuge with fixed-angle rotor for 30–40 ml tubes

For liver: Potter-Elvehjem (Teflon and glass) homogenizer (30–40 ml) attached to a high-torque overhead motor (thyristor controlled) ③

For cells: Ball-bearing homogenizer or syringe with 23- or 25-gauge needle ③

Dounce homogenizer (loose fitting, Wheaton type B)

Syringe (20 ml) with metal cannula (internal diameter approx. 0.8 mm) ④

Procedure

Carry out all operations at 0–4 °C.

1. *For liver:* Mince the tissue very finely with scissors (or with a tissue chopper) and transfer to the Potter-Elvehjem homogenizer with HM (use 10 ml medium for every 2.5 g tissue). Homogenize using approx. six strokes of the pestle (500–700 rpm). ⑤

2. *For cells:* Wash $1–3 \times 10^8$ cells in 5 ml of phosphate buffered saline and again with 5 ml of HM. Suspend the cells in 3 ml of HM and homogenize in a ball-bearing homogenizer using five passes. ⑥ ⑦

3. Pellet the nuclei, cell debris and any unbroken cells by centrifuging at 800g for 10 min.

4. Decant or aspirate (use a syringe and cannula) the supernatant and retain on ice.

5. Resuspend the pellet in 10 ml (5 ml for cells) of HM using two to three gentle strokes of the pestle of a loose-fitting Dounce homogenizer. ⑧

6. Repeat the centrifugation and combine the supernatants.

7. Centrifuge the combined supernatants at 3000g for 10 min. ⑨

8. Centrifuge the supernatant from step 7 at 15 000g for 10 min. ⑩

9. Resuspend the light mitochondrial pellet in 10 ml (5 ml for cells) of HM using two to three gentle strokes of the pestle of a loose-fitting Dounce homogenizer. ⑧

10. Centrifuge at 15 000 *g* for 10 min.

11. Resuspend the pellet in a suitable medium for analysis or further processing. Use this for *Protocols 4.10, 4.12, 4.15* and *4.17*.

Notes

This procedure will take approx. 2 h.

① Any suitable buffered isoosmotic 0.25 M sucrose solution may be used, with or without EDTA. If the aim is to isolate peroxisomes, include 0.1% ethanol.

② The pelleting of the nuclei, etc. (step 3), although often carried out in a low-speed centrifuge, may alternatively be performed in the high-speed fixed-angle rotor.

③ For more information on homogenizing cells and tissues, see ref. 2.

④ Metal 'filling' cannulas can be obtained from any surgical equipment supplies company.

⑤ Most soft tissues can be homogenized in a Potter-Elvehjem homogenizer, but alternatives such as the Polytron homogenizer have been used.

⑥ For the isolation of organelles from cultured cells, the ball-bearing homogenizer is considered to offer the gentlest means of disruption [14]. Alternatives are passage through a syringe needle or a tight-fitting Dounce homogenizer.

⑦ Always monitor the efficacy of the homogenization by phase contrast microscopy.

⑧ Resuspension of the pellet must be carried out under the mildest of conditions.

⑨ This step to remove the heavy mitochondria may be omitted.

⑩ A variety of centrifugation conditions have been used for isolation of the light mitochondrial fraction; RCFs of 12 000–20 000g and times of 10–20 min are the most common. Check the recovery of the required organelle and modify as required. Always use the minimum *g*-force to give an adequate recovery.

Purification of yeast mitochondria in a discontinuous Nycodenz® gradient [38]

Reagents

Yeast spheroplast preparation in sorbitol buffer ①

Nycoprep™ Universal (60% w/v Nycodenz®)

Nycodenz® working solution: 50% w/v Nycodenz®

Sorbitol buffer: 0.6 M sorbitol, 20 mM Mes-KOH, pH 6.0

Sorbitol buffer (2×): 1.2 M sorbitol, 40 mM Mes-KOH, pH 6.0

Suspension buffer (SB): 0.6 M sorbitol, 20 mM Hepes-KOH, pH 7.4

Add protease inhibitors as required to the buffers

Equipment

High-speed centrifuge with 8 × 50 ml fixed-angle rotor

Ultracentrifuge with swinging-bucket rotor (approx. 12–13 ml tubes)

Dounce homogenizer (approx. 40 ml tight-fitting, Wheaton type A)

Dounce homogenizer (approx. 40 ml loose-fitting, Wheaton type B)

Syringe (20 ml) with metal cannula (internal diameter approx. 0.8 mm) ②

Procedure

Carry out all operations at 0–4 °C.

1. Homogenize the spheroplasts using 15 strokes of the pestle of the tight-fitting Dounce homogenizer.

2. Dilute with an equal volume of sorbitol buffer and centrifuge in the fixed-angle rotor at 1500*g* for 5 min.

3. Decant the supernatants and retain.

4. Resuspend the pellets to the original volume with sorbitol buffer and repeat steps 1 and 2.

5. Combine all the supernatants and centrifuge at 12 000*g* for 10 min.

6. Resuspend the crude mitochondrial pellets in approx 40 ml sorbitol buffer using a few gentle strokes of the pestle of the loose-fitting Dounce homogenizer.

7. Remove any residual nuclei and debris by centrifugation at 1500*g* for 5 min.

8. Remove the supernatants using the syringe and cannula, taking care not to disturb the loosely packed pellet. ③

9. Centrifuge the supernatants at 12 000*g* for 10 min and resuspend the pellets (loose-fitting Dounce homogenizer) in approx 4 ml of sorbitol buffer. ③

10. Prepare two Nycodenz® solutions of 18 and 14.5% (w/v) from 25 ml of sorbitol buffer (2×) and 18 ml (and 14.5 ml) of the 50% Nycodenz®

working solution; make up each to 50 ml with water.

11. In tubes for the swinging-bucket rotor load 5 ml of each of the Nycodenz® solutions and approx. 2 ml of the crude mitochondrial suspension (top up with sorbitol buffer if necessary). ④

12. Centrifuge at approx $200\,000g_{av}$ for 30 min. The purified mitochondria band at the interface of the two Nycodenz® layers.

13. Harvest the band with the syringe and metal cannula; dilute with 4 vol. of SB; centrifuge at $12\,000g$ for 15 min and resuspend the pellet in SB.

Notes

This procedure will take approx. 2 h.

① The spheroplast suspension from approx 25 g of packed yeast cells (wet weight) should be approx. 35 ml.

② Metal 'filling' cannulas can be obtained from any surgical equipment supplies company.

③ Decantation of the supernatant may lead to contamination from the pellet material.

④ See ref. 18 regarding the preparation of discontinuous gradients.

Purification of mitochondria from mammalian liver or cultured cells in a median-loaded discontinuous Nycodenz® gradient [41, 42]

Reagents

Homogenization medium (HM): 0.25 M sucrose, 1 mM EDTA, 10 mM Hepes-NaOH, pH 7.4 ①

Nycoprep™ Universal

Nycoprep™ Universal diluent (NUD): 6 mM EDTA, 60 mM Hepes-NaOH, pH 7.4

Nycodenz® working solution (50%, w/v): Mix 5 vol. of Nycoprep™ Universal with 1 vol. of NUD.

Light mitochondrial fraction (see *Protocol 4.8*)

Include protease inhibitors in these solutions as required

Equipment

Ultracentrifuge and swinging-bucket rotor with a tube capacity of approx. 38 ml, smaller volume 17 ml or 13 ml tubes are also acceptable

Syringe (10 ml) with metal cannula (internal diameter approx. 0.8 mm) ②

Procedure

1. Dilute the Nycodenz® working solution with HM to produce gradient solutions with Nycodenz® concentrations of 40, 34, 30, 23 and 20% (w/v). Keep these solutions at 0–4 °C. ③

2. Resuspend the washed light mitochondrial pellet (LMP) in HM (4–10 ml) and mix with an equal volume of Nycodenz® working solution.

3. For 36–38 ml tubes, use a syringe with metal cannula to underlayer (low density end first) the following Nycodenz® solutions: 2 ml of 20%, 7 ml of 23%, 10 ml of LMP in 25%, 7 ml of 30%, and 5 ml each of 34 and 40%. ④ ⑤ ⑥

4. After centrifugation at 95 000 *g* for 2 h harvest the mitochondria from the 25/30% Nycodenz® interface or unload the entire gradient in a series of equal volume fractions prior to analysis. ⑦

Notes

The procedure will take approx. 4.5 h, including preparation of the LMP.

① Use the same HM as used in *Protocol 4.8.*

② Metal 'filling' cannulas can be obtained from any surgical equipment supplies company.

③ The low-density end of the gradient is probably ineffective at resolving any Golgi membranes from the lysosomes. For this an additional layer of 15% Nycodenz® might be layered on top.

④ For methods on the best ways of constructing discontinuous gradients, see ref. 18.

⑤ Top up the tubes, if necessary, with either 20% Nycodenz® or HM, according to the manufacturer's recommendations for tube filling. If using the additional layer of 15%

Nycodenz®, use 2 ml of this and reduce the layer of 40% Nycodenz® to 3 ml.

⑥ Scale down the volumes proportionally for smaller tubes.

⑦ See ref. 18 for more information on harvesting and analysing gradients.

Reagents

Ethyl acetate/ethanol/trichloroacetic acid (EET) 5 : 5 : 1 (v/v/w)

p-Iodonitrotetrazolium violet (2.5 mg/ml in phosphate buffer)

Phosphate buffer pH 7.4 (50 mM)

Substrate: 0.01 M succinate in phosphate buffer

Equipment

Microcentrifuge with 1.5–2.0 ml tubes

Spectrophotometer (visible wavelength) with 1 ml cuvettes (glass or EET-resistant plastic)

Procedure [73]

1. Dilute fractions from a gradient with an equal volume phosphate buffer and sediment in a microcentrifuge at 4 °C for 5–10 min.

2. Resuspend the pellets by vortex mixing in 0.3 ml of substrate. ①

3. After incubation at 37 °C for 10 min add 0.1 ml of INT and incubate for a further 10 min. ②

4. Stop the reaction by addition of 1 ml of EET.

5. Centrifuge any precipitate in a microcentrifuge for 2 min and measure the absorbance of the supernatant at 490 nm against a suitable control. ③ ④

Notes

The procedure will take approx. 1 h.

① Rather than centrifuging and resuspending a gradient fraction in substrate solution, small volumes of gradient fraction (<50 μl) may be added directly to the substrate solution. Nycodenz® does not interfere with the enzyme assay.

② This incubation time may be extended if the concentration of enzyme is low.

③ Ideally, for each test incubation, a control incubation should be set up in which the substrate solution is replaced by phosphate buffer. In most cases, however, a single control containing substrate solution, buffer, INT and EET is sufficient.

④ The molar extinction coefficient of reduced INT is 19 300.

PROTOCOL 4.12

Isolation of lysosomes in a discontinuous Nycodenz® gradient

Reagents

Homogenization medium (HM): 0.25 M sucrose, 1 mM EDTA, 10 mM Hepes-NaOH, pH 7.4 ①

Nycoprep™ Universal

Nycoprep™ Universal diluent (NUD): 6 mM EDTA, 60 mM Hepes-NaOH, pH 7.4

Nycodenz® working solution (50%, w/v): Mix 5 vol. of Nycoprep™ Universal with 1 vol. of NUD

Add protease inhibitors to solutions as required

Light mitochondrial fraction (see *Protocol 4.8*)

Equipment

Ultracentrifuge with swinging-bucket rotor with a tube capacity of approx. 38 ml (smaller volume 17 ml or 13 ml tubes are also acceptable).

Syringe (10 ml) with metal cannula (internal diameter approx. 0.8 mm) ②

Procedure [45]

Carry out all operations at 0–4 °C.

1. Dilute the Nycodenz® working solution with HM to produce gradient solutions with Nycodenz® concentrations of 30, 26, 24 and 19% (w/v). Keep these solutions at 0–4 °C.

2. Resuspend the washed light mitochondrial pellet in HM (4–10 ml) and mix with 2 vol. of Nycodenz® working solution.

3. For 36–38 ml tubes: Transfer 8 ml of 19% Nycodenz® into each tube and underlayer, using a syringe and metal cannula, 7 ml each of the 24, 26 and 30% Nycodenz® and 8–9 ml of the light mitochondrial pellet suspension. ③ ④ ⑤

4. Centrifuge at 95 000*g* for 2 h and then harvest the lysosomes from the 19/24% Nycodenz® interface or unload the entire gradient in a series of fractions prior to marker analysis. ⑥

Notes

The procedure will take approx. 4.5 h, including preparation of the LMP.

① Use the same HM as used in *Protocol 4.8*.

② Metal 'filling' cannulas can be obtained from any surgical equipment supplies company.

③ For methods on the best ways of constructing discontinuous gradients, see ref. 18.

④ Top up the tubes, if necessary, with either 29% Nycodenz® or HM, according to the manufacturer's recommendations for tube filling.

⑤ Scale down the volumes proportionally for smaller tubes.

⑥ See ref. 18 for more information on harvesting and analysing gradients.

β-Galactosidase (spectrophotometric assay)

Reagents

Reaction buffer: 0.5% (w/v) Triton X-100, 0.05 M sodium-citrate-phosphate buffer, pH 4.3

Stop buffer: 0.25 M glycine-NaOH, pH 10 ①

Substrate: 6 mM *p*-nitrophenyl β-D-galactopyranoside in reaction buffer

Equipment

Microcentrifuge with 1.5–2.0 ml tubes

Spectrophotometer (visible wavelength) with 1 ml cuvettes

Procedure [73]

1. For each test, mix 0.5 ml of substrate with a membrane fraction pelleted in a microcentrifuge tube (see step 1 of *Protocol 4.11*) or with up to 50 μl of a membrane suspension. ②

2. Set up controls as in step 1 and add 1 ml of stop buffer immediately. ③

3. Incubate tests at 37 °C for 30 min.

4. Add 1 ml of stop buffer to tests.

5. Remove any sedimentable material in a microcentrifuge (1–2 min).

6. Read the absorbance of tests vs controls at 410 nm.

Notes

The procedure will take approx. 1 h.

① Do not allow the pH of the stop buffer to rise above 10 as this will cause hydrolysis of the substrate.

② If the sample is sufficiently concentrated, up to 50 μl of sample can be added directly to the substrate solution. Dilute samples should first be concentrated by pelleting in a microcentrifuge.

③ If none of the samples contributes to the overall absorbance, it is permissible to use a single control without membrane suspension. There is negligible hydrolysis of the substrate at pH 4.3 in the absence of enzyme.

PROTOCOL 4.14

β-Galactosidase (fluorometric assay)

Reagents

Reaction buffer: 100 mM sodium acetate-acetic acid, pH 4.3, 100 mM KCl, 2% Triton X-100, 5 mM DTT

Substrate: 1.7 mM 4-methylumbelliferyl-β-D-galactoside (4-MUG) in reaction buffer

Equipment

Fluorimeter, 355 nm emission, 460 nm excitation with cuvettes to take 200 μl

Microcentrifuge

Water bath at 37 °C

Procedure [74]

1. After dilution with two volumes of buffer, pellet the gradient fraction in the microcentrifuge for 10 min at 0–4 °C.

2. Suspend the pellet in 0.25 ml of reaction buffer. ①

3. Mix with an equal volume of substrate and incubate at 37 °C for 1 h. Set up sample blanks using 0.25 ml of reaction buffer instead of substrate. ①

4. Set up also a substrate blank.

5. After 1 h measure the fluorescence. ②

Notes

This procedure will take approx. 2 h.

① It will be necessary to run a few trial assays to optimize the amount of membrane required; determine that the rate of reaction is linear with time and membrane concentration. The assay is approx. 5–10 times more sensitive than the spectrophotometric assay (see *Protocol 4.13*).

② The substrate blank should read no more than 10 fluorescence units against a value of at least 500 units for the test.

Isolation of mammalian peroxisomes in an iodixanol gradient

Reagents

OptiPrep™ (60% iodixanol)

Homogenization medium (HM): 0.25 M sucrose, 1 mM EDTA, 0.1% (v/v) ethanol, 5 mM Mops pH 7.2. ①

Diluent: 6 mM EDTA, 0.6% ethanol, 30 mM Mops, pH 7.2

1 M sucrose

Add protease inhibitors to HM and diluent as required

Light mitochondrial fraction prepared using HM (see *Protocol 4.8*)

Equipment

Dounce homogenizer (10 ml, loose-fitting, Wheaton type B)

Two-chamber gradient maker or Gradient Master™

Syringe (5 ml) with metal cannula (internal diameter approx. 0.8 mm) ②

Ultracentrifuge and 30–40 ml fixed-angle rotor capable of approx. 100 000g, with thick-walled tubes of approx 30 ml capacity

Gradient collector (e.g. Beckman Fraction Recovery System)

Procedure [75]

Carry out all operations at 0–4 °C.

1. Make up the following gradient solutions from OptiPrep™, diluent, 1 M sucrose and water using respectively, these ratios by volume: 50% (w/v) iodixanol (5 + 0.6 + 0.4 + 0.0), 40% iodixanol (4 + 0.6 + 0.7 + 0.7) and 20% iodixanol (2 + 0.6 + 1.1 + 2.3). ③ ④ ⑤

2. Resuspend the light mitochondrial pellet in HM using a loose-fitting Dounce homogenizer (two to three strokes of the pestle). Adjust to a volume of about 0.5 ml per g of tissue.

3. Use a two-chamber gradient maker or a Gradient Master to prepare a linear gradient from 9 ml each of 20 and 40% iodixanol solutions in thick-walled polycarbonate tubes for the fixed-angle rotor and, using the syringe and metal cannula, underlayer each gradient with 2 ml of 50% iodixanol. ⑥ ⑦

4. Layer 3 ml of the suspension over each gradient and centrifuge at 105 000g_{av} for 1 h.

5. Allow the rotor to decelerate from 1000 rpm without the brake and collect the gradient in 1 ml fractions, dense end first. Alternatively, collect the peroxisomes that band approx. two-thirds of the way down the gradient, using the syringe and metal cannula. ⑧

Notes

① Use this HM to prepare the light mitochondrial fraction.

② Metal 'filling' cannulas can be obtained from any surgical equipment supplies company.

③ The use of the four components to prepare the gradient solutions assures that each is isoosmotic.

④ Van Veldhoven *et al.* [75] developed this method for rat liver. Iodixanol gradients (15–45%) have also been used in vertical rotors for mouse [76] and human liver [77]. The shorter sedimentation path length of a vertical rotor permits the use of considerably lower *g*-forces ($31\,000g$ for 75 min). This may improve the retention of functional integrity of the organelles. On the other hand, a 4–30% iodixanol gradient at $100\,000g$ for 17 h has been used for CV1 cells [78].

⑤ Gradients covering a similar density range are used with Nycodenz®:

for yeast, 15–42.5% at $174\,000g$ for 75 min [79] and 15–36% at $100\,000g$ for 1 h [80]; and for rat liver, gradients vary from 25–50% [81] to 15–48% [82] and centrifugation conditions from $60\,000g$ for 75 min [81] to $35\,000g$ for 75 min [83]. Many of these methods employ a vertical rotor.

⑥ If neither of these devices is available, mix equal volumes of the 20 and 40% iodixanol solutions to produce an additional 30% iodixanol solution. Create a discontinuous gradient from 6 ml each of the three solutions and allow a continuous gradient to be formed by diffusion overnight at 0–4 °C.

⑦ See ref. 18 for more information about making gradients.

⑧ Use either tube puncture or aspiration from the bottom of the tube to collect the gradient. See ref. 18 for more information about harvesting and analysing gradients.

Catalase assay

Reagents

Peroxide stock: 30 mM H_2O_2 in T-BSA ①

Sample buffer: 2 vols. T-BSA + 1 vol. 2% Triton X-100

Titanium oxysulfate (2.25 g/l) in 1 M H_2SO_4 Δ ②

T-BSA: 20 mM Tris-HCl, pH 7.0 containing 0.1% (w/v) bovine serum albumin

Equipment

Microcentrifuge with 2.0 ml microcentrifuge tubes

Spectrophotometer

Procedure [84]

Carry out all procedures on ice in 2 ml microcentrifuge tubes and keep all solutions on ice.

1. Make up the assay mixture freshly by diluting 8.5 ml of stock peroxide to 100 ml with T-BSA.

2. Add 1.0 ml of titanium oxysulfate to 0.5 ml of assay mixture and measure the absorbance at 405 nm; it should be ~1.5. If it is not, then adjust the H_2O_2 concentration accordingly.

3. Mix 10 μl of sample with 30 μl of sample buffer; for the reagent control, use 40 μl of sample buffer. ③ ④

4. Add 0.5 ml of assay mixture to all tubes; do this in batches of *six* tubes only.

5. After exactly 1 min, add 1.0 ml of titanium oxysulfate and microcentrifuge (~13 500g) for 2 min.

6. Allow all tubes to reach room temperature and measure the absorbance at 405 nm against a blank containing 0.54 ml of T-BSA and 1 ml of titanium oxysulfate. ⑤

Notes

This procedure will take approx. 1 h.

① Keep the stock peroxide at 4 °C.

② Titanium oxysulfate is highly corrosive.

③ The enzyme is very active in mammalian liver fractions: peroxisome-rich fractions may have to be diluted up to 100× with sample buffer.

④ It is usually unnecessary to use a blank for each fraction; if it is deemed necessary, then set up duplicate samples (step 4) and add the titanium oxysulfate before the assay mixture to one of each pair.

⑤ The titanium oxysulfate reacts with the H_2O_2 remaining after the incubation; thus if all the substrate had been used up, the absorbance would be close to zero. The reagent control absorbance measures the substrate concentration at zero time.

Analysis of major organelles in a preformed iodixanol gradient

Reagents

OptiPrep™ (60% w/v iodixanol)

Homogenization medium (HM): 0.25 M sucrose, 1 mM EDTA, 20 mM Hepes-NaOH, pH 7.4

OptiPrep™ Diluent (OD): 0.25 M sucrose, 6 mM EDTA, 120 mM Hepes-NaOH, pH 7.4

Iodixanol (50%, w/v) working solution (IWS): 5 vols. OptiPrep™ + 1 vol. OD ①

Add protease inhibitors to HM and OD as required

Light mitochondrial fraction prepared using HM (see *Protocol 4.8*)

Equipment

Dounce homogenizer (loose-fitting, Wheaton type B)

Two-chamber gradient maker or Gradient Master™

Syringe (5 ml) with metal cannula (internal diameter approx. 0.8 mm) ②

Ultracentrifuge with swinging-bucket rotor (tube capacity of approx. 17 ml)

Gradient collection device for harvesting gradient low-density or high-density end first ③

Procedure

Carry out all operations at 0–4 °C.

1. Dilute IWS with HM to produce gradient solutions with iodixanol concentrations of 19 and 27% (w/v). Keep these solutions on ice until required. ④

2. Resuspend the light mitochondrial pellet in approx 3.0 ml of HM using the loose-fitting Dounce homogenizer (two to three strokes of the pestle).

3. Adjust the suspension to 30% iodixanol by mixing with IWS.

4. Use a two-chamber gradient maker or a Gradient Master™ to prepare a linear gradient from 6 ml each of the two gradient solutions in tubes for the swinging-bucket rotor. ⑤

5. Using the syringe and metal cannula layer 3–4 ml of the suspension (step 3) beneath the gradient.

6. Layer 1–2 ml of HM on top of the gradient to fill the tube to about 3 mm from its top and centrifuge in a suitable swinging-bucket rotor at approx $70\,000g_{av}$ for 1.5–2 h.

7. Collect the gradient in 1 ml fractions low-density end first by upward displacement with a dense medium or by aspiration from the meniscus. Alternatively, collect dense end first by tube puncture or by carefully introducing a narrow metal cannula (connected to a peristaltic pump) to the bottom of the tube. ③ ⑥ ⑦

Table 4.2 Fractionation of major organelles in iodixanol gradients: variations of sample and gradient conditions

Tissue/cell	Gradient input[1]	Gradient[2]	RCF (time)	Ref.
Brain (mouse)	17 000*g*/10 min pellet from PNS	10–30	52 000*g* (90 min)	85
Glioblastoma	15 000*g*/10 min pellet from PNS	10–30	52 000*g* (90 min)	86
Glomerular epithelial	100 000*g*/1 h pellet from PNS	2.5–25 (d)	100 000*g* (3 h)	87
H4 neuroglioma	17 000*g*/10 min pellet from PNS	10–30	52 000*g* (90 min)	88
HEK293	3000*g* PNS	10–30	30 000*g* (16 h)	89
HEK293	PNS	10–30(d)	100 000*g* (3 h)	90
Human carcinoma	PNS underlaid in 35% iodixanol	1–30	100 000*g* (1.5 h)	91
Human liver	Light mitochondrial pellet	13–45	33 000*g* (75 min)	92
Human epithelial	17 000*g*/10 min pellet from PNS	10–30	52 000*g* (90 min)	93
MDCK	17 000*g* pellet from PNS	10–30	100 000*g* (1 h)	94
Vero	3000*g* supernatant	2.5–30 (d)	126 000*g* (25 min)	95

[1] PNS = post-nuclear supernatant; unless stated all samples layered on top of gradient.
[2] Figures are % (w/v) iodixanol; d = discontinuous gradient.
RCF = relative centrifugal force.

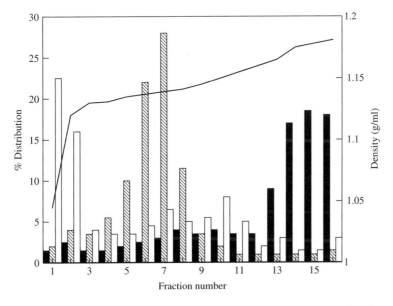

Figure 4.1 Fractionation of a mammalian liver light mitochondrial fraction by flotation: percentage distribution of enzyme markers. Light mitochondrial fraction layered in 30% iodixanol beneath a 19–27% iodixanol gradient; centrifuged at 70 000*g* for 1.5 h. White bars, β-galactosidase; hatched bars, succinate dehydrogenase; black bars, catalase

Notes

① This solution is approx. isoosmotic.

② Metal 'filling' cannulas can be obtained from any surgical equipment supplies company.

③ The most satisfactory method for collecting low-density end first is by aspiration from the meniscus using the Labconco Auto Densi-flow gradient unloader, but upward displacement with a dense medium such as Maxidens™, introduced to the bottom of the gradient by tube puncture, is satisfactory. The Beckman Fraction Recovery System incorporates a good tube puncture device. This can also be used for collection dense end first.

④ There are variations to the density range of the gradient used in this protocol. The 19–27% iodixanol gradient is optimal for rat liver; some other examples are summarized in Table 4.2. In nearly all cases, the sample was layered on top of the gradient rather than beneath it, even though in the latter format the resolution is probably greater.

⑤ If neither of these devices is available create a discontinuous gradient from 3 ml each of 19, 22, 25 and 27% iodixanol and allow a continuous gradient to be formed by diffusion overnight at 0–4 °C. For more information on the formation of gradients, see ref. 18.

⑥ For more information about the harvesting and analysis of gradients, see ref. 18.

⑦ A typical result with rat liver is shown in Figure 4.1.

Separation of smooth and rough ER in preformed sucrose gradients [48]

Reagents

Homogenization medium (HM): 0.25 M sucrose, 10 mM Hepes-NaOH, pH 7.4

Gradient solutions: 13.6, 17.5, 50 and 51.5% (w/v) sucrose in 10 mM Hepes-NaOH, pH 7.4

Rough ER suspension medium (RSM) 70% (w/v) sucrose, 3 mM imidazole-HCl, pH 8.2

Stripping solution: 68% (w/v) sucrose, 40 mM pyrophosphate, 3 mM imidazole-HCl, pH 8.2

Light mitochondrial supernatant: *Protocol 4.8 (using the above HM) as far as step 8*

Include protease inhibitors in the solutions as required

Equipment

Dounce homogenizer (20–30 ml loose-fitting, Wheaton type B)

Two-chamber gradient maker or Gradient Master™

Syringe (5 or 10 ml) and metal cannula ①

Ultracentrifuge with 8 × 39 ml fixed-angle rotor (with open-topped thick-walled tubes) and 14–17 ml swinging-bucket rotor

Procedure

The method is designed for approx 10 g of rat liver. Carry out all operations at 0–4 °C. ②

1. Centrifuge the 15 000g supernatant at 100 000g for 40 min.

2. Discard the supernatant and resuspend the microsomal pellet in 20 ml of the homogenization medium using the Dounce homogenizer. ③

3. Using a two-chamber gradient maker or Gradient Master™, prepare 17.5–50% (w/v) sucrose gradients in tubes for the swinging-bucket rotor (for 17 ml tubes make a 5 ml gradient). ④

4. Underlayer the gradient with 2 ml of 51.5% sucrose using the syringe and metal cannula.

5. Layer 10 ml of the microsomal suspension on top and centrifuge at 100 000g for 16 h.

6. Harvest the smooth ER band by removing the top 15 ml of the gradient using a syringe and metal cannula and concentrate (after dilution with an equal volume of HM) by centrifugation at 100 000g for 1 h.

7. Resuspend the pellet of rough microsomes (from the gradient tube) in the residual medium (2 ml) and add 12 ml of the RSM plus 5 ml of stripping solution.

8. Prepare 13.6–50% (w/v) linear sucrose gradients (12 ml in 17 ml tubes) and underlay each with the 5 ml of sample. ⑤

9. Centrifuge at $200\,000g$ for 2 h.

10. The stripped RER vesicles band in the bottom half of the gradient (27–50% sucrose) while peroxisomal and mitochondrial contaminants are located below this band.

11. Collect the band using a syringe and metal cannula.

12. Dilute with 2 vols. of HM and pellet at $100\,000g$ for 1 h.

Notes

This procedure will take approx. 1.5 days.

① Metal 'filling' cannulas can be obtained from any surgical equipment supplies company.

② Although developed for rat liver the method may be used for any tissue.

③ The rough ER in the $100\,000g$ pellet tends to make the latter gelatinous; make sure this is well dispersed.

④ If neither of these devices is available then allow a discontinuous gradient (1 ml each of 17.5, 25.5, 32.5, 40.5 and 50% sucrose) to diffuse for approx 6 h. See ref. 18 for more information about making gradients.

⑤ If neither a two-chamber mixer nor a Gradient Master™ is available, allow a discontinuous gradient (2.5 ml each of 13.6, 22.6, 31.6, 41.6 and 50% sucrose) to diffuse overnight. See ref. 18 for more information about making gradients.

Separation of smooth and rough ER in a self-generated iodixanol gradient

Reagents

OptiPrep™ (60% iodixanol)

Homogenization medium (HM): 0.25 M sucrose, 10 mM Tris-HCl

Light mitochondrial supernatant: *Protocol 4.8 (using the above HM) as far as step 8*

Include protease inhibitors in the HM as required

Equipment

Ultracentrifuge with 8×39 ml fixed-angle (with open-topped thick-walled tubes) and a vertical or near-vertical rotor (e.g. Beckman VTi65.1 or equivalent) and suitable sealed tubes (approx 11 ml) ①

Dounce homogenizer (20–30 ml loose-fitting, Wheaton type B)

Gradient collection device for unloading low-density end first ②

Procedure [96]

The method is designed for approx. 10 g of liver. Carry out all procedures at 0–4 °C. ③

1. Centrifuge the light mitochondrial (15 000g) supernatant at 100 000g for 40 min.

2. Discard the supernatant and resuspend the microsomal pellet in 30 ml of HM using the Dounce homogenizer. ④

3. Mix the microsome suspension with half its volume of OptiPrep™ (final iodixanol concentration 20%, w/v) and transfer to 11.2 ml Optiseal tubes for the Beckman VTi65.1 vertical rotor.

4. Centrifugation at 350 000g_{av} for 2 h at 4 °C. Turn the brake off, or use a slow deceleration program, during below 2000 rpm.

5. Harvest the gradients by upward displacement with a dense medium in 0.5 ml fractions for analysis. ⑤ ⑥ ⑦

Notes

① Beckman Optiseal™ tubes are the most convenient sealed tubes to use and are the recommended ones for this procedure. They are sealed by a central plastic plug that can be removed after centrifugation to allow easy access to the gradient.

② The tube puncture part of a Beckman Fraction Recovery System or the Labconco Auto Densi-flow is the recommended device for unloading Optiseal™ tubes. Crimp or heat-sealed tubes are less easy to unload and must be collected dense end first.

③ Although developed for rat liver the method may be used for any tissue.

④ The rough ER in the 100 000g pellet tends to make the latter gelatinous; make sure this is well dispersed.

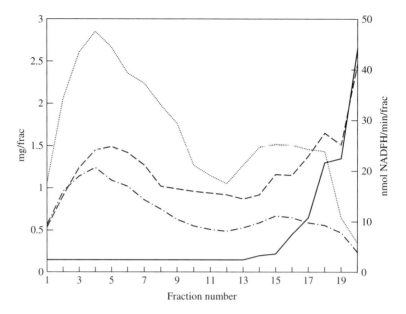

Figure 4.2 Fractionation of rat liver microsomes in a self-generated iodixanol gradient. Starting concentration of iodixanol = 20%; Beckman VTi65.1 vertical rotor at 350 000g for 2 h. Protein (------) mg/fraction, phospholipids (-·-·-·-·) mg/fraction, RNA (———) mg/fraction (×5), NADPH cytochrome c reductase (·············)

⑤ The bottom of the gradient is extremely viscous, due to the high concentration of iodixanol and the presence of RNA, so collection of the gradient dense end first is not really the best option.

⑥ For more information on the harvesting and analysis of gradients, see ref. 18.

⑦ A typical profile of major markers for the SER and RER is given in Figure 4.2. Protein, lipid and NADH

cytochrome c reductase profiles of the gradient reveal two broad bands of material. RNA increases gradually in the lower third of the gradient, the very sharp increase in the bottom fraction being accompanied by a rapid fall in phospholipid and NADH cytochrome c reductase. The SER is thus distributed broadly in the top half of the gradient, the RER in the bottom third and ribosomes band in the last fraction.

NADPH-cytochrome c reductase assay [50]

Reagents

Buffer: 50 mM phosphate, pH 7.7, containing 0.1 mM EDTA

Cytochrome c (25 mg/ml) in buffer (make up fresh; keep on ice)

NADPH (2 mg/ml) in buffer (make up fresh; keep away from light; keep on ice)

Rotenone: 1 mg/ml in ethanol

Equipment

Recording spectrophotometer (visible wavelength) with 1 ml cuvettes. ①

Procedure

The procedure is given for a single-beam spectrophotometer linked up to a chart recorder (0.2 absorbance units full-scale deflection).

1. Bring the buffer to room temperature and carry out all operations at this temperature.

2. In a 1 ml cuvette add 50 μl of cytochrome c and 10 μl of rotenone to 1 ml of buffer, then mix well with up to 50 μl of sample.

3. When the absorbance at 550 nm is steady, add 0.1 ml of NADPH.

4. Mix well and continue to record the absorbance until a linear increase in value can be measured over a period of 1–2 min. ②

Notes

① A dual-beam spectrophotometer is perhaps the best instrument; if one is available, replace the NADPH with buffer in the blank cuvette. Single beam instruments linked up to a chart recorder (used in this procedure) will probably be the most widely available. Even non-recording spectrophotometers can be used, although the measurement is rather more tedious.

② Calculate the enzyme activity in terms of μmoles cytochrome c reduced; the molar extinction coefficient of reduced cytochrome c is 27 000.

Glucose-6-phosphatase assay [49]

Reagents

2.5% (w/v) Ammonium molybdate in 2.5 M H_2SO_4

Fiske-Subbarow reducing solution (prepare and store as per manufacturer's instructions)

Substrate solution: 0.1 M glucose-6-phosphate

Buffer: 5 mM EDTA, 20 mM histidine-HCl buffer pH 6.5

8% (w/v) Trichloroacetic acid (TCA), keep ice-cold

Equipment

Centrifuge (low-speed refrigerated) with swinging-bucket rotor to hold the assay tubes

Spectrophotometer (visible wavelength)

Tubes, approx. 4 ml ①

Water baths set at 37 and 100 °C

Procedure

1. Prepare the test solution: mix substrate and buffer solution in the ratio 2.5 : 2 (v/v).

2. Add 50 µl of sample to 0.45 ml of the test solution and mix. For each sample also set up a control tube containing only 0.45 ml of test solution.

3. Incubate the tubes at 37 °C for 30 min.

4. Transfer the tubes to an ice/water bath; add 50 µl of sample to the control tubes.

5. Add 2.5 ml of ice-cold TCA to all tubes.

6. Keep at 0–4 °C for 20 min. ②

7. Centrifuge all the TCA containing tubes at 1000g for 15 min at 4 °C.

8. Remove 1 ml of supernatant to a new tube.

9. Add 1.15 ml water, 0.25 ml of ammonium molybdate solution and 0.1 ml of reducing solution.

10. Heat all tubes at 100 °C for 10 min.

11. Cool and read absorbance of all test solutions at 820 nm against the appropriate controls.

Notes

This procedure takes 2–3 h.

① Enzyme assays based on the measurement of released phosphate can be a problem if care is not taken to eliminate, as far as possible, phosphate contamination from the tubes and bottles used for the reagents. Glass tubes and glass containers must be acid-washed and then rinsed three times in distilled water before use. Plastic tubes used for the incubation at 37 °C need to be checked for absence of phosphate.

② Longer times are not harmful.

RNA analysis [97]

Reagents

Orcinol reagent: 0.5 g orcinol, 0.25 g FeCl$_3$.6H$_2$O in 50 ml concentrated HCl (make up freshly and keep at 4 °C until required)

Trichloroacetic acid (TCA) solution: 20% (w/v) in water

RNA standard: 250 μg/ml in water

Equipment

Microcentrifuge and 1.5–2.0 ml microcentrifuge tubes

Water bath

Spectrophotometer

Procedure

1. Dilute RNA stock to give 50, 100, 150 and 200 μg/ml solutions.

2. Transfer 0.5 ml of each RNA standard (50–250 μg/ml) and a water blank to microcentrifuge tubes.

3. Dilute gradient samples in microcentrifuge tubes to 0.5 ml with water.

4. Add TCA to all tubes so that the final concentration is 5%.

5. Heat in a water bath at 90 °C for 20 min.

6. Microcentrifuge for 2 min.

7. Transfer the supernatant to another tube and add an equal volume of the orcinol reagent.

8. After 20 min at 100 °C, cool and measure the absorbance at 660 nm. ①

Note

The procedure will take approx. 1.5 h.

① Neither Nycodenz® nor iodixanol interfere with this assay.

Isolation of Golgi membranes from liver [52]

Reagents

Homogenization buffer (HM): 0.5 M sucrose, 1% (w/v) dextran ($M_r = 225\,000$) in 0.05 M Tris-maleate, pH 6.4

Density barrier: 1.2 M sucrose, 0.05 M Tris-maleate, pH 6.4

Include protease inhibitors in these two solutions as required

Equipment

High-speed centrifuge with swinging-bucket rotor for 30–50 ml clear plastic tubes. ①

Phase contrast microscope

Polytron homogenizer

Syringe (10 or 20 ml) and metal cannula ②

Ultracentrifuge with swinging-bucket rotor for 17 ml tubes

Glass rod

Procedure

Carry out all operations at 0–4 °C. The procedure is for a single rat liver (approx. 10 g).

1. Mince the liver finely using a razor blade and suspend in HM (about 10 g liver per 20 ml buffer).

2. Homogenize in a Polytron homogenizer set at 10 000 rpm (setting 1) for 30–50 s. ③

3. Centrifuge for 15 min at 5000g and very carefully aspirate the supernatant using the syringe and cannula. ④

4. The Golgi membranes are contained in the upper (yellow-brown) portion of the bipartite pellet; resuspend them in some of the residual supernatant by very gentle stirring with a glass rod. Be very careful to avoid resuspension of the lower part of the pellet, which contains nuclei, some heavy mitochondria and whole cells.

5. Transfer the resuspended Golgi material to a new tube using the syringe and cannula.

6. Adjust the concentration of the suspension with HM to 6 ml per 10 g liver and layer over 2 volumes of 1.2 M sucrose in a tube for the swinging-bucket rotor.

7. Centrifuge at 120 000g for 30 min.

8. Remove the Golgi membranes that collect at the interface with a syringe and cannula. ⑤

Notes

This procedure will take approx. 1.5 h.

① A fixed-angle rotor is not an option because of the difficulty such a rotor would pose for step 4.

② Metal 'filling' cannulas can be obtained from any surgical equipment supplies company.

③ Use the Polytron in 10 s 'bursts', interposed by 20 s 'rests' to avoid any heating. Monitor the progress of the homogenization by phase contrast microscopy.

④ Keep the tip of the cannula close to the meniscus.

⑤ The dextran in the HM maintains the stacking of the Golgi tubules. If it is necessary to isolate the various domains from the Golgi, the stacks must be disaggregated by hydrolysis of the dextran. This is achieved by incubating the isolated Golgi in 4–5 ml of HM with 3 mg each of α-amylase Type X-A from *Aspergillus oryzae* and α-amylase Type III-A from barley for 45 min at 4 °C. Destacking is completed by gentle liquid shearing through a Pasteur pipette (tip i.d. = 1 mm) or through a syringe with metal cannula [98].

Assay of UDP-galactose galactosyl transferase [54]

Reagents

Keep all solutions on ice.

Acceptor solution: 10 mM ATP, 5% (w/v) ovalbumin in assay buffer (make up fresh)

Assay buffer: 10 mM $MnCl_2$, 30 mM 2-mercaptoethanol, 0.1 M cacodylate buffer, pH 6.2, containing 0.2% (w/v) Triton-X100

TCA (10%, w/v)

UDP-gal: 12.5 mM UDP-galactose containing UDP-[6-^{3}H]-galactose (70 kBq/ml)

Scintillant ①

Equipment

Filter paper discs (2.4 cm)

Ice-water bath

Plastic assay tubes (0.3 ml)

Polystyrene board with pins

Scintillation counter

Water bath at 37 °C

Procedure

1. Prepare two filter paper discs (2.4 cm) per sample (numbered lightly with a pencil).

2. Pin the discs to a polystyrene board, making sure that the discs do not touch the board.

3. Place 250 ml of 10% TCA in a beaker in ice.

4. In 0.3 ml assay tubes (in an ice-water bath), mix 50 μl of acceptor solution with 50 μl sample and add 10 μl radiolabelled UDP-gal solution.

5. At zero time and after a 20 min incubation at 37 °C transfer 50 μl of the incubation mixture to a filter disc.

6. Plunge the discs into the ice-cold TCA. ②

7. Leave for at least 2 h, swirling the discs occasionally.

8. Wash the discs in several changes of distilled water; leave to dry overnight at room temperature.

9. Count in a suitable scintillant.

Notes

This procedure will take 18–20 h.

① Any commercial scintillant for non-aqueous samples will suffice.

② Do not allow the discs to float on the TCA; the pencil mark is liable to come off.

Purification of human erythrocyte 'ghosts' [55]

Reagents

Human blood (0.38% citrate or 1 mM EDTA as anticoagulant)

Hypotonic buffer: dilute isotonic buffer $15.5\times$ with distilled water

Isotonic buffer: mix 0.155 M NaH_2PO_4 and 0.103 M Na_2HPO_4 to pH 7.4

Equipment

High-speed centrifuge with fixed-angle rotor (8×50 ml)

Low-speed (refrigerated) centrifuge with swinging-bucket rotor (50 ml tubes)

Wide-bore pipette attached to automatic pipette-filler

Procedure

Carry out all operations at $0-4\,^{\circ}C$. The procedure is for 30–40 ml of blood.

1. Centrifuge the blood in a capped tube at $800g$ for 15 min. ①

2. Aspirate the plasma and buffy coat into a trap containing disinfectant.

3. Add isotonic buffer to the original volume and gently resuspend the erythrocyte pellet using the wide-bore pipette.

4. Centrifuge the erythrocyte suspension at $800g$ for 15 min.

5. Repeat steps 2–4 three more times. ②

6. After removing the buffer from the final wash, resuspend the erythrocytes in isotonic buffer (32 ml final volume).

7. Place 28 ml of hypotonic buffer in each of 16 tubes for the high-speed centrifuge and from an automatic pipette quickly add 2 ml of the washed erythrocyte suspension. ③

8. Centrifuge at $20\,000g$ for 20 min.

9. Aspirate the supernatant and resuspend the erythrocyte 'ghost' pellet in the hypotonic buffer by gentle pipetting, avoiding any hard-packed buttons. ④

10. Repeat the centrifugation and washing procedure three times.

11. Finally, resuspend the washed 'ghosts' in a suitable buffer.

Notes

The procedure will take approx. 3 h.

① Operators need to be properly trained in the handling of human blood.

② Significant numbers of leukocytes in the final erythrocyte suspension can cause serious aggregation due to lysis of their nuclei during the hypotonic buffer steps. It is therefore important that as much of the leukocyte fraction, which forms the buffy coat, is removed, even at the expense of losing some of the erythrocytes.

③ Effective lysis of the erythrocytes only occurs if the cells are rapidly exposed to the hypotonic buffer.

④ The hard-packed button contains aggregated material (debris, partially disrupted leukocytes and leukocyte nuclei). This must be avoided in the resuspension of the erythrocyte ghosts.

Isolation of plasma membrane sheets from rat liver [56]

Reagents

Homogenization medium (HM): 0.25 M sucrose, 1 mM $MgCl_2$, 10 mM Tris-HCl, pH 7.4

Dense sucrose solution (DSS): 2.0 M sucrose, 1 mM $MgCl_2$, 10 mM Tris-HCl, pH 7.4

Include protease inhibitors in these two solutions as required

Equipment

Dounce homogenizer (loose-fitting Wheaton type B, clearance 0.1–0.3 mm)

Low-speed refrigerated centrifuge with swinging-bucket rotor for 50 ml tubes

Nylon mesh (750 μm pore size)

Refractometer

Syringe (20 ml) with metal filling cannula

Ultracentrifuge with 37–39 ml swinging-bucket rotor and suitable tubes

Procedure

Carry out all operations at 4 °C. The procedure is given for one liver from an adult male rat.

1. Blanch the liver of an anaesthetized rat by perfusing about 20 ml of HM through the portal vein. ①

2. Rapidly remove and weigh the liver, then chop finely with scissors in a beaker on ice. ②

3. Suspend the mince in the HM (approx. 4 ml/g of liver).

4. Homogenize the liver (in two batches) in the Dounce homogenizer using 10 strokes of the pestle. ③

5. Filter the homogenate through the nylon mesh to remove connective tissue and unbroken cells; facilitate filtration by gentle stirring with a glass rod. ④

6. Dilute the filtrate with HM to 5 ml/g liver and divide between two 50 ml tubes for centrifugation at 280g for 5 min.

7. Aspirate the supernatant using a syringe and metal cannula. ⑤ ⑥

8. Centrifuge the 280g supernatant at 1500g for 10 min; decant and discard the supernatant.

9. Resuspend the pellet in 25 ml of HM using the Dounce homogenizer (five strokes of the pestle) and mix well with 2 volumes of DSS. Adjust the density if necessary to 1.18 g/ml by checking the refractive index ($\eta = 1.4106$).

10. Divide the suspension between two tubes for a Beckman SW 28 swinging-bucket rotor and fill by overlaying with about 2 ml of HM and centrifuge at 113 000g for 1 h.

11. Collect the plasma membrane sheets, which band at the interface, using a syringe and metal cannula.

12. Dilute the suspension to 40 ml with HM and harvest the membranes by centrifugation at 3000g for 10 min.

13. Resuspend the pellets in the required volume (approx 1 mg protein/ml) of a suitable medium.

Notes

This procedure will take approx 3 h.

① Cannulation of blood vessels of an experimental animal under anaesthesia must be carried out by licensed and trained operators.

② Failure to mince the liver adequately will result in difficult and irreproducible homogenization.

③ Avoid frothing during homogenization.

④ Do not squeeze the homogenate through the filter.

⑤ Metal 'filling' cannulas can be obtained from any surgical equipment supplies company.

⑥ The pellet is very loosely packed; do not decant the supernatant.

Reagents

Ammonium molybdate: 2.5%, w/v in 2.5 M H_2SO_4

Buffer: 0.1 M Tris-HCl, pH 8.5

Fiske-Subbarow reducing agent (make up according to manufacturer's recommendations)

Levamisole (1 mM): 240 µg/ml ①

Substrate: 20 mM adenosine monophosphate (AMP); 20 mM $MgCl_2$, 0.2 M Tris-HCl, pH 8.5

Trichloroacetic acid (TCA): 10%, w/v

Equipment

Microcentrifuge with 1.5 ml tubes

Plastic assay tubes (1.5 and 2.0 ml)

Spectrophotometer (visible wavelength)

Vortex mixer

Water bath set at 37 °C

Procedure ②

1. Dispense 0.7 ml of TCA in microcentrifuge tubes (two per sample) and keep on ice.

2. On ice, adjust a resuspended membrane pellet or gradient fraction (0.1 ml or less) to 0.5 ml with buffer in assay tubes and mix with 25 µl of levamisole if required.

3. Add 0.5 ml of substrate to all tubes.

4. Transfer 0.3 ml from each tube to one of the prepared TCA containing tubes (zero time reading) and mix.

5. Incubate the remaining samples at 37 °C for 30 min.

6. Transfer 0.3 ml of the suspension to the remaining TCA containing tubes and mix.

7. After a further 30 min on ice, remove the precipitate in a microcentrifuge (1 min).

8. Transfer 0.5 ml of the supernatant to a 2 ml plastic assay tube.

9. For a reagent blank use 0.5 ml from a mixture of 0.7 ml of TCA and 0.2 ml buffer.

10. To each assay tube, add 0.6 ml of water, 125 µl molybdate solution and 50 µl of reducing agent, mixing well after each addition.

11. Heat at 37 °C for 20 min and then measure the absorbance at 820 nm against the appropriate reagent blank. ③

Notes

① Levamisole inhibits the hydrolysis of AMP by a non-specific alkaline phosphatase.

② Enzyme assays based on the measurement of released phosphate can be a problem if care is not taken to eliminate, as far as possible, phosphate contamination from the tubes and bottles used for the reagents. Glass tubes and glass containers must be acid-washed and then rinsed three times in distilled water before use. Plastic tubes used for the incubations need to be checked for absence of phosphate.

③ Inorganic phosphate standards in the range 20–200 nmol should be used in calibrating the assay.

PROTOCOL 4.28

Assay for alkaline phosphodiesterase [100]

Reagents

Buffer: 50 mM glycine-NaOH, pH 9.8

NaOH (0.25 M)

$MgCl_2$ (1.0 M)

Substrate: p-nitrophenylthymidine 5′-phosphate (5 mM) (make up fresh)

Equipment

Microcentrifuge

Plastic assay tubes (1.5 ml)

Vortex mixer

Water bath set at 37 °C

Procedure

1. Prepare the assay solution by mixing equal volumes of buffer and substrate and to every 10 ml add 0.1 ml of 1 M $MgCl_2$.

2. Add 0.5 ml assay solution to a resuspended pellet or gradient fraction (no more than 50 µl). ①

3. Incubate at 37 °C for 20 min and stop the reaction by adding 0.5 ml of NaOH solution.

4. Set up a blank for each sample in which the NaOH is added before the assay reagent (zero time sample) and a reagent blank containing assay solution and NaOH. ②

5. Microcentrifuge (15 000 rpm) for 2 min.

6. Measure the absorbance due to the p-nitrophenol at 410 nm against the reagent blank. The molar extinction coefficient of p-nitrophenol is 9620.

Notes

① If the membrane is at a sufficiently high concentration, gradient samples may be used directly in the assay without removal of solutes such as sucrose, Nycodenz® or iodixanol.

② It may be permissible to omit a blank for each sample. The membrane should contribute no significant light scattering or absorbance at 410 nm in NaOH.

Assay for ouabain-sensitive Na$^+$/K$^+$-ATPase [101]

Reagents

Phosphate reagent A: to 0.75 g ascorbic acid in 25 ml 0.5 M HCl, add 1.25 ml 10% (w/v) ammonium heptamolybdate with vigorous agitation

Phosphate reagent B: 3% (w/v) sodium-*m*-arsenite, 2% (w/v) trisodium citrate (dihydrate) and 2% (v/v) acetic acid

Substrate A: 3 mM MgCl$_2$, 3 mM ATP, 130 mM NaCl, 20 mM KCl, 30 mM histidine buffer, pH 7.5

Substrate B: as medium A plus 1 mM ouabain (0.73 mg/ml) ①

Equipment

Microcentrifuge

Plastic assay tubes (4 ml)

Spectrophotometer (visible wavelength)

Vortex mixer

Water bath set at 37 °C

Procedure ②

1. For each sample set up two tubes containing 1 ml each of substrates A and B.

2. Bring the tubes to 37 °C; add 25 μl of sample to each tube and incubate for 5–10 min. ③

3. Add 0.5 ml of phosphate reagent A to each tube and transfer to an ice-water bath for 6 min.

4. Return to 37 °C and add 1.5 ml phosphate reagent B.

5. After 10 min, transfer 1.5 ml to a microcentrifuge tube and centrifuge for 1 min.

6. Read absorbance of each supernatant at 850 nm. ④

Notes

① Ouabain inhibits specifically the Na$^+$/K$^+$-ATPase.

② Enzyme assays based on the measurement of released phosphate can be a problem if care is not taken to eliminate, as far as possible, phosphate contamination from the tubes and bottles used for the reagents. Glass tubes and glass containers must be acid-washed and then rinsed three times in distilled water before use. Plastic tubes used for the incubations need to be checked for absence of phosphate.

③ The activity of the enzyme is very variable; depending on the source of the membrane, use a time that provides a measurable release of phosphate, while the substrate concentration remains non-rate-limiting.

④ The difference in the phosphate concentration between the two incubations (with and without ouabain) provides that due to the Na$^+$/K$^+$-ATPase.

Isolation of chloroplasts from green leaves or pea seedlings [58]

Reagents

Homogenization buffer (spinach): 330 mM sorbitol, 10 mM $Na_2P_4O_7$ (pyrophosphate), 5 mM $MgCl_2$, 2 mM sodium isoascorbate, pH 6.5 ①

Homogenization buffer (pea): 330 mM glucose, 5 mM $MgCl_2$, 0.1% (w/v) bovine serum albumin (BSA), 0.1% (w/v) NaCl, 0.2% (w/v) sodium iso-ascorbate, 50 mM phosphate buffer, pH 6.5 ①

Wash medium: 330 mM sorbitol, 2 mM EDTA, 1 mM $MnCl_2$, 1 mM $MgCl_2$, 50 mM Hepes-KOH, pH 7.6

40% (v/v) Percoll in wash medium

Equipment

Centrifuge (high speed) with swinging-bucket rotor or fixed angle rotor (250 ml polycarbonate tubes) and swinging-bucket rotor for 15 ml polycarbonate tubes ②

Cheesecloth

Cotton wool

Safety razor blade

Muslin

Rotating blades homogenizer: e.g. Polytron, Ultra-Turrax or Waring Blender

Procedure

Pre-chill all equipment and solutions. Carry out all operations at 0–4 °C and use approx. 50 g of chosen plant material. ③

1. Harvest the material at the start of the light period and clean off any dirt adhering to the plant material in cold water and dry.

2. Devein the spinach leaves.

3. Chop the green leaves or aerial shoots of peas very finely using razor blades and then suspend in the appropriate homogenization buffer (approx. 150 ml).

4. Homogenize in a Polytron at top speed for approx 5–10 s. ④

5. Prepare two filters (A) from two layers of muslin and (B) from a sandwich of cotton wool between four layers of cheesecloth. ⑤

6. Position filter A over a 250 ml beaker on ice.

7. Pre-wet the filter with the homogenization buffer and squeeze the homogenate through the filter.

8. Pass the filtrate through filter B (do not squeeze).

9. Centrifuge the filtrate at 1500–6000g for 0.5–1.0 min. ⑥

10. Decant the supernatant and remove any material adhering to the tubes by wiping with a tissue.

11. Slowly pour a little wash medium down the side of the tube and gently swirl the contents to resuspend any broken chloroplasts, which form a loosely packed layer at the top of the pellet.

12. Decant the liquid and resuspend the remaining pellet in a few ml of wash medium, by very gentle agitation.

13. Dilute to the original volume of homogenate; repeat the centrifugation and resuspend the pellet in 5 ml of wash medium.

14. To remove damaged chloroplasts, layer the suspension over 6 ml of the Percoll solution and centrifuge at $2000–3000g_{av}$ for 1 min. (6)

15. Resuspend the pelleted intact chloroplasts in wash medium.

Notes

The procedure will take approx. 1 h.

(1) There is a great variety of homogenization media and choice of the appropriate medium will depend not only on the material but also on subsequent requirements. For more information, see ref. 58.

(2) It is important for chloroplast intactness that the inner surfaces of the tubes are not scratched.

(3) Some varieties of spinach, which have high levels of oxalate and phenolics, should be avoided (see ref. 58 for more information).

(4) Because of the delicacy of the organelles, homogenization times should be kept to a minimum; it is preferable to accept lower yields in the interest of improved chloroplast integrity.

(5) The muslin filter is critical as this retains the nuclei and broken chloroplasts.

(6) The centrifugation conditions need to be optimized by the user.

PROTOCOL 4.31

Measurement of chloroplast chlorophyll [58]

Reagent

Acetone

Equipment

Glass centrifuge tubes, screw-capped (6 ml)

High-speed centrifuge with swinging-bucket rotor for 6 ml tubes

Spectrophotometer (visible wavelength) with glass cuvettes (1 cm)

Procedure

1. To 10–100 µl of chloroplast suspension, add water to a final volume of 1 ml.
2. Add 4 ml of acetone and mix well.
3. Centrifuge at $3000g$ for 2–3 min.
4. Measure absorbance of supernatant at 652 nm. $A_{652} \times 27.8 = \mu g$ chlorophyll per ml of acetone extract.

PROTOCOL 4.32

Assessment of chloroplast integrity [58]

Reagents

Suspending medium (SM): 660 mM sorbitol, 4 mM EDTA, 2 mM $MnCl_2$, 2 mM $MgCl_2$, 100 mM Hepes-KOH, pH 7.6

Potassium ferricyanide (500 mM): make up fresh

DL-Glyceraldehyde (2 M)

NH_4Cl (500 mM)

Equipment

Illuminator (150 W bulb) with heat filter ①

Oxygen electrode ②

Procedure

1. To the electrode chamber add in the following order: 1 ml SM, 0.9 ml distilled water, 0.1 ml of chloroplast suspension (*Protocol 4.30*), 10 µl ferricyanide and 10 µl glyceraldehyde.

2. Illuminate the mixture in the electrode chamber.

3. After 1 min, uncouple electron flow from proton gradient development by adding 10 µl NH_4Cl.

4. Measure the rate of oxygen evolution (rate 1). ③

5. Repeat steps 1–4, but in step 1 add water to the chloroplast suspension first, followed by SM and the other reagents (rate 2). ④

Notes

① The heat filter may be a glass water bottle.

② Consult the manufacturer's handbook re operating instructions.

③ This is a measure of the rate of ferricyanide reduction by intact chloroplasts plus any broken organelles.

④ This is a measure of the rate of ferricyanide reduction by totally broken chloroplasts. The percentage of intact chloroplasts = [(rate 2 − rate 1)/ rate 2] × 100.

References

1. Graham, J. M. (1993) In: *Methods in Molecular Biology*, Vol. 19 (J. M. Graham and J. A. Higgins, eds), pp. 29–40. Humana Press, Totowa, NJ.
2. Graham, J. M. (1997) In: *Subcellular Fractionation – a Practical Approach* (J. M. Graham and D. Rickwood, eds), pp. 1–29. IRL Press at Oxford University Press.
3. Chappel, J. B. and Perry, S. V (1954) *Nature*, **173**, 1094–1095.
4. Bhattacharya, S. K. Thakar, J. H. Johnson, P. L. and Shanklin, D. R. (1991) *Anal. Biochem.*, **192**, 344–349.
5. Bullock, G. Carter, E. E. and White, A. M. (1970) *FEBS Lett.*, **8**, 109–111.
6. Volkl, A. and Fahimi, H. D. (1985) *Eur. J. Biochem.*, **149**, 257–265.
7. Blobel, G. and Potter, V. R. (1966) *Science*, **154**, 1662–1665.
8. Marsh, M., Schmid, S., Kern, H., Harms, E, Male, P., Mellman, I. and Helenius, A. (1987) *J. Cell Biol.*, **104**, 875–886.

9. Goldberg, D. E. and Kornfeld, S. (1983) *J. Biol. Chem.*, **258**, 3159–3165.

10. Balch, W. E. and Rothman, J. E. (1985) *Arch. Biochem. Biophys.*, **240**, 413–425.

11. Jones, D. R., D'Santos, C. S., Merida, I. and Divecha, N. (2002) *Int. J. Biochem. Cell Biol.*, **34**, 158–168.

12. Anyatonwu, G. I., Buck, E. D. and Ehrlich, B. E. (2003) *J. Biol. Chem.*, **278**,45528–45538.

13. Cai, X. and Lytton, J. (2004) *J. Biol. Chem.*, **279**, 5867–5876.

14. Carlsson, S. R. and Fukuda, M. (1992) *Arch. Biochem. Biophys.*, **296**, 630–639.

15. Graham, J. M. (2001) In: *Biological Centrifugation*, pp. 1–12. Taylor and Francis Books Ltd, Oxford, UK.

16. Graham, J. M. (2001) In: *Biological Centrifugation*, pp. 15–42. Taylor and Francis Books Ltd, Oxford, UK.

17. Graham, J. M. (2001) In: *Biological Centrifugation*, pp. 43–60. Taylor and Francis Books Ltd, Oxford, UK.

18. Graham, J. M. (2001) In: *Biological Centrifugation*, pp. 61–84. Taylor and Francis Books Ltd, Oxford, UK.

19. Osmundsen, H. (1982) *Int. J. Biochem.*, **14**, 905–914.

20. Graham, J., Ford, T. and Rickwood, D. (1994) *Anal. Biochem.*, **220**, 367–373.

21. Provost, J. J., Fudge, J., Israelit, S., Siddiqi, A. R. and Exton, J. H. (1996) *Biochem. J.*, **319**, 285–291.

22. Hofer, T. and Moler, L. (2002) *Chem. Res. Toxicol.*, **15**, 426–432.

23. Yamamoto, Y., Jones, K. A., Mak, B. C., Muehlenbachs, A. and Yeung, R. S. (2002) *Arch. Biochem. Biophys.*, **404**, 210–217.

24. German, D. C., Ng, M. C., Liang, C-L., McMahon, A. and Iacopino, A. M. (1997) *Neuroscience*, **81**, 735–743.

25. Merritt, S. E., Mata, M., Nihalani, D., Zhu,C., Hu, X. and Holzman, L. B. (1999) *J. Biol. Chem.*, **274**, 10195–10202.

26. Iuchi, S., Hoffner, G., Verbeke, P., Dijan, P. and Green, H. (2003) *Proc. Nat. Acad. Sci. USA*, **100**, 2409–2414.

27. Valenzuela, S. M., Martin, D. K., Por, S. B., Robbins, J. M., Warton, K., Bootcov, M. R., Schofield, P. R., Campbell, T. J. and Breit,S. N. (1997) *J. Biol. Chem.*, **272**, 12575–12582.

28. Zippin, J. H., Farrell, J., Huron, D., Kamenetsky, M., Hess, K. C., Fischman, D. A., Levin, L. R. and Buck, J. (2004) *J. Cell Biol.*, **164**, 527–534.

29. Morrison, J. A., Klingelhutz, A. J. and Raab-Traub, N. (2003) *J. Virol.*, **77**, 12276–12284.

30. Sinzger, C., Kahl, M., Laib, K., Klingel, K., Rieger, P., Plachter, B. and Jahn, G. (2000) *J. Gen. Virol.*, **81**, 3021–3035.

31. Qin, Z-H., Wang, Y., Kikly, K. K., Sapp,E., Kegel, K. B., Aronin, N. and DiFiglia, M. (2001) *J. Biol. Chem.*, **276**, 8079–8086.

32. Kegel, K. B., Meloni, A. R., Yi, Y., Kim, Y. J., Doyle, E., Cuiffo, B. G., Sapp, E., Wang, Y., Qin, Z-H., Chen, J. D., Nevins, J. R., Aronin, N. and DiFiglia, M. (2002) *J. Biol. Chem.*, **277**, 7466–7476.

33. Lu, Z., Ghosh, S., Wang, Z. and Hunter, T. (2003) *Cancer Cell*, **4**, 499–515.

34. Shaw, J. P., Large, A. T., Chipman, J. K., Livingstone, D. R. and Peters, L. D. (2000) *Marine Envir. Res.*, **50**, 405–409.

35. Ford, T. C., Baldwin, J. P. and Lambert,S. J. (1998) *Plant Club Ann. Symp. York, UK.*, Abstr. 54.

36. Hovius, R., Lambrechts, H., Nicolay, K. and de Kruijff, B. (1990) *Biochim. Biophys. Acta*, **1021**, 217–226.

37. Graham, J. M., Ford, T. and Rickwood, D. (1990) *Anal. Biochem.*, **187**, 318–323.

38. Glick, B. J. and Pon, L. A. (1995) *Meth. Enzymol.*, **260**, 213–223.

39. Meeusen, S., Tieu, Q., Wong, E., Weiss, E., Schieltz, D., Yates, J. R. and Nunnari, J. (1999) *J. Cell Biol.*, **145**, 291–304.

40. Wattiaux, R. and Wattiaux-De Coninck, S. (1983) In: *Iodinated Density Gradient Media – a Practical Approach* (D. Rickwood, ed.), pp. 119–137. IRL Press at Oxford University Press, Oxford, UK.

41. Okado-Matsumoto, A. and Fridovich, I. (2001) *J. Biol. Chem.*, **276**, 38388–38393.

42. Teoh, M. L. T., Walasek, P. J. and Evans, D. H. (2003) *J. Biol. Chem.*, **278**, 33175–33184.

43. Leighton, F., Poole, B., Beaufay, H., Baudhuin, P., Coffey, J. W., Fowler, S. and de Duve, C. (1968) *J. Cell Biol.*, **37**, 482–513.

44. Symons, L. J. and Jonas, A. J. (1987) *Anal. Biochem.*, **164**, 382–390.

45. Olsson, G. M., Svensson. I., Zdolsek, J. M. and Brunk, U. T. (1989) *Virchows Archiv B Cell Pathol.*, **56**, 385–391.

46. Engstrom, L. (1961) *Biochim. Biophys. Acta*, **52**, 36–48.

47. Mannaerts, G. P., Van Veldhoven, P., Van Broekhoven, A., Vanderbroek, G. and Debeer, L. J. (1982) *Biochem. J.*, **204**, 124–134.

48. Amar-Costesec, A., Godelaine, D. and Hortsch, M. (1988) In: *Cell Free Analysis of Membrane Traffic* (D. J. Morré, K. E. Howell, G. M. C. Cook and W. H. Evans, eds), pp. 211–225. Alan R Liss Inc., New York.

49. Aronson, N. N. and Touster O. (1974) *Meth. Enzymol.*, **31**, 90–102.

50. Williams, C. H. and Kamin, H. (1962) *J. Biol. Chem.*, **237**, 587–595.

51. Balch, W. E., Dunphy, W. G., Braell, W. A. and Rothman, J. E. (1984) *Cell*, **39**, 405–416.

52. Morré, D. J., Cheetham, R. D. and Nyquist, S. E. (1972) *Prep. Biochem.*, **2**, 61–66.

53. Fleischer, B., Fleischer, S. and Ozawa, H. (1969) *J. Cell Biol.*, **43**, 59–79.

54. Beaufay, H., Amar-Costesec, A., Feytmans, E., Thines-Sempoux, D., Wibo, M., Robbi, M. and Berthet, J. (1974) *J. Cell Biol.*, **61**, 188–200.

55. Dodge, J. T., Mitchell, C. and Hanahan, D. J. (1963) *Arch. Biochem. Biophys.*, **100**, 119–130.

56. Poole, R. K. (1993) In: *Methods in Molecular Biology*, Vol. 19 (J. M. Graham and J. A. Higgins, eds), pp. 109–121. Humana Press, Totowa, NJ.

57. Scott, L., Schell, M. J. and Hubbard, A. L. (1993) In *Methods in Molecular Biology*, Vol. 19 (J. M. Graham and J. A. Higgins, eds), pp. 59–69. Humana Press, Totowa, NJ.

58. Whitehouse, D. G. and Moore, A. L. (1997) in *Subcellular Fractionation – a Practical Approach* (J. M. Graham and D. Rickwood, eds), pp. 243–270. IRL Press at Oxford University Press.

59. Kegel, K. B., Kim, M., Sapp, E., McIntyre, C., Castano, J. G., Aronin, N. and DiFiglia, M. (2000) *J. Neurosci.*, **20**, 7268–7278.

60. Wray, W. (1976) In: *Biological Separations in Iodinated Density Gradient Media* (D. Rickwood, ed.), pp. 57–69. IRL Press at Oxford University Press, Oxford.

61. Comings, D. E. (1978) In: *Methods in Cell Biology* (G. Stein, J. Stein and L. J. Kleinsmith, eds), Vol XVII, pp. 115–132. Academic Press, London.

62. Harris, J. R. (1985) *Micron Microscop. Acta*, **16**, 89–108.

63. Harris, J. R. and Milne, J. F. (1974) *Biochem. Soc. Trans.*, **2**, 1251–1253.

64. Harris, J. R. (1978) *Biochim. Biophys. Acta*, **515**, 55–104.

65. Rickwood, D., Messent, A. and Patel, D. (1997) In: *Subcellular Fractionation – a Practical Approach* (J. M. Graham and D. Rickwood, eds), pp. 71–105. IRL Press at Oxford University Press.

66. Stubbs, G. and Harris, J. R. (1978) *Biochem. Soc. Trans.*, **6**, 1172–1174.

67. Marshall, P. and Harris, J. R. (1979) *Biochem Soc. Trans.*, **7**, 928–929.

68. Dwyer, N. and Blobel, G. (1976) *J. Cell Biol.*, **70**, 581–591.

69. Rout, M. P. and Blobel, G. (1993) *J. Cell Biol.*, **123**, 771–783.

70. Belgrader, P., Siegal, A. J. and Berezney, R. (1991) *J. Cell Sci.*, **98**, 281–291.

71. Higashinakagawa, T., Muramatsu, M. and Sugano, H. (1972) *Expt. Cell Res.*, **71**, 65–74.

72. Rice, J. E. and Lindsay, J. G. (1997) In: *Subcellular Fractionation – a Practical Approach* (J. M. Graham and D. Rickwood, eds), pp. 107–142. IRL Press at Oxford University Press.

73. Graham, J. M. (1993) In: *Methods in Molecular Biology*, Vol. 19 (J. M. Graham and J. A.Higgins, eds), pp. 1–18. Humana Press, Totowa, NJ.

74. Young, D. C., Kingsley, S. D., Ryan, K. A. and Dutko, F. J. (1993) *Anal. Biochem.*, **215**, 24–30.

75. Van Veldhoven, P. P., Baumgart, E. and Mannaerts, G. P. (1996) *Anal. Biochem.*, **237**, 17–23.

76. Hunt, M. C., Solaas, K., Kase, B. F. and Alexson, S. E. H. (2002) *J. Biol. Chem.*, **277**, 1128–1138.

77. Solaas, K., Sletta, R. J., Soreide, O. and Kase, B. F. (2000) *Scand. J. Clin. Lab. Invest.*, **60**, 91–102.

78. Dunphy, J. T., Schroeder, H., Leventis, R., Greentree, W. K., Knudsen, J. K., Silvius, J. R. and Linder, M. E. (2000) *Biochim. Biophys. Acta*, **1485**, 185–198.

79. Watkins, P. A., Lu, J-F., Steinberg, S. J., Gould, S. J., Smith, K. D. and Braiterman, L. T. (1998) *J. Biol. Chem.*, **273**, 18210–18219.

80. Thieringer, R., Shio, H., Han, Y., Cohen, G. and Lazarow, P. B. (1991) *Mol. Cell Biol.*, **11**, 510–522.

81. Wilcke, M. and Alexson, S. E. H. (1994) *Eur. J. Biochem.*, **222**, 803–811.

82. Prydz, K., Kase, B. F., Bjorkhem, I. and Pedersen, J. I. (1988) *J. Lipid Res.*, **29**, 997–1004.

83. Appelqvist, E-L., Reinhart, M., Fischer, R., Billheimer, J. and Dallner, G. (1990) *Arch. Biochem. Biophys.*, **282**, 318–325.

84. Baudhuin, P. (1974) *Meth. Enzymol.*, **31**, 17–23.

85. Murata, Y., Sun-Wada, G-H., Yoshimizu, T., Yamamoto, A., Wada, Y. and Futai, M (2002) *J. Biol. Chem.*, **277**, 36296–36303.

86. Sharer, J. D., Shern, J. F., Van Valkenburgh, H., Wallace, D. C. and Kahn, R. A. (2002) *Mol. Biol. Cell*, **13**, 71–83.

87. Liu, J., Takano, T., Papillon, J., Khadir, A. and Cybulsky, A. V. (2001) *Biochem. J.*, **353**, 79–90.

88. Greeve, I., Hermans-Borgmeyer, I., Brellinger, C., Kasper, D., Gomez-Isla, T., Behl, C., Levkau, B. and Nitsch, R. M. (2000) *J. Neurosci.*, **20**, 7345–7352.

89. Simonova, M., Wall, A., Weissleder, R. and Bogdanov, A. (2000) *Cancer Res.*, **60**, 6656–6662.

90. Miyazaki, T., Neff, L., Tanaka, S., Horne, W. C. and Baron, R. (2003) *J. Cell Biol.*, **160**, 709–718.

91. Weissleder, R., Tung, C-H., Mahmood, U. and Bogdanov, A. (1999) *Nature Biotech.*, **17**, 375–378.

92. Solaas, K., Ulvestad, A., Soreide, O. and Kase, B. F. (2000) *J. Lipid Res.*, **41**, 1154–1162.

93. Glunde, K., Guggino, S. E., Ichikawa, Y. and Bhujwalla, Z. M. (2003) *Mol. Imaging*, **2**, 24–36.

94. Sabo, S. L., Lanier, L. M., Ikin, A. F., Khorkova, O., Sahasrabudhe, S., Greengard, P. and Buxbaum, J. D. (1999) *J. Biol. Chem.*, **274**, 7952–7957.

95. Majoul, I. V., Bastiaens, P. I. H. and Soling, H-D. (1996) *J. Cell Biol.*, **133**, 777–789.

96. Cartwright, I. J., Higgins, J. A., Wilkinson, J., Bellavia, S., Kendrick, J. S. and Graham, J. M. (1997) *J. Lipid Res.* **38**, 531–545.

97. Schneider, W. C. (1957) *Meth. Enzymol.*, **3**, 680–684.

98. Hartel-Schenk, S., Minnifield, N., Reuter, W., Hanski, C., Bauer, C. and Morre, D. J. (1991) *Biochim. Biophys. Acta*, **1115**, 108–122.

99. Widnell, C. C. (1974) *Meth. Enzymol.*, **32**, 368–374.

100. Brightwell, R. and Tappel, A. L. (1968) *Arch. Biochem. Biophys.*, **124**, 325–332.

101. Esmann, M. (1988) *Meth. Enzymol.*, **156**, 105–115.

5

Fractionation of Subcellular Membranes in Studies on Membrane Trafficking and Cell Signalling

John Graham

Protocol 5.1	Separation of basolateral and bile canalicular plasma membrane domains from mammalian liver in sucrose gradients	160
Protocol 5.2	Isolation of rat liver sinusoidal domain using antibody-bound beads	162
Protocol 5.3	Fractionation of apical and basolateral domains from Caco-2 cells in a sucrose gradient	163
Protocol 5.4	Fractionation of apical and basolateral domains from MDCK cells in an iodixanol gradient	165
Protocol 5.5	Isolation of lipid rafts	167
Protocol 5.6	Isolation of caveolae	170
Protocol 5.7	Analysis of Golgi and ER subfractions from cultured cells using discontinuous sucrose–D_2O density gradients	172
Protocol 5.8	Analysis of Golgi, ER, ERGIC and other membrane compartments from cultured cells using continuous iodixanol density gradients	174
Protocol 5.9	Analysis of Golgi, ER, TGN and other membrane compartments in sedimentation velocity iodixanol density gradients (continuous or discontinuous)	177
Protocol 5.10	SDS-PAGE of membrane proteins	180
Protocol 5.11	Semi-dry blotting	182
Protocol 5.12	Detection of blotted proteins by enhanced chemiluminescence (ECL)	183
Protocol 5.13	Separation of membranes and cytosolic fractions from (a) mammalian cells and (b) bacteria	185

Cell Biology Protocols. Edited by J. Robin Harris, John Graham, David Rickwood
© 2006 John Wiley & Sons, Ltd

Protocol 5.14   Analysis of early and recycling endosomes in preformed
iodixanol gradients; endocytosis of transferrin in
transfected MDCK cells   188

Protocol 5.15   Analysis of clathrin-coated vesicle processing in
self-generated iodixanol gradients; endocytosis of
asialoglycoprotein by rat liver   191

Protocol 5.16   Polysucrose–Nycodenz® gradients for the analysis of
dense endosome–lysosome events in mammalian liver   194

Introduction

In conjunction with confocal microscopy, density gradients are often used to analyse processes that might be broadly defined as membrane trafficking and cell signalling. They are relevant to a diverse range of cellular events such as secretion, biosynthesis, endocytosis and virus processing. They involve the use of gradients to fractionate not only the principal subcellular organelles such as smooth and rough endoplasmic reticulum (ER) and the Golgi, but also subcompartments such as the *trans*-Golgi network (TGN), the ER-Golgi Intermediate Compartment (ERGIC), early and late endosomes, transport and secretory vesicles and plasma membrane domains such as caveolae and lipid rafts. This rich area of cell biology research also involves studies into the transfer of proteins from the cytosol to a membrane, which might occur for example during cell differentiation.

Methods available

Frequently, a simple post-nuclear supernatant is used as a gradient input rather than a partially purified fraction such as a light mitochondrial pellet or microsomes. Since the gradient is often used not so much to purify one particular type of membrane (see Chapter 4), but rather to analyse the translocation of proteins or other molecules from compartment to compartment, it is often considered more important that none of the potentially involved compartments gets lost in some pre-gradient treatment. Such a simple treatment of the homogenate also reduces the experimentation time and number of manipulations to a minimum.

For many years, sucrose gradients were the principal vehicle for analysing these membrane compartments and a number of the most well-established protocols using this solute are presented in this chapter. Increasingly, however, the use of iodinated density gradient media, notably Nycodenz® and iodixanol, are being used for this purpose and a number of examples of the use of these gradient solutes are given. They have several advantages over sucrose; they are commercially available as dense solutions, making gradient solutions easy to prepare, and their lower osmolality compared to sucrose solutions of similar density (particularly iodixanol solutions) often results in a greater resolving power than that of sucrose.

The viscosity of Nycodenz® and iodixanol solutions is also lower than that of sucrose solutions of similar density and there are many examples that take advantage of the consequent more rapid movement of membrane particles through the gradient. Gradients of Nycodenz® and iodixanol are frequently run for 2–4 h, rather than overnight, which is the norm for sucrose gradients. However, there is a school of thought that maintains that to get true equilibrium density banding it is necessary to centrifuge for long periods (> 12 h) at relatively low centrifugation speeds, irrespective of the gradient medium. A number of iodixanol gradient methods support this contention (see *Protocol 5.8*). There are also some examples of the use of very short centrifugation times, which tend to separate particles principally on the basis of sedimentation velocity (see *Protocol 5. 9*).

One property of iodixanol that is not available for sucrose is its ability to form self-generated gradients, given a sufficiently high *g*-force and the right sort of rotor (vertical or near-vertical) [1]. There are several advantages of using self-generated gradients over the more standard preformed gradients: they are very easy to prepare (the sample is simply adjusted to a certain iodixanol concentration); they are very reproducible and very shallow gradients may be generated (under the appropriate centrifugation conditions) – such gradients are not easily preformed. *Protocol 5.15* is an example of the use a self-generated gradient.

Thus, the protocols in this chapter are generally not aimed at the isolation of an individual membrane compartment; rather they describe the use of a variety of gradient systems that might be used for analysing some aspect of membrane trafficking or cell signalling. An exception is the group of protocols at the start of this chapter that describes the isolation of various types of plasma membrane domain. Because plasma membrane domains have a unique composition and function, they are of particular importance in studies on the means by which the cell directs and controls the flow of molecules to and from the surface.

Plasma membrane domains

Bile canalicular and basolateral domains of rat liver plasma membrane are prepared from the plasma membrane sheets, isolated in *Protocol 4.26*, after vesiculation, which is usually produced either by sonication [2, 3] or liquid shear [4–6]. The two domains are then separated in a continuous or discontinuous sucrose gradient (*Protocol 5.1*). Sometimes Nycodenz® gradients are used to provide additional purification [7, 8]. The sinusoidal domain is rather less easy to purify by density gradients and it is more usually obtained by density perturbation with protein A-sepharose beads (*Protocol 5.2*). The beads are bound to an antibody to the polymeric IgA receptor, often called the secretory component (SC), which is a specific marker for the sinusoidal plasma membrane domain [9]. Once bound to the beads, the sinusoidal membrane vesicles can be harvested in a microcentrifuge. The methodology has been widely reviewed [10, 11]. *Protocols 5.3 and 5.4* provide methods for the fractionation of apical and basolateral domains from polarized cells such Caco-2 [12] and MDCK cells [13, 14] in sucrose and iodixanol gradients respectively.

Lipid rafts, the specialized cholesterol-, sphingolipid- and caveolin-rich microdomains at the surface of a variety of cell types, are involved in an important range of signal-transduction events, virus processing and lipid transport. They are routinely isolated as detergent-resistant membranes (DRMs) by flotation through a discontinuous iodixanol gradient (*Protocol 5.5*). The method was initially worked out for a line of mouse mammary epithelial cells [15] and for MDCK cells [16], but has since been extended to a huge range of cell types and the details of the gradient centrifugation have been varied considerably. In contrast, the flotation gradient for the purification of caveolae (*Protocol 5.6*), which has also been extended from human skin fibroblasts [17] to the use of a vast range of tissue and cell types, has hardly been modified at all [18]. It relies on the sonication of a purified plasma membrane preparation, followed by two iodixanol flotation gradients: the first being a purification in a continuous gradient, the second being a concentration in a discontinuous gradient [17, 18].

Analysis of membrane compartments in the endoplasmic reticulum – Golgi – plasma membrane pathway

Sucrose – D₂O gradients

Before iodinated density gradient media became widely used for analysis of membrane trafficking, one particular strategy to reduce the concentration of sucrose solutions necessary to band subcompartments of the ER and Golgi membrane systems was to dissolve the sucrose in D_2O (*Protocol 5.7*). This allowed gradients of lower osmolality and viscosity to be used [19]. The method was used primarily to study the transfer of proteins from the rough endoplasmic reticulum to the *cis*-Golgi. The gradient showed that this intercompartmental transport involved a low-density vesicle quite distinct from its source and destination membranes. Part of the resolution of such gradients is achieved by collection of the gradients in small fraction volumes (12 ml gradients in 32 fractions). The gradients are also capable of a high degree of resolution of ER-Golgi-*trans* Golgi network (TGN) events [19, 20] and subfractionation of the Golgi membranes in the analysis of N-linked oligosaccharide processing [21].

Buoyant density iodixanol gradients (preformed)

Continuous iodixanol gradients were first used by Yang *et al.* [22] and Zhang *et al.* [23] to fractionate the ER and Golgi from COS-7 and CHO cells (*Protocol 5.8*). The density of these membranes generally decreases in the order ER > Golgi > PM. Most centrifugations are carried out for approx. 3 h, but the separation of ER and Golgi is enhanced by longer periods (16 h) and relatively low *g*-forces ($<100\,000g$). This strategy has permitted the clear resolution of ERGIC in the middle of the gradient between the Golgi and the ER [24]. Long period centrifugations have also permitted the resolution of perinuclear ER from that ER more further from the nuclear region [25]. Most gradients are in the region of 1–25% iodixanol, and depending on the cell type, other compartments such as early and late endosomes and TGN may also be at least partially resolved (see Table 5.2 in *Protocol 5.8* for a summary of some of the variations).

Continuous linear gradients may be constructed by diffusion of the solute between layers of the same volume, whose density increases regularly from one layer to the other. Thus a 10–30% iodixanol gradient might be formed from equal volumes of 10, 20 and 30% iodixanol. Sometimes increased resolution may be obtained in convex or concave gradients. Although there are devices to create such gradients [26] they are not particularly common, nor are they easy to use; an alternative is to construct the gradient from steps of increasing (or decreasing) volume and/or increasing (or decreasing) concentration interval. Table 5.3 in *Protocol 5.8* describes some of the variants and their use.

Sedimentation velocity iodixanol gradients (preformed)

There are several examples of the use of preformed iodixanol gradients for the separation of membrane compartments, principally on the basis of their sedimentation rate (*Protocol 5.9*). They have the advantage of the use of relatively low *g*-forces and also short centrifugation times (never more than 90 min and sometimes as short as 25 min). Their disadvantage is that small changes in gradient and centrifugation conditions, which are unlikely to affect buoyant density separations, may have serious effects on their resolution. Nevertheless it seems that such separations are able to provide useful fractionations: for example, of *cis*-medial Golgi, TGN, ER and early endosomes [27]. Two examples are given, one using a continuous gradient, the other using a discontinuous gradient.

Analysis by SDS-PAGE and electroblotting (Western blotting)

Traditionally, subcellular membranes have been characterized by measuring marker enzymes (usually spectrophotometrically), whose location has been established by histochemistry or by association with a structure identified unambiguously by electron microscopy. Protocols of enzyme assays for all of the major subcellular organelles are given in the relevant sections of Chapter 4. Modern analytical techniques rely more on the antibody probing of Western blots from SDS-PAGE gels. This approach is frequently used to identify proteins in membrane compartments involved in trafficking and cell signalling (see refs 28–30).

The standard Laemmli gel electrophoresis system is commonly used for fractionating membrane proteins. As long as the protein in the gradient fraction is sufficiently concentrated for the detection system, then the non-ionic true solutes such as sucrose, Nycodenz® and iodixanol do not need to be removed before the sample is applied to the gel. These solutes can thus replace the glycerol normally added to the sample to render it dense enough to layer beneath the running buffer. Percoll® however does require removal as the silica particles interfere with the uniform movement of the protein molecules into the gel.

The semi-dry Western blotting apparatus that uses small volumes of buffer and short inter-electrode distances is the most widely used one. Nitrocellulose blotting membranes are the most popular, but polypropylene or polyvinylidine difluoride are more robust and have higher binding capacities. Visualization of the proteins on the blotting membrane can be performed by general methods using Amido Black or the more sensitive silver staining, but only probing the blot with antibodies to specific proteins will provide the means of identifying the subcellular particle. The antibody itself may be tagged with a suitable radiolabel for detection, but it is far more common (and safer) to use a secondary reagent conjugated with an enzyme that is subsequently monitored by an enhanced colorimetric or fluorometric assay. For more information see refs 29 and 30 and *Protocols 5.10–5.12*.

Separation of membrane vesicles from cytosolic proteins

In addition to the frequent requirement for establishing whether a protein has a cytosolic or membrane location, the separation of these two compartments is also often necessary following experiments with permeabilized mammalian cells and also those involving vesicle budding. The strategy for separating membranes and cytosol, which involves flotation of the membranes through an iodixanol barrier from a dense sample, has been developed for both eukaryotes and bacteria (*Protocol 5.13*). This strategy allows maximum resolution of the two components, since the proteins, which are much denser than the membranes, tend to sediment rather than float. It removes any ambiguity in the results from top-loaded sucrose gradients, in which the membranes sediment and the proteins move more slowly in the same direction due to sedimentation and diffusion. The difference in density between the proteins and the membrane vesicles is also much greater in iodixanol than in sucrose, primarily due to the much lower osmolality of the iodixanol gradients. Similar methods are used for the separation of liposomes from denser proteoliposomes.

Endocytosis

Ligand labelling

In most instances the endocytic process is followed experimentally by using a ligand that can be identified by (i) a radiolabel, (ii) a covalently bound enzyme, subsequently detected by provision of a substrate, or (iii) colloidal gold. Although such modifications to the ligand may modulate its behaviour, unless the ligand itself can be easily identified, they are difficult to avoid. The basic strategy is to present the labelled ligand to the tissue or cells in short pulse, after which it is chased through the various intracellular compartments.

Tissue systems

Establishment of a perfused rat liver the system requires an anaesthetized animal; it must be performed by a trained and licensed person. After allowing the isolated perfused liver to equilibrate with the perfusion medium, using a continuous recycling system, the labelled ligand is presented in a single-pass perfusion for 1 min, followed by unlabelled medium for varying times. The process is arrested by perfusion with ice-cold homogenization medium. The perfused liver provides a situation as close to that *in vivo* as possible, but it is difficult to use with very short chase times or when studying large numbers of variables.

Cell systems

A cell monolayer culture provides the most easily managed system for investigating incubation conditions and exposure to perturbants or use of more than one ligand; cultured hepatocytes or MDCK cells are popular choices. The ligand is allowed to bind to the cells at $0\,^\circ$C; the ligand-containing medium is then removed and the monolayer washed with culture medium. Endocytosis is then allowed to proceed at $37\,^\circ$C (in the presence or absence of perturbants) and then rapidly cooled to $0\,^\circ$C (with ice-cold homogenization medium) to arrest further ligand processing. Detailed information on strategies for studying endocytosis has been published elsewhere [31].

Analysis

Sucrose gradients can be used to separate the endocytic compartments on the basis of density [32] but unless the ligand is attached to a marker that can perturb the density of the endosomes, the resolving power is not great. The two most commonly employed methods use either colloidal gold [33] or horseradish peroxidase (HRP) [34]. In the latter case the density perturbation is achieved prior to gradient by incubation of the membrane fraction with H_2O_2 and benzamidine, which is polymerized in the HRP-containing vesicles. There are, however, several gradient techniques which do not require these sometimes inconvenient perturbation techniques and they have been reviewed in ref. 31. Three examples are provided: a preformed iodixanol gradient for analysing early and recycling endosomes (*Protocol 5.14*), a self-generated iodixanol gradient for analysing the processing of clathrin-coated vesicles (*Protocol 5.15*) and a polysucrose-Nycodenz® gradient for analysis of late endosomes/lysosome events (*Protocol 5.16*).

Separation of basolateral and bile canalicular plasma membrane domains from mammalian liver in sucrose gradients [5]

Reagents

Suspension buffer (SB): 0.25 M sucrose, 10 mM Tris-HCl, pH 7.4

Gradient solutions: 35, 39 and 44% (w/v) sucrose in SB

Rat liver plasma membrane sheets (see *Protocol 4.26*)

Include protease inhibitors in these solutions as required

Equipment

Dounce homogenizer (5–10 ml tight-fitting, Wheaton type A)

Phase contrast microscope

Syringe (10 ml) with metal filling cannula ①

Ultracentrifuge with 14–17 ml swinging-bucket rotor and 13 ml fixed-angle rotor (with open-topped tubes)

Procedure

Carry out all operations at 0–4 °C.

1. Suspend the plasma membrane sheets in 2–5 ml of SB.

2. Disrupt the membranes using at least 30 vigorous strokes of the pestle of the Dounce homogenizer. ②

3. Monitor the procedure using a phase contrast microscope (25× objective); the large sheets of membrane should be converted principally to vesicles. ③

4. In tubes for the swinging-bucket rotor, use the syringe and metal cannula to prepare discontinuous gradients from 3–5 ml each of the gradient solutions and layer 1–2 ml of the suspension on top. ④

5. Centrifuge at 196 000g for 3 h.

6. The canalicular membrane vesicles band at the sample/gradient interface; the basolateral (contiguous) domain at the 39/44% sucrose interface.

7. Collect each band with the syringe and cannula; dilute each with two volumes of SB and pellet at 200 000g for 30 min in the fixed-angle rotor.

8. Resuspend the pellet in SB for analysis.

Notes

This procedure will take approx. 4 h.

① Metal 'filling' cannulas can be obtained from any surgical equipment supplies company.

② A bath sonicator (80 W output) or a Polytron homogenizer (setting 8) is an alternative device for disrupting the membrane sheets [2, 3]. Whatever method is used, the disruption must be regularly monitored. Start monitoring the preparation, for example, after 15 strokes of the pestle (Dounce homogenizer) or after five bursts of sonication.

③ The sheets of membrane, which are observed as characteristic dark Y-shaped structures (15–20 μm), should

be replaced by mass of black dots when disruption is complete.

④ The precise volumes will depend on the size of the centrifuge tube. For more information about the preparation of gradients see ref. 26.

Isolation of rat liver sinusoidal domain using antibody-bound beads [9]

Reagents

Anti-SC serum (or other primary antibody)

Medium A: 0.25 M sucrose, 50 mM Tris-HCl, pH 7.4

Medium B: 2% (w/v) Triton X-100, 150 mM NaCl, 2 mM EDTA, 30 mM Tris-HCl, pH, 7.4

Medium C: 1% (w/v) Triton X-100, 0.2% (w/v) SDS, 150 mM NaCl, 5 mM EDTA, 10 mM Tris-HCl, pH 8.0

Medium D: as medium C minus the detergents

Non-immune serum

Protein–sepharose beads

Light mitochondrial supernatant: *Protocol 4.8 as far as step 8 (use medium A for the homogenization medium)* ①

Include protease inhibitors in solutions as required

Equipment

Dounce homogenizer (10 ml loose-fitting, Wheaton type B)

Shaking water bath at 24 °C

Microcentrifuge and microcentrifuge tubes

Ultracentrifuge with fixed-angle rotor for approx 13 ml tubes

Vortex mixer

Procedure

Carry out the isolation of the microsome fraction at 0–4 °C.

1. Centrifuge the supernatant (*Protocol 4.8*, Step 8) at 120 000g for 90 min to pellet the microsomes.

2. Resuspend the microsomes (using the Dounce homogenizer) in medium A (at 2 mg microsomal protein/ml).

3. Incubate 200 µl of a 60% (v/v) suspension of protein A–sepharose beads in medium B at 24 °C with 50 µl of anti-SC serum (or non-immune serum) for 60 min.

4. Pellet the beads in a microcentrifuge and wash them four times in medium C, and a further two times in the medium D (use the vortex mixer to resuspend the beads in the media).

5. Incubate the beads in 500 µl of the microsome fraction for 2 h at room temperature.

6. Sediment the beads in a microcentrifuge and wash four times in medium A and then finally resuspend in this medium for further analysis.

Note

This procedure will take approx. 6 h (including preparation of the light mitochondrial supernatant).

① In *Protocol 4.8* do not omit the first centrifugation step at 800g. Failure to include this step will cause a significant loss of microsomal material due to entrapment by large, rapidly sedimenting particles and debris.

Fractionation of apical and basolateral domains from Caco-2 cells in a sucrose gradient [12]

Reagents

Homogenization buffer (HM): 0.25 M sucrose, 10 mM Tris-HCl, pH 7.4

Suspension buffer B (SB): as HM plus 5 mM EDTA

Hepes-buffered saline (HBS)

1 mM $MgCl_2$

Gradient solutions: 34, 40, 54% (w/v) sucrose in 10 mM Tris-HCl, pH 7.4

Include protease inhibitors in solutions as required

Equipment

Low-speed refrigerated centrifuge with swinging-bucket rotor (10–15 ml tubes)

Nitrogen pressure vessel (Artisan Industries, Waltham, Mass, USA or Baskerville, Manchester, UK)

Syringe and metal filling cannula ①

Ultracentrifuge with (a) fixed-angle rotor for 13.5 ml open-topped tubes and (b) swinging-bucket rotor for approx. 5 ml tubes

Procedure

Carry out all operations at 0–4 °C.

1. Rinse the cell monolayers once in HBS and once in HM.

2. Scrape the cells off the substratum with a rubber policeman into 5 ml of HM.

3. Homogenize the cells by nitrogen cavitation using nitrogen at 3795 kPa (550 psi) for 10 min and collect the homogenate in a beaker (50 ml) covered with Parafilm™, through which the delivery tube is sealed. ② ③

4. Allow the foam to subside by gentle stirring and then centrifuge the homogenate at 270g for 10 min.

5. Recentrifuge the supernatant at 920g for 10 min and carefully decant the supernatant.

6. Add 1 M $MgCl_2$ to the supernatant to a final concentration of 10 mM.

7. Stir for 15 min before centrifuging at 2300g for 15 min.

8. Carefully decant the supernatant and centrifuge it at 170 000g for 45 min.

9. Resuspend the pellet in 0.5 ml of SB.

10. Using the syringe and cannula prepare a discontinuous gradient from 34, 40 and 54% (w/v) sucrose, 10 mM Tris-HCl, pH 7.4 (1.4, 1.7 and 1.4 ml respectively) in tubes for the swinging-bucket rotor. ④

11. Layer the suspension over the gradient and centrifuge at 68 000g for 4.5 h.

12. Remove the material at the interfaces with the syringe. The top band is enriched in Golgi membranes, the middle, in basolateral membranes and the bottom, in apical membranes.

13. Dilute the fractions with at least two volumes of HM and harvest the membranes by centrifugation at about 40 000g for 1 h.

Notes

This procedure will take approx. 9–10 h.

① Metal 'filling' cannulas can be obtained from any surgical equipment supplies company.

② Assemble and operate the nitrogen cavitation vessel in accordance with the manufacturer's recommendations.

Note that when the homogenate is collected, the valve on the delivery tube should be released only very slowly. For more information on the homogenization procedure see ref. 35.

③ It may be permissible to use other cell homogenization media, for example, the ball-bearing homogenizer or if neither of these machines is available, several passages through a gauge 23 or 25 syringe needle may suffice. The gradient fractionation is not known to be homogenization method sensitive. See ref. 35 for more information on the homogenization of cultured cells.

④ For more information on the preparation of gradients see ref. 26.

Fractionation of apical and basolateral domains from MDCK cells in an iodixanol gradient [13, 14]

Reagents

OptiPrep™

Homogenization medium (HM): 0.25 M sucrose, 90 mM KOAc, 2 mM Mg(OAc)$_2$, 20 mM Hepes-KOH, pH 8.0 ①

Diluent: 540 mM KOAc, 12 mM Mg(OAc)$_2$, 120 mM Hepes-KOH, pH 8.0

Working solution (50% iodixanol): mix 5 vol. of OptiPrep™ and 1 vol. of diluent

Include protease inhibitors in HM and the diluent as required

Equipment

Ball-bearing homogenizer ②

Gradient collector ③

Low-speed refrigerated centrifuge with swinging-bucket rotor (10–15 ml tubes)

Phase contrast microscope

Refractometer (Abbé)

Syringe and metal filling cannula ④

Ultracentrifuge with vertical rotor or near-vertical for 11–13 ml tubes ⑤

Procedure

Carry out all operations at 0–4 °C.

1. Homogenize the cells in HM in the ball-bearing homogenizer using four to six passages. Monitor the efficacy of the homogenization by phase contrast microscopy.

2. Centrifuge the homogenate at 1000g for 5 min to pellet the nuclei. Decant and retain the post-nuclear supernatant (PNS).

3. Mix the working solution with the PNS to produce a final iodixanol concentration of 30% (format A) or 35% (format B). ⑥

4. Dilute the working solution with HM to produce solutions of 10 and 20% iodixanol (Format A) or 10, 15, 20, 25 and 30 (Format B). ⑥

5. Layer equal volumes of the dense PNS and the chosen iodixanol solutions in tubes for the vertical or near-vertical rotor. ⑦ ⑧

6. Centrifuge at 353 000g_{av} for 3 h using a slow acceleration program.

7. Allow the centrifuge to decelerate to rest from 2000 rpm without the brake or use a slow deceleration program.

8. Unload the gradient by tube puncture, upward displacement or aspiration from the meniscus in 0.5 ml fractions and assay for membrane markers [13, 14]. Measure the density profile of a blank gradient using a refractometer. ⑨ ⑩

Notes

(1) Homogenization media often have to be tailored to a particular cell type; some alternatives are given in ref. 35.

(2) Other homogenization methods may be permissible, such as Dounce homogenization or repeated passage through a gauge 23 or 25 syringe needle; for more information see ref. 35.

(3) The type of unloading system that is permissible will depend on the tube type. Heat-sealed or crimp-sealed tubes must be unloaded by tube puncture, or possibly aspiration from the bottom of the gradient. Beckman Optiseal™ tubes, which are sealed with a central plastic plug, may also be unloaded by upward displacement with Maxidens™ or by aspiration from the meniscus using the Labconco Auto Densi-flow machine. See ref. 26 for more information.

(4) Metal 'filling' cannulas can be obtained from any surgical equipment supplies company.

(5) The most convenient sealed tubes are the Optiseal™ for Beckman rotors. Some small volume fixed-angle rotors with a sedimentation path length <25 mm may be suitable for making self-generated gradients.

(6) Choose whichever format is most suitable; format A was used by Amieva *et al.* [14]; format B was used by Yeaman [13]. Both will lead to the formation of a gradient that is more or less linear after the centrifugation period.

(7) A small volume of HM may be added to fill the tube to the required level.

(8) Use the syringe and metal cannula to introduce the layers into the tube; see ref. 26 for more information.

(9) Measurement of the density is not an absolute requirement but it makes comparison of banding patterns of basolateral and apical domain markers with published data easier. It also confirms the density profile, which should be linear over most of the gradient, becoming more steep in the densest region.

(10) The peak density of the basolateral domain is approx 1.17 g/ml (about two-thirds of the way down the gradient), while that of the apical domain is approx. 1.08 g/ml (about a third of the way down the gradient).

Isolation of lipid rafts [15, 16]

Reagents

Homogenization medium (HM): 150 mM NaCl, 5 mM dithiothreitol (DTT), 5 mM EDTA, 25 mM Tris-HCl, pH 7.4 ①

Lysis medium (LM): HM + 1% Triton X100 ②

OptiPrep™

Phosphate-buffered saline (PBS)

Triton X-100

Add protease inhibitors to HM as required

Equipment

Dounce homogenizer (tight-fitting, Wheaton type A) or syringe with 23 or 25 gauge needle ③

Gradient collector (optional) ④

Low-speed refrigerated centrifuge with swinging-bucket rotor (approx. 15 ml tubes)

Phase contrast microscope (for isolation from a post-nuclear supernatant only)

Syringe (5 ml) with metal cannula ⑤

Ultracentrifuge with swinging-bucket rotor (5 or 13 ml tubes) ⑥

Procedure

Carry out all operations at 0–4 °C.

Isolation from a total cell lysate

The procedure can be applied to any confluent cells from a single 3 cm culture dish. Scale up all volumes proportionally for larger amounts of cells.

1. Wash the cell monolayer twice with PBS and scrape into this medium using a rubber policeman.

2. Pellet the cells and resuspend in 0.2 ml of LM; then leave on ice for 30 min.

Isolation from a post-nuclear supernatant

The procedure is more convenient with larger numbers of cells (from one or two 9 cm dishes) because of the difficulty in homogenizing very small volumes.

1. Homogenize the cells in HM (up to 2 ml) using up to 20 strokes of the pestle of the Dounce homogenizer or up to the same number of passes through the syringe needle. ⑦

2. Centrifuge the homogenate at 1000*g* for 10 min.

3. Decant the supernatant and adjust it to 1% Triton X-100 and leave on ice for 30 min.

Gradient separation

1. Add 2 vol. of OptiPrep™ to 1 vol. of either the homogenate or 1000*g* supernatant.

2. Dilute OptiPrep™ with LM to give 35, 30, 25 and 20% (w/v) iodixanol. ⑧

3. In tubes for the swinging-bucket rotor layer 0.6 ml each of the sample, the four

Table 5.1 Some methodological variations for the isolation of lipid rafts

Source material	EDTA; DTT; TX100[1]	Iodixanol gradient[2]	RCF (time)	Ref.
MDCK TGN vesicles	5 mM; 5 mM; 0.1%	30%, 20%, 10%, 5%	160 000g (4 h)	16
MDCK cells	2 mM; 2 mM; 2.0%	43%, 35%, 30%, 25%, 20%, 0%	250 000g (2.5 h)	36
	1 mM; none; 2.0%	40%, 30%, 5%	78 000g (4 h)	37
	5 mM[3]; 5 mM; 1.0%	40%, 25%, 0%	160 000g (4 h)	38
	5 mM[3]; 1 mM; 1.0%	40%, 30%, 25%, 5%[4]	100 000g (4 h)	39
	none; 1 mM; 1.0%	40%, 35%, 30%, 25%, 20%, 0%	120 000g (12 h)	40
Drosophila	0.2 mM[3]; none; 2.0%	24%, 21%, 15%, 6%	130 000g (5 h)	41
Human breast carcinoma	5 mM; none; 0.2%	35%, 30%, 0%	170 000g (4 h)	42
HEK293	5 mM; none; 0.1%	35%; 30%	160 000g (4 h)	43
	5 mM; 1 mM; 2.0%[5]	45%, 35%, 30%, 20%, 0%	180 000g (4 h)	44
Yeast	5 mM; none; 1.0%	40%, 30%, 0% (0.1% TX100)	200 000g (2 h)	45
	5 mM; none; 1.0%	35%, 30%, 0%	147 000g (16 h)	46
Jurkat	5 mM; none; 1.0%	40%, 35%, 30%, 25%, 0%	150 000g (7 h)	47
Oligodendrocytes	5 mM; none; 1.0%[6]	40%, 30%, 0%	200 000g (2 h)	48
Neurons	2 mM; 5 mM; 1%	40%, 30%, 5%	150 000g (5 h)	49
Neuroblastoma cells	5 mM, none, 0.1%	35%, 30%, 0%	200 000g (4 h)	50
Ciliary ganglion[7]	none; 1 mM; 0.1%	35%, 30%, 0%	285 000g (4 h)	51
COS-1	5 mM; none; 0.5%	50%, 40%, 30%, 20%, 10%	170 000g (4 h)	52

[1] Most cell lysis media also contained NaCl and a Tris buffer, DTT = dithiothreitol, TX100 = Triton X100.
[2] The first figure in each series gives the % iodixanol in the sample.
[3] EGTA was used in place of EDTA.
[4] All solutions contained 10% sucrose, the density layers did not contain DTT.
[5] Density solutions contained Ca^{2+} and no DTT.
[6] 20 mM CHAPS was also used in place of triton X-100.
[7] Membrane fraction first obtained in preliminary OptiPrep™ gradient.

gradient solutions and LM to fill the tube. Modify or scale up the gradient for larger sample volumes ⑧

4. Centrifuge at 160 000g_{av} for 4 h. ⑨

5. Collect the lipid rafts from the top interface or harvest the gradient in a number of equal volume fractions and analyse as required. ⑩

Notes

The procedure will take approx. 5.5 h, using a 4 h centrifugation.

① The isolation media used by both Oliferenko *et al.* [15] and Lafont *et al.* [16] were similar, although the level of DTT used by Oliferenko

et al. [15] was 1 mM rather than 5 mM and EDTA was omitted. Some proteins, which associate with lipid rafts, exhibit a Ca^{2+}-dependence, so inclusion of a chelating agent may be detrimental to the study. DTT is often now omitted and sometimes the level of Triton X-100 in the discontinuous gradient is lower than in the sample layer. Protease inhibitors such as PMSF, leupeptin, antipain, aprotinin, etc. should be included in all of the media. Some of the variations in isolation media are given in Table 5.1 (column 2)

② The level of Triton X100 is sometimes increased to 2% or reduced as low as 0.1% (see Table 5.1). Occasionally Triton X-100 is substituted by CHAPS [48, 53]; sometimes the detergent is in the sample but omitted from the density gradient solutions.

③ Use whatever device is known to be suitable for the cell type; a ball-bearing homogenizer might be a useful alternative. For more information on the homogenization of cultured cells see ref. 35.

④ For gradient collection low-density end first use either the tube puncture device of the Beckman Fraction Recovery System to deliver a dense liquid such as Fluorinert™ (Sigma Aldrich) or Flutec-Blue™ (F2 chemicals Ltd, Preston UK) to the bottom of the tube or the Labconco Auto Densi-flow device to aspirate from the meniscus. For collection dense-end first use tube-puncture. More information about gradient collection can be found in ref. 26.

⑤ Metal 'filling' cannulas (i.d. 0.8–1.0 mm) can be obtained from any surgical equipment supplies company.

⑥ The smaller volume tubes are generally sufficient if a total cell lysate is to be used for the gradient input, but a larger volume tube may be easier for scale-ups or if a post-nuclear supernatant is to be prepared.

⑦ The details of the homogenization will have to be worked out for the specific cell type used. Use phase contrast microscopy to monitor the homogenization.

⑧ Some of the variations in gradient format are given in Table 5.1 (column 3). Some workers have reduced the total number of layers to three (see Table 5.1); in this format the 5 ml tubes are capable of accommodating larger sample volumes. Lipid raft subdomains are best analysed using a different gradient format (see ref. 52 and Table 5.1). It may even be better to use a continuous gradient rather than a discontinuous one.

⑨ Oliferenko et al. [15] used a longer centrifugation time of 12 h at a slightly lower RCF ($120\,000g_{av}$). Because of the relatively short sediment path length of the rotor, 4 h at the higher RCF is probably satisfactory, but the centrifugation conditions may vary with the mode of preparation (see Table 5.1).

⑩ Depending on the resolution that is required it may be sufficient to use an automatic pipette to collect the gradient in four or five broad zones. Alternatively for higher resolution the gradient should be unloaded either by tube puncture, upward displacement or automatic aspiration from the meniscus (see note 4 and ref. 26). Always check on the distribution of raft and non-raft markers in the gradient to check that the centrifugation has achieved a satisfactory resolution and recovery of rafts.

Protocol 5.6

Isolation of caveolae [17, 18]

Reagents

OptiPrep™

Diluent: 0.25 M sucrose, 6 mM EDTA, 120 mM Tricine-NaOH, pH 7.8

Working solution (50% iodixanol, $\rho = 1.282$ g/ml): mix 5 vol. of OptiPrep™ with 1 vol. of diluent

Suspension medium (SM): 0.25 M sucrose, 1 mM EDTA, 20 mM Tricine-NaOH, pH 7.6

Add protease inhibitors to diluent and SM as required

Equipment

Gradient maker (two-chamber) or Gradient Master™

Gradient collector (optional)

Sonicator

Syringe and metal cannula (i.d. 0.8–1.0 mm) ①

Ultracentrifuge with swinging bucket (14–17 ml tube size)

Procedure

Carry out all operations at 0–4 °C.

1. Prepare a plasma membrane fraction by a suitable technique. ②

2. Suspend the plasma membrane in 2 ml of SM in a suitable tube (approx. 1.5 cm diameter).

3. Place the tube in ice and introduce the tip of a sonicator probe (approx. diameter 3 mm) to a point equidistant from the top and bottom of the suspension.

4. Sonicate twice for 6 s at a total power of 50 J/W per second, then allow to rest for 2 min before repeating the sonication procedure twice more, i.e. a total of six sonication bursts. ③

5. Add 1.84 ml of 50% iodixanol working solution and 0.16 ml of SM. The final iodixanol concentration is 23% (w/v).

6. Produce two gradient solutions of 10% (w/v) iodixanol ($\rho = 1.076$ g/ml) and 20% (w/v) iodixanol ($\rho = 1.125$ g/ml) by diluting the working solution with SM (1 + 4 and 2 + 3 v/v respectively). ④

7. Using a standard two-chamber gradient former or a Gradient Master™ produce a 9 ml linear 10–20% iodixanol gradient in an approx. 13 ml tube for the swinging-bucket rotor. ④ ⑤

8. Using the syringe and metal cannula, underlayer the gradient with 4 ml of the sample and centrifuge at $53\,000g_{max}$ for 90 min.

9. Collect the top 5 ml of the gradient; transfer to a new centrifuge tube and mix with 4 ml of working solution. ⑥ ⑦

10. Produce two new gradient solutions of 5 and 15% (w/v iodixanol by diluting working solution with SM (1 + 9 and 3 + 7 v/v respectively)

11. Layer 1.0 ml of 15% and 0.5 ml of 5% iodixanol over the sample and centrifuge at 52 000*g* for 90 min. ⑧

12. Collect the caveolae-rich opaque layer that forms above the 15% iodixanol layer.

Notes

The procedure, excluding the preparation of the plasma membrane, will take approx. 4.5 h.

① Metal 'filling' cannulas (i.d. 0.8–1.0 mm) can be obtained from any surgical equipment supplies company.

② The method used to prepare the plasma membrane will vary with the type of tissue or cell. As the source material may be a 'routine' cultured cell line such as human fibroblasts [54, 55], COS 7 [56, 57] or NIH 3T3 [18], a genetically modified line such as CHO-ldlA-7 [58, 59] or tissue-derived cells such as lung microvascular endothelial cells [60, 61] or brain astrocytes [62], the operator should refer to the appropriate literature for a plasma membrane isolation technique. However, one of the methods described in *Protocols 5.8 and 5.9* may be relevant.

③ If the sonicator does not conform to these parameters, it may be necessary to modulate the conditions in order to achieve the necessary fragmentation of the membrane. This can only be monitored by determining the caveolin and protein distribution in the first gradient (step 9).

④ The volume of the gradient needs to be calculated to fill the tube when the 4 ml sample is underlayered. An approximately 9 ml gradient was used by Smart *et al.* [17]. The method can be scaled up or down as required, adjusting the volumes of sample and gradient proportionately.

⑤ If a gradient-forming device is not available, then prepare a discontinuous gradient from equal volumes of 10, 15 and 20% iodixanol and allow them to diffuse overnight at 4 °C. See ref. 26 for more information about making gradients.

⑥ If the volume of the gradient is increased, the volume occupied by the caveolae-containing fractions may also increase. As an approximation, the top third-to-half of the gradient should be removed. If the relative volumes of gradient and sample are significantly different it may be advisable to check the distribution of the caveolae by assaying fractions for caveolin by electroblotting (see *Protocols 5.10–5.12*).

⑦ Until the banding of the caveolin-rich, protein-poor material in the gradient is well established, it might be pertinent to collect the gradient by upward displacement with a dense medium; for more information on unloading gradients see ref. 26. Once the banding position of the caveolae has been established, the relevant part of the gradient can be removed using a syringe and cannula.

⑧ The volumes used in the second gradient may also have to be modified for larger volume tubes; it is probably good practice to increase the volumes of the sample and the two iodixanol layers, proportionally.

Analysis of Golgi and ER subfractions from cultured cells using discontinuous sucrose–D$_2$O density gradients [19]

Reagents

Homogenization buffer (HB): 0.25 M sucrose, 10 mM Hepes-NaOH, pH 7.4

Gradient solutions: 10, 12.5, 15, 17.5, 20, 22.5, 25, 27.5, 30 and 50% (w/v) sucrose in 10 mM Hepes-NaOH, pD 7.4 in D$_2$O

Add protease inhibitors to all solutions as required

Equipment

Dounce homogenizer (10 ml tight-fitting, Wheaton type A) or ball-bearing homogenizer or syringe fitted with 25 gauge needle

Gradient collector (upward displacement or collection from the meniscus) ①

Low-speed refrigerated centrifuge with swinging-bucket rotor for 10 ml plastic tubes

Phase contrast microscope

Syringe (5 ml) and metal filling cannula ②

Ultracentrifuge with swinging-bucket rotor (approx. 13 ml tubes)

Procedure

Carry out all operations at 0–4 °C.

1. Suspend the cells in HB (no more than 5 ml) and homogenize using up to 20 strokes of the pestle of the Dounce homogenizer, or up to 10 passes through the ball-bearing homogenizer or syringe. ③

2. Pellet the nuclei by centrifugation at 1000g for 10 min.

3. Using the syringe and cannula, prepare a discontinuous gradient from 1 ml of each of the gradient solutions in tubes for the ultracentrifuge rotor. ④

4. Layer 2 ml of the post-nuclear supernatant over the gradient and centrifuge at 160 000g for 3 h.

5. Unload the gradient by upward displacement (or aspiration from the meniscus) into approx 0.3 ml fractions. ⑤

Notes

① For gradient collection low-density end first use either the tube puncture device of the Beckman Fraction Recovery System to deliver a dense liquid such as Fluorinert™ (Sigma Aldrich) or Flutec-Blue™ (F2 chemicals Ltd, Preston UK) to the bottom of the tube or the Labconco Auto Densi-flow device to aspirate from the meniscus. For collection dense-end first use tube puncture. More information about gradient collection can be found in ref. 26.

② Metal 'filling' cannulas (i.d. 0.8–1.0 mm) can be obtained from any surgical equipment supplies company.

③ Monitor the progress of the homogenization by phase contrast microscopy; ideally no less than 90% cell lysis should be obtained, while the nuclei remain intact. More information about cell homogenization can be found in ref. 35.

④ Ref. 26 provides more information about the preparation of gradients.

⑤ Low density ER-Golgi intermediates band in the region 12.5–17.5% sucrose, while the ER itself bands around 22% sucrose.

Analysis of Golgi, ER, ERGIC and other membrane compartments from cultured cells using continuous iodixanol density gradients [22, 23]

Reagents

OptiPrep™

Homogenization medium (HM): 0.25 M sucrose, 1 mM EDTA 10 mM Hepes-NaOH, pH 7.4 ①

Diluent: 0.25 M sucrose, 6 mM EDTA, 60 mM Hepes-NaOH, pH 7.4

50% (w/v) Iodixanol working solution ($\rho =$ 1.272 g/ml): 5 vol. of OptiPrep™ + 1 vol. of diluent

Phosphate-buffered saline

Add protease inhibitors to HM and diluent as required

Equipment

Ball-bearing homogenizer, Dounce homogenizer (5 ml tight-fitting, Wheaton type A) or syringe with 25-gauge needle ②

Gradient maker (two-chamber or Gradient Master™)

Gradient collector ③

Low-speed refrigerated and/or high-speed centrifuge with swinging bucket rotor for 15 ml tubes

Phase contrast microscope

Ultracentrifuge with swinging bucket rotor for approx. 13 ml tubes and fixed-angle rotor for 4–10 ml open-topped tubes (optional, see step 5 below) ④

Procedure

Carry out all operations at 0–4 °C.

1. Wash the cells twice in phosphate-buffered saline to remove the culture medium, and then once in the homogenization medium before resuspending in this medium. ⑤

2. Suspend the cells in a small volume of HM (0.5–5.0 ml) and disrupt them in a ball-bearing homogenizer, Dounce homogenizer or repeated passages through a fine syringe needle. ⑥

3. Centrifuge the homogenate at 1000–3000g for 10 min; decant and retain the supernatant.

4. Resuspend the pellet in 0.5–2.0 ml of HM; repeat step 3 and combine the two supernatants. ⑦

5. **Optional step:** Centrifuge the supernatant(s) at 100 000g for 40 min and then resuspend the pellet in 1–2 ml of HM. ⑧

6. Prepare a 25% (w/v) iodixanol solution by mixing equal volumes of HM and the 50% iodixanol working solution.

7. Prepare 12 ml gradients in tubes for the swinging-bucket rotor from equal volumes of HM and the 25% iodixanol

Table 5.2 Preformed continuous iodixanol gradients (2–3 h) for buoyant density separation of ER, Golgi, PM and other membrane compartments: variations in gradient and centrifugation conditions

Cell type	Input[1]	Grad[2]	g	Time (h)	Membrane compartments[3]	Ref.
BHK	pns	2.5–25	130 000	3.0	ER; G; PM; early endosomes	63
CV1	pns	8–22	95 000	2.0	ER sub; *cis*-G	64
	pns	2.5–25	100 000	3.0	ER sub; G; PM	65
HEK293	pns	0–26	288 000	2.0	ER; G; PM	66
	pns	10–24	170 000	1.5	ER; *cis*-medial G	67
LLC-PK₁	pns	2.5–25	100 000	3.0	ER; G; PM	68
3T3	pns	2.5–25	100 000	3.0	ER sub; G; PM	69
M7 transgenic	ppns	0–15	200 000	2.0	ER; G; PM	70
Neuroblastoma	lmp	0–26	160 000	2.0	ER; light vesicles	71
RBL-2H3[4]	ppns	1–20	200 000	3.0	ER; G; PM; granules	72

[1] pns = post-nuclear supernatant, ppns = 80–100 000g pellet from a pns, lmp = light mitochondrial pellet from pns.
[2] Gradient density range in % iodixanol.
[3] Membrane compartments analysed: ER = endoplasmic reticulum, G = Golgi, PM = plasma membrane, ERGIC = ER Golgi intermediate compartment, sub = subfraction.
[4] RBL = rat basophilic leukaemia cells.

Table 5.3 Preformed continuous iodixanol gradients (16–18 h) for buoyant density separation of ER, Golgi, PM and other membrane compartments: variations in gradient conditions

Cell type	Input[1]	Grad[2]	g	Time (h)	Membrane compartments[3]	Ref.
CHO	pns	8–34	100 000	18	ER; G; early endosomes	73
H4	pns	10–28	100 000	18	ER; G; ERGIC	24
LLC-PK₁	lms	8–34	100 000	18	ER; G; PM	74
PC12	pns	5–25	88 000	16	PM, early endosomes	75
3T3		10–40	48 000	18	ER sub; end; coated vesicles	25

[1] pns = post-nuclear supernatant, lms = light mitochondrial supernatant.
[2] Gradient density range in % iodixanol.
[3] Membrane compartments analysed: ER = endoplasmic reticulum, G = Golgi, PM = plasma membrane, ERGIC = ER Golgi intermediate compartment, sub = subfraction.

solution using a two-chamber gradient maker or a Gradient Master. ⑨ ⑩

8. Layer the vesicle suspension on top of the gradient and centrifuge at 200 000g for 2–3 h. ⑩

9. Collect the gradient in 0.5 ml fractions by tube puncture or upward displacement and analyse by SDS-PAGE and electroblotting, probing with the appropriate antibodies to proteins in the membranes of interest (see *Protocols 5.10–5.12*). Prior to analysis, it may be necessary to concentrate the fraction either by precipitation with trichloroacetic acid or by diluting it with 2 vol. of HM and pelleting by centrifugation at 100 000–150 000g for 40 min before resuspending in a small volume of HM. See the references in Tables 5.2 and 5.3 for details of the banding of the membrane compartments from a variety of cell types.

Notes

This procedure will take approx. 2 h.

① With cultured cells the homogenization medium often has to be tailored to the cell type. Sometimes an alternative buffer such as 10 mM triethanolamine-acetic acid or the HM used in *Protocol 5.4* achieves a better cell disruption. For more information see ref. 35.

② The device of choice is the ball-bearing homogenizer. Sometimes a syringe needle with a lower gauge number (wider orifice) will work satisfactorily. As with the homogenization medium, the homogenization protocol will need to be tailored to the cell type. See ref. 35 for more information.

③ For gradient collection low-density end first use either the tube puncture device of the Beckman Fraction Recovery System to deliver a dense liquid such as Fluorinert™ (Sigma Aldrich) or Flutec-Blue™ (F2 Chemicals Ltd, Preston UK) to the bottom of the tube or the Labconco Auto Densi-flow device to aspirate from the meniscus. For collection dense-end first use tube puncture. More information about gradient collection can be found in ref. 26.

④ Other swinging-bucket rotors or even vertical rotors may be used. Larger volume swinging-bucket rotors may require longer centrifugation times but vertical rotors will need shorter times. Most of the gradients have been run in 13 ml tubes but the gradients and sample volume can be scaled down as required.

⑤ Washing the cells with saline may be carried out at room temperature rather than at 4 °C if preferred.

⑥ Up to 10 cycles with the ball-bearing homogenizer, up to 15 strokes of

the pestle of a Dounce homogenization or up to 20 passages through a syringe needle should be sufficient. Always monitor the progress of homogenization by phase contrast microscopy.

⑦ Washing the low-speed pellet will release some trapped vesicles and improve yields.

⑧ The advantage of using a low-speed supernatant for the gradient input is that the procedure is quick and that vesicles in the supernatant are not further exposed to the shearing forces that are required to resuspend a $100\,000g$ pellet. On the other hand, if it is important to remove any soluble cytosolic proteins then preparing a $100\,000g$ pellet may be a useful step.

⑨ If neither of these devices is available then a continuous gradient can be prepared by diffusion of a discontinuous gradient [26].

⑩ The actual density range of the gradient may require modulation to suit the operator's requirements. If the main interest, for example, is the Golgi and lower density fractions then a 0–20% or 0–15% iodixanol gradient may be more useful. Table 5.2 summarizes some of the variations for the 2–3 h centrifugations. Table 5.3 summarizes some of the 16–18 h centrifugation separations. A commonly used non-linear gradient is generated from layering the following iodixanol solutions: 1 ml of 2.5%, 2 ml each of 5, 7.5 and 10%, 0.5 ml of 12.5%, 2 ml of 15%, and 0.5 ml each of 17.5, 20 and 30%. These gradients appear particularly good for resolving the TGN [76–80].

PROTOCOL 5.9

Analysis of Golgi, ER, TGN and other membrane compartments in sedimentation velocity iodixanol density gradients (continuous or discontinuous)

Reagents

OptiPrep™

Homogenization medium I (HMI): 0.25 M sucrose, 1 mM EDTA 10 mM Hepes-NaOH, pH 7.4 ①

Diluent I: 0.25 M sucrose, 6 mM EDTA, 60 mM Hepes-NaOH, pH 7.4 ①

Homogenization medium II (HMII): 130 mM KCl, 25 mM NaCl, 1 mM EGTA, 25 mM Tris-Cl, pH 7.4 ②

Diluent II: 130 mM KCl, 25 mM NaCl, 6 mM EGTA, 150 mM Tris-Cl, pH 7.4 ②

Working solution of 50% (w/v) iodixanol: 5 vol. of HM +1 vol. of diluent I or II ③

Equipment

Ball-bearing homogenizer, Dounce homogenizer (5 ml tight-fitting, Wheaton type A) or syringe with 25-gauge needle ④

Gradient collector ⑤

Gradient maker (two-chamber or Gradient Master™), for continuous gradient only ⑥

Low-speed refrigerated and/or high-speed centrifuge with swinging bucket rotor for 15 ml tubes

Syringe (5 ml) and metal cannula, for discontinuous gradient only ⑦

Ultracentrifuge with swinging-bucket rotor for approx. 13 ml tubes ⑧

Procedure

Carry out all operations at 0–4 °C.

1. Produce a cell homogenate from the chosen tissue or cells using HM (I or II) using steps 1–2 of *Protocol 5.8*.

2. Centrifuge the homogenate at 1000g for 10 min.

3. Centrifuge the supernatant from step 2 at 3000g for 10 min; decant and retain the supernatant.

4. **For a discontinuous gradient**: Prepare 10 ml each of 2.5, 5, 7.5, 10, 12.5, 15, 17.5, 20 and 30% (w/v) iodixanol solution by mixing HM with the 50% iodixanol working solution containing the appropriate diluent (I or II).

5. **For a continuous gradient**: Prepare 20 ml of 10 and 30% iodixanol solutions as in step 4.

6. In 13 ml tubes for the swinging-bucket rotor **(discontinuous gradient)**: layer 1.2 ml each of the nine gradient solutions using the syringe and metal

Table 5.4 Preformed iodixanol gradients for sedimentation velocity separation of ER, Golgi and other membrane compartments

Cell/tissue	Input[1]	Gradient	g	Min	Membrane compartments[2]	Ref.
CHO	pns	10–30% (or 25%) continuous	54 000	90	ER; G; PM	82
HEK93	pns	2.5–30% discontinuous	126 000	30	ER sub, G sub	83, 84, 85
MDCK	lms	0–40% continuous	85 000	45	ER, G, end, TGN	86
Mouse brain	hms	2.5–30% discontinuous	120 000	32	ER sub, G sub	87
Mouse brain and fibroblasts	hms	2.5–30% discontinuous	126 000	30	ER sub; G sub	81
Neuroblastoma	hms	2.5–30% discontinuous	126 000	30	ER; *cis*- med-G, TGN, EE	27
Vero	hms	2.5–30% discontinuous	126 000	25	ER, G, mit	88, 89
Xenopus	hms	10–30% (or 25%) continuous	26 000	90	ER, G, ERGIC, post-TGN	90

[1] pns = post-nuclear supernatant; lms = light mitochondrial supernatant; hms = heavy mitochondrial supernatant.
[2] Membrane compartments analysed: ER = endoplasmic reticulum, G = Golgi, sub = subfractions, EE = early endosome, TGN = *trans*-Golgi network, ERGIC = ER Golgi intermediate compartment.

cannula; (**continuous gradient**): use a two-chamber gradient maker or Gradient Master™ to make an approx. 12 ml gradient from equal volumes of the 10 and 30% iodixanol solutions. ⑨

7. Layer the supernatant from step 3 on top of the gradient.

8. Centrifuge at $126\,000g_{\mathrm{av}}$ for 30 min, $85\,000g$ for 45 min or $26\,000g$ for 90 min. ⑩

9. Collect the gradient in 0.5 ml fractions by tube puncture or upward displacement and analyse by SDS-PAGE and electroblotting, probing with the appropriate antibodies to proteins in the membranes of interest (see *Protocols 5.10–5.12*). Prior to analysis, it may be necessary to concentrate the fraction either by precipitation with trichloroacetic acid or by diluting it with 2 vol. of HM and pelleting by centrifugation at $100\,000$–$150\,000g$ for

40 min before resuspending in a small volume of HM. See the references in Table 5.4 for details of the banding of the membrane compartments from a variety of cell types.

Notes

This procedure will take approx. 4 h.

① This is general purpose HM and diluent and has been used for many types of cultured cells and tissues.

② This salt-containing HM and diluent was used for neuroblastoma cells [27] and brain tissue [81] but may be particularly applicable to any cells and tissues that release proteins that may cause agglutination of membranes. See ref. 35 for other homogenization media.

③ Use the appropriate diluent for the chosen HM.

④ The device of choice is the ball-bearing homogenizer. Sometimes a syringe needle with a lower gauge number (wider orifice) will work satisfactorily. As with the homogenization medium, the homogenization protocol will need to be tailored to the cell type. See ref. 35 for more information.

⑤ For gradient collection low-density end first use either the tube puncture device of the Beckman Fraction Recovery System to deliver a dense liquid such as Fluorinert™ (Sigma Aldrich) or Flutec-Blue™ (F2 Chemicals Ltd, Preston UK) to the bottom of the tube or the Labconco Auto Densiflow device to aspirate from the meniscus. For collection dense-end first use tube puncture. More information about gradient collection can be found in ref. 26.

⑥ If neither of these devices is available then a continuous gradient can be prepared by diffusion of a discontinuous gradient [26].

⑦ Metal 'filling' cannulas (i.d. 0.8–1.0 mm) can be obtained from any surgical equipment supplies company.

⑧ Other swinging-bucket rotors or even vertical rotors may be used. Larger volume swinging-bucket rotors may require longer centrifugation times but vertical rotors will need shorter times. Most of the gradients have been run in 13 ml tubes but the gradients and sample volume can be scaled down as required.

⑨ See ref. 26 for information on making discontinuous gradients. Continuous gradients can be prepared by allowing a discontinuous gradient of equal volumes of 10, 20 and 30% iodixanol to diffuse overnight.

⑩ These are the most widely reported centrifugation conditions; see Table 5.4 for examples of separations using the various conditions described in this protocol.

SDS-PAGE of membrane proteins

Reagents

Ammonium persulfate (10% w/v, freshly made up)

Reservoir buffer: 0.192 M glycine, 25 mM Tris, 0.1% (w/v) SDS

10% (w/v) SDS

Sample buffer (SB): 4% SDS, 0.002% bromophenol blue, 20% glycerol, 10% 2-mercaptoethanol, 125 mM Tris-HCl, pH 6.5 ①

Separating gel buffer: 1.5 M Tris-HCl, pH 8.8

Stacking gel buffer: 1.0 M Tris-HCl, pH 8.8

Stock acrylamide (30% T, 2.7% C): 20.2 g acrylamide, 0.8 g bisacrylamide in 100 ml distilled water Deionize with 1 g Amberlite MB-1 monobed resin per 100 ml for 2 h. Store in an amber bottle at 4 °C after filtering

N,N,N',N'-tetramethylenediamine (TEMED)

20% (w/v) trichloroacetic acid (TCA)

Water-saturated isobutanol

Equipment

Commercial gel cassette system, power pack, etc.

Narrow-ended tips for automatic pipette

Vacuum pump and side-arm flask

Procedure [28, 30]

This procedure will take approx. 21–36 h.

1. Assemble the gel mould cassette according to the manufacturer's instructions.

2. Make a 10% T separating gel by mixing: 33.3 ml of acrylamide stock, 25 ml of 1.5 M Tris-HCl, pH 8.8, 40.5 ml of distilled water and 150 μl of ammonium persulfate. ②

3. Degas the solution using the vacuum pump.

4. Carefully add 1.0 ml of SDS and 50 μl of TEMED; mix by gentle swirling.

5. Pour the gel solution carefully into the cassette to about 1 cm below the position of the comb.

6. Overlayer the gel with a small volume of the water saturated isobutanol.

7. Allow the gel to polymerize for 2–16 h at room temperature. ③

8. Prepare the stacking gel solution from: 3 ml of acrylamide stock, 3.75 ml of 1.5 M Tris-HCl, pH 6.8, 22.8 ml of distilled water and 135 μl of ammonium persulfate.

9. After degassing, carefully add 375 μl of SDS and 30 μl of TEMED; mix by gentle swirling.

10. Fill the cassette with the stacking gel solution and carefully insert the comb.

11. Polymerize for 2 h at room temperature.

12. Remove the comb and set up the cassette in the electrophoresis apparatus according to the manufacturer's instructions.

13. Add 1 vol. of sample buffer to 1 vol. of membrane suspension and heat at 100 °C for 3 min.

14. Apply the samples into the wells of the stacking gel with the micropipette; 20–100 μl of the sample is commonly used, depending on the well size and the protein concentration. ④

15. Run the gels at constant current. Standard 1.5 mm thick gels can be run at 15 mA per gel overnight.

16. Remove the gels from the cassette, fix in TCA if necessary and stain as required, unless the gels are going to be electroblotted (see *Protocol 5.11*). ⑤ ⑥ ⑦

Notes

① The glycerol may be omitted if dense gradient fractions containing sucrose, Nycodenz® or iodixanol are to be used directly. For membrane fractions in 0.25 M sucrose include the glycerol.

② This 10% T gel is a useful general purpose gel for resolving most proteins in the M_r range 20 000–150 000; for resolution of higher M_r proteins use a 5% T gel, for lower M_r values use a 15 or 20% T gel.

③ Polymerization overnight is often convenient.

④ The amount of sample loaded on to each track of a standard SDS-PAGE gel will vary with the sample and will depend on the staining technique used. With Coomassie blue staining the load should be 50–100 μg of protein; with silver staining, 10–50 μg of protein is normally sufficient.

⑤ Coomassie blue stock: 0.125 g Coomassie brilliant blue in methanol-water (405 : 520 v/v); add 7 ml of glacial acetic acid in 100 ml before use. Destain in the same solvent mixture.

⑥ Staining is enhanced if the gels are first rinsed with water to remove some of the SDS.

⑦ For glycoproteins use periodic acid-Schiff or Alcian blue [28, 30].

Semi-dry blotting

Reagents

Transfer buffer: 20 mM Tris, 150 mM glycine, pH 8.3. ①

Equipment

Filter paper (Whatman 3 mM) cut to size of gel

Rocker

Semi-dry apparatus with high current capability power supply (e.g. Bio-Rad PowerPac HC)

Transfer membrane (nitrocellulose) cut to size of gel

Procedure [29, 30] ②

1. Place gel from cassette (see *Protocol 5.10*) in transfer buffer for 30 min. ③

2. Soak 12 sheets of filter paper and the nitrocellulose membrane in transfer buffer. ③

3. After allowing excess buffer to drain off each paper and gel, place six sheets of the filter paper, the membrane, the gel and six more sheets of filter paper on the anode plate of the semi-dry apparatus.

4. Place the cathode in position.

5. Transfer the proteins at ~ 0.8 mA/cm^2 for 40 min.

6. Blotted proteins are normally detected by antibody binding using a primary antibody to a membrane protein, peroxidase-conjugated rabbit IgG as secondary antibody followed by enhanced chemiluminescence using one of a variety of commercially available kits. See *Protocol 5.12* and refs 29 and 30 for more information.

Notes

The procedure will take approx. 2 h.

① An alternative is to use an anode buffer of 60 mM Tris-HCl, pH 9.6, 40 mM CAPS, 15% methanol and a cathode buffer of 60 mM Tris-HCl, pH 9.6, 40 mM CAPS, 0.1% SDS.

② Carry out all reactions at room temperature and handle all materials with gloved hands.

③ If the two-buffer system is used, soak the gel and six sheets of filter paper in the anode buffer and the other six sheets in the cathode buffer.

Detection of blotted proteins by enhanced chemiluminescence (ECL)

Reagents

Diaminobenzidine (DAB): 0.05% (w/v) DAB, 0.01% (v/v) H_2O_2 in PBS

ECL electroblotting detection reagents; a variety of kits are commercially available and very convenient to use

Phosphate buffered saline (PBS): 8 g NaCl, 0.2 g KCl, 1.15 g Na_2HPO_4, 0.2 g KH_2PO_4 in 1 litre of water

PBS-T: 0.05% Tween 20 in PBS

PBS-TM: 3% (w/v) non-fat dried milk in PBS-T

Primary (monoclonal) antibody for detecting membrane antigens

Secondary antibody reagent: e.g. peroxidase-conjugated rabbit IgG; dilute this 1 : 1000 in PBS-TM

Equipment

Cling-film

Detection film (blue-light sensitive) and suitable photoradiography cassette ① ②

Glass plate

Plastic bags ③

Rocker

Procedure [29, 30]

Carry out all operations at room temperature and use gloved hands.

1. Remover the membrane carefully from the semi-dry blotter and incubate in a plastic bag with PBS-TM for 1 h to block non-specific binding sites on the membrane.

2. Transfer the membrane to a suitable dilution (e.g. 1 : 1000–1:10 000) of the monoclonal antibody and incubate for 1 h. ④

3. Wash the membrane three times in PBS-T.

4. Incubate for 1 h with the peroxidase-conjugated secondary antibody.

5. Wash the membrane three times in PBS-T.

6. Mix the ECL reagents as recommended by the manufacturer and incubate with the membrane (0.125 ml/cm^2 membrane) for 1 min.

7. Drain excess solution from the membrane; place on a glass plate (protein side uppermost) and cover with cling-film to prevent drying out. ⑤

8. Expose the membrane to the film in the dark. ⑥

9. Develop the film as recommended by the manufacturer.

10. To visualize the proteins after completion of the ECL exposure, wash the membrane twice for 15 min in PBS-T and stain with the DAB solution for 10 min.

Notes

① Consult local manufacturer's instructions and ref. 30 for details.

② Rather than use the traditional film method, the blot can also be detected via a gel documentation system such as a Bio-Rad ChemiDoc XRS or a VersaDoc imager. The convenience of this method and the production of a digital image make this an increasingly popular choice.

③ Compared to plastic boxes, small volume plastic bags make more economical use of often expensive or sparingly available antibodies.

④ Use the minimum volumes permitted by the bag.

⑤ Avoid all air bubbles and creasing of the membrane.

⑥ Times between 15 s and 60 min are common. Check the manufacturer's instructions.

Separation of membranes and cytosolic fractions from (a) mammalian cells and (b) bacteria

(a) Mammalian cells [91, 92]

Reagents

OptiPrep™

OptiPrep™ diluent (OD): 140 mM KCl, 60 mM Hepes-KOH, pH 7.2 containing *either* 15 mM Mg(OAc)$_2$ *or* 12 mM EGTA and 6 mM DTT ① ②

Working solution (WS) of 50% iodixanol: mix 5 vol. of OptiPrep™ with 1 vol. of OD

WS diluent: 140 mM KCl, 10 mM Hepes-KOH, pH 7.2 containing *either* 2.5 mM Mg(OAc)$_2$ *or* 2 mM EGTA and 1 mM DTT ① ②

Include protease inhibitors in solutions as required

Equipment

Gradient collector (optional)

Low-speed refrigerated centrifuge or microcentrifuge (in cold room)

Ultracentrifuge with swinging-bucket rotor for approx 5 ml tubes

Procedure

Carry out all operations at 0–4 °C.

1. Remove cells from the vesicle-containing suspension either in a microcentrifuge or by centrifugation at 1000g for 5 min in a low-speed centrifuge. ③

2. Aspirate the supernatant and adjust it to 30% (w/v) iodixanol by thorough mixing with WS. ④

3. Prepare solutions containing 25 and 5% (w/v) iodixanol by diluting WS with WS diluent. ⑤

4. In tubes for the swinging-bucket rotor layer 2 ml each of the crude vesicle fraction in 30% iodixanol and the 25% iodixanol. ⑥

5. Fill the tube by overlaying with 5% iodixanol. ⑦

6. Centrifuge at approx. 250 000g for 3 h; collect the vesicles which band at the top interface and the bottom layer containing cytosolic proteins. Alternatively the gradient may be collected in a number of equal volume fractions. ⑧

(b) Bacteria [93] ⑨

Reagents

OptiPrep™

OptiPrep™ diluent (OD): 3.0 M KOAc, 30 mM Mg(OAc)$_2$, 300 mM Hepes-KOH, pH 7.6 ⑩

Working solution (WS) of 50% iodixanol: mix 5 vol. of OptiPrep™ with 1 vol. of OD

Suspension buffer (SB): 0.5 M KOAc, 5 mM Mg(OAc)$_2$, 50 mM Hepes-KOH, pH 7.6

WS diluent: 0.125 M sucrose in SB ⑩

Include protease inhibitors in solutions as required

Equipment

Ultracentrifuge with swinging-bucket rotor for approx. 5 ml tubes

Procedure

1. Suspend the crude vesicle preparation in SB.

2. Adjust the suspension to 44% iodixanol by addition of WS.

3. Transfer to tubes for the chosen swinging-bucket rotor.

4. Dilute WS with WS diluent to give a 30% iodixanol solution.

5. Overlayer the sample with approx. 5 vol. of 30% iodixanol. ⑥

6. Fill the tubes by overlayering with SB and centrifuge at approx $170\,000g$ for 3 h. ⑦

7. Either collect the gradient in three or four fractions or harvest the vesicle band at the top interface and the soluble protein fraction (original sample zone at the bottom of the tube). ⑧

Notes

① The preparation of a 50% iodixanol working solution using OptiPrep™ and OD maintains the concentration of buffer and $Mg(OAc)_2$ (or EGTA and DTT) constant through the gradient when WS is diluted with WS diluent. The buffer and additives may be changed to suit the operator's requirements; as long as their concentration in the gradient is no more than approx. 20 mM, they can be present in the OD at $6\times$ the required concentration.

The KCl concentration in the OD is not similarly raised; if it were, gradient solutions would be highly hyperosmotic and affect the density of the vesicles.

② The MgOAc was used by Love *et al.* [92] in the separation of vesicles and cytosol from permeabilized cells and the DTT/EGTA used by Scheiffele *et al.* [91] in the separation of budded vesicles and cytosol.

③ If the requirement is to separate the cytosol and total microsomes from a homogenate it is probably advisable to centrifuge the homogenate at $15\,000g$ for 15 min to remove all of the larger organelles (nuclei, mitochondria, lysosomes, peroxisomes) before adjusting the density of the supernatant.

④ Rather than adjusting the crude vesicle fraction to 30% iodixanol by the addition of a 50% iodixanol WS, Love *et al.* [92] first pelleted the vesicles and then suspended them directly in 30% iodixanol. Either strategy should be effective.

⑤ Most published protocols use very similar discontinuous gradient strategies; it is worth noting, however, that Joglekar *et al.*[94] used a continuous 5–25% iodixanol gradient to separate COP-coated vesicles (budded from the ER/Golgi) from the cytosol also by flotation.

⑥ The procedure can be scaled down or up as required to suit the sample volume; change the volume of the solution above the sample proportionally; the small volume of low-density solution on top need not be increased.

⑦ As with all flotation methods a small volume of buffer or 5–10% iodixanol is always layered on top of the low-density barrier to prevent banding of the vesicles at an air/liquid interface.

⑧ See ref. 26 for more information on gradient fractionation. Small volume gradients may be divided into three of four zones simply by very careful aspiration using an automatic pipette.

⑨ The same flotation strategy has been applied to lysed yeast spheroplasts; the lysate is adjusted to 37% (w/v) iodixanol; layered beneath 30, 25, 19 and 0% iodixanol; and centrifuged at $75\,000g_{av}$ for 4 h to separate cytosolic and vacuolar fractions [95].

⑩ The preparation of a 50% iodixanol working solution (WS) using Opti-Prep™ and OD, maintains the concentration of buffer KOAc and Mg(OAc)$_2$ constant through the gradient when WS is diluted with WS diluent. The buffer and additives may be changed to suit the operator's requirements; they can be present in OD at 6× the required concentration as long as their final concentration in the gradient does not significantly alter its osmolality.

Analysis of early and recycling endosomes in preformed iodixanol gradients; endocytosis of transferrin in transfected MDCK cells [96]

Reagents

OptiPrep™

Optiprep™ diluent (OD): 235 mM KCl, 12 mM MgCl$_2$, 25 mM CaCl$_2$, 30 mM EGTA, 150 mM Hepes-NaOH pH 7.0

40% Iodixanol working solution (WS): 2 vol. of OptiPrep™ + 1 vol. of OD

WS diluent: 78 mM KCl, 4 mM MgCl$_2$, 8.4 mM CaCl$_2$, 10 mM EGTA, 50 mM Hepes-NaOH pH 7.0

Homogenization medium (HM): 0.25 M sucrose, 78 mM KCl, 4 mM MgCl$_2$, 8.4 mM CaCl$_2$, 10 mM EGTA, 50 mM Hepcs-NaOII pH 7.0 ①

Protease inhibitors should be added to the HM and WS diluent as required

Equipment

Ball-bearing homogenizer, syringe with 25-gauge needle or tight-fitting Dounce homogenizer (Wheaton type A) or other suitable device ②

Gradient collector ③

Gradient maker (two-chamber) or Gradient Master™

Low-speed refrigerated centrifuge with swinging-bucket rotor for 15 ml tubes

Phase contrast microscope

Ultracentrifuge with swinging-bucket rotor for 13–17 ml tubes

Procedure

Following any experimental procedures to allow binding, uptake and processing of a ligand or other functional manipulations, all operations must be carried out at 4 °C.

1. Remove any unbound ligand or other components from the cell surface by washing the cell monolayer twice in any solution compatible with the study and then scrape the cells in 1 ml of HM. ④

2. Homogenize the cells using a ball-bearing homogenizer (cell cracker); four passes of the cell suspension should be sufficient. Check for adequate homogenization by phase contrast microscopy. If such a device is not available then use several passages through a fine gauge syringe needle or a Dounce homogenizer. ⑤

3. Centrifuge the homogenate in a swinging-bucket rotor at 1000g for 5 min to pellet the nuclei and any cell debris. The pellet may be washed with HM if necessary and the two supernatants combined.

4. Prepare the low- and high-density gradient solutions of 5 and 20% (w/v) iodixanol by diluting WS with the WS diluent.

5. In tubes for the swinging-bucket rotor prepare 11–14 ml 5–20% iodixanol gradients using either a two-chamber gradient maker or a Gradient Master. ⑥

6. Layer the 1000g supernatant(s) on top of the gradient and centrifuge at 90 000g_{av} for 18–20 h. ⑦ ⑧

7. Collect the gradient in approx. 0.25 ml fractions either by upward displacement with a dense liquid or by tube puncture or by aspiration from the meniscus. The delivery of the gradient into a multi-well plate via a Gilson Fraction Collector is a very convenient mode for analysis. ⑨ ⑩

8. If it is necessary to remove the iodixanol, fractions can be pelleted at 200 000g for 20 min after dilution with 3 vol. of HM.

Notes

① Other homogenization media may be permissible and may need modulating if other cell types are used; see ref. 35 for more details.

② The device of choice is the ball-bearing homogenizer. Sometimes a syringe needle with a lower gauge number (wider orifice) will work satisfactorily. As with the homogenization medium, the homogenization protocol will need to be tailored to

the cell type. See ref. 35 for more information.

③ Either the Beckman Fraction Recovery or the Labconco Auto Densi-flow is suitable. For more information see ref. 26.

④ Surface-bound transferrin, for example, is removed by washing the cells twice in 280 mM sucrose, 25 mM citric acid, 24.5 mM sodium citrate, pH 4.6 [96].

⑤ Up to 10 cycles with the ball-bearing homogenizer may be needed (depending on cell type and/or HM composition), up to 15 strokes of the pestle of a Dounce homogenization or up to 20 passages through a syringe needle should be sufficient. Always monitor the progress of homogenization by phase contrast microscopy.

⑥ Alternatively allow a discontinuous gradient of equal volumes of 5, 10, 15 and 20% iodixanol to diffuse overnight. See ref. 26 for information on making continuous gradients.

⑦ Satisfactory separation of endosomal compartments from other cells or tissues may require modulation of either the density gradient or the

Table 5.5 Preformed iodixanol gradients for analysis of endocytosis

Cell/tissue	Input[1]	Grad[2]	g	hours	Fractionation[3]	Analysis	Ref.
CHO (25RA)	pns	5–20	90 000	20.0	EE, LE, TGN, PM	LDL-derived cholesterol trafficking	97
HeLa	pns	5–20	125 000	20.0	EE, RE, PM, lys	Degradation of EGF receptor	98
	pns	5–20	90 000	18.0	EE, RE, PM	Rab11 endocytic targeting	99
MDCK	pns	5–20	90 000	18.0	EE, RE, PM, lys	Transferrin endocytosis	96
NRK	pns	12.5–30[d]	113 000	1.5	EE	Caveolin recruitment to early endosomes	100
Renal BB[4]	mic	15–25	100 000	3.0	CP, endosomes	Myosin localization	101

[1] pns = post-nuclear supernatant; mic = microsomal fraction.
[2] Gradient density range in % iodixanol; [d] = discontinuous.
[3] EE = early endosomes; LE = late endosomes, RE = recycling endosomes, TGN = *trans* Golgi network, PM = plasma membrane; CP = coated pit; lys = lysosomes.
[4] BB = brush border.

centrifugation conditions. This can only be determined by trial and error.

⑧ Use of other centrifugation conditions (shorter times at higher RCFs for example) should be checked against the recommended parameters for their efficacy.

⑨ For gradient collection low-density end first, use either the tube puncture device of the Beckman Fraction Recovery System to deliver a dense liquid such as Maxidens™ to the bottom of the tube or the Labconco Auto Densi-flow device to aspirate from the meniscus. For collection dense-end first, use tube puncture. More information about gradient collection can be found in ref. 26.

⑩ In this gradient the density of plasma membrane, lysosomes, recycling endosomes and early endosomes increases in that order[96]. Some other published papers that have reported the use of similar gradients are given in Table 5.5.

Analysis of clathrin-coated vesicle processing in self-generated iodixanol gradients; endocytosis of asialoglycoprotein by rat liver [102, 103]

Reagents

Homogenization medium (HM): 0.25 M sucrose, 1 mM EDTA, 10 mM Tris-HCl, pH 7.4 ①

OptiPrep™

OptiPrep™ diluent (OD): 0.25 M sucrose, 6 mM EDTA, 60 mM Tris-HCl, pH 7.4

Working solution (WS) of 50% iodixanol: 5 vol. of OptiPrep™ + 1 vol. of OD.

Add protease inhibitors to HM and OD as required

Equipment

Gradient collector ②

High-speed centrifuge with swinging-bucket for 40–50 ml tubes

Syringe (2 ml) with metal cannula ③

Ultracentrifuge with a vertical or near vertical rotor capable of approx. $350\,000g$ (e.g. Beckman VTi65.1 or NVT65) with 11.2 ml Optiseal™ tubes ④

Procedure ⑤

Carry out ligand binding, uptake and processing, as required. Subsequently all operations must be carried out at $4\,^{\circ}$C.

1. Homogenize the liver in HM (approx. 4 ml per g of tissue) as described in steps 4–6 of *Protocol 4.6*.

2. Centrifuge the homogenate in a swinging-bucket rotor at $3000g_{av}$ for 10 min. The pellet may be washed with HM if necessary and the two supernatants combined. ⑥

3. Make a 20% iodixanol solution ($\rho = 1.127$ g/ml) by diluting WS with HM.

4. Dilute the $3000g$ supernatant with WS $3:1$ (v/v), so the final iodixanol concentration is 12.5% (w/v).

5. Transfer approx. 9 ml of the suspension to the Optiseal™ tube for the vertical or near-vertical rotor.

6. Underlay with 1.5 ml 20% iodixanol and overlay with HM to fill the tube. ⑦

7. Centrifuge at approx. $350\,000g_{av}$ for 1.5 h (slow acceleration to 800 rpm). Use slow deceleration (or no brake) from 800 rpm. ⑧

8. Collect the gradient, low-density end first, by upward displacement or aspiration from the meniscus or dense end first by tube puncture, in approx. 0.5 ml fractions and analyse as required. ⑨

9. If it is necessary to remove cytosolic proteins from the fractions and/or to

concentrate them, dilute with an equal volume of buffer and sediment the membranes at approx. 350 000g for 15 min. ⑩

Notes

This procedure will take approx. 3–4 h (excluding the operations prior to homogenization).

① The inclusion of EDTA in this buffer removes ligand bound to the plasma membrane so that only intracellular compartments are labelled.

② Either the Beckman Fraction Recovery System or the Labconco Auto Densiflow is suitable. For more information see ref. 26.

③ Metal 'filling' cannulas (i.d. 0.8–1.0 mm) can be obtained from any surgical equipment supplies company.

④ The sedimentation path length of the tube should be <24 mm. The protocol provides centrifugation times and g-forces for 11 ml Optiseal™ tubes for the Beckman VTi65.1; they may need to be optimized to produce the required iodixanol density gradient in other rotors. Smaller volume rotors can be used with little or no modification to the protocol, but larger volumes may require significantly longer centrifugation times. For more information on the formation of self-generated gradients see ref. 26. Although tubes other than Optiseal tubes may be used, the latter are the easiest of any sealed tube type for gradient unloading.

⑤ Although this fractionation protocol was used to study the endocytosis of neogalactosylalbumin by the perfused rat liver, it may also be used to analyse

the internalization of other ligands by, for example, isolated hepatocytes.

⑥ Although washing of the pellet will maximize the recovery of endosomes, it may make the volume of the post-heavy mitochondrial supernatant inconveniently large.

⑦ If a near-vertical rotor is used, the 20% iodixanol cushion may be omitted, it is present in tubes for vertical rotors to prevent any dense particle reaching the wall of the centrifuge tube; likewise the overlay may also be omitted in near-vertical rotors.

⑧ Slow acceleration is not needed if the dense cushion and overlay are omitted with near-vertical rotors.

⑨ For gradient collection low-density end first, use either the tube puncture device of the Beckman Fraction Recovery System to deliver a dense liquid such as Fluorinert™ (Sigma Aldrich) or Flutec-Blue™ (F2 chemicals Ltd, Preston UK) to the bottom of the tube or the Labconco Auto Densi-flow device to aspirate from the meniscus. For collection dense-end first, use tube puncture. More information about gradient collection can be found in ref. 26. Under the recommended conditions, the gradient contains a central shallow region to separate light and dense endosomes. Lysosomes band in the sharp gradient formed at the bottom of the tube, while mitochondria and peroxisomes band below the lysosomes. For a more linear gradient use longer centrifugation times. To subfractionate the early clathrin-coated vesicles and the plasma membrane, which tend to have a high density, the starting concentration of iodixanol should be increased to 15 or 17.5%. To analyse more effectively the low-density endosomes, the starting concentration of iodixanol might be reduced to 10%.

⑩ By removing cytosolic proteins <u>after</u> fractionation, endosomes need not be pelleted and resuspended prior to separation. Small volume open-topped thick-walled tubes for a micro-ultra centrifuge are a convenient way of recovering sedimented membrane fractions. Do not use more than 15 min at $350\,000g$ or $1-1.5$ h at $100\,000g$, otherwise sedimentation of the iodixanol molecules themselves may interfere with pellet formation.

Polysucrose–Nycodenz® gradients for the analysis of dense endosome–lysosome events in mammalian liver [104–108]

Reagents

Nycoprep™ Universal

Homogenization medium (HM): 0.25 M sucrose, 1 mM $MgCl_2$, 10 mM TES-NaOH, pH 7.4

Nycoprep™ Universal diluent (NUD): 6 mM EDTA, 60 mM TES-NaOH, pH 7.4

Gradient diluent (GD): 0.25 M sucrose, 1 mM EDTA, 10 mM TES-NaOH, pH 7.4

Nycodenz® working solution (50%, w/v): mix 5 vol. of Nycoprep™ Universal with 1 vol. of NUD

45% (w/v) Nycodenz®: mix 4.5 vol. of Nycodenz® working solution with 0.5 vol. of 1 mM EDTA, 10 mM TES-NaOH, pH 7.4 (use NUD diluted with 5 vol. of water)

20% (w/v) Nycodenz®: mix 2 vol. of 45% Nycodenz® with 2.5 vol. of GD

20% (w/v) polysucrose in GD ① ②

Equipment

Gradient collector ③

High-speed or refrigerated low-speed centrifuge with swinging bucket for 40–50 ml tubes

Syringe (2 ml) with metal cannula ④

Ultracentrifuge and vertical rotor with a tube volume of approx. 36 ml (e.g. Beckman VTi50 rotor) ⑤

Procedure ⑥

Carry out ligand binding, uptake and processing, as required. Carry out all operations at 0–4 °C.

1. Perfuse with the liver with HM until the lobes are well blanched. ⑦

2. Homogenize the liver in HM (approx. 4 ml per g of tissue) as described in steps 3 and 4 of Protocol 4.7.

3. Centrifuge the homogenate at 2000g for 10 min to sediment cell debris, nuclei and most of the heavy mitochondria.

4. Transfer 12.5 ml of the 20% polysucrose to 36 ml Optiseal tubes for the VTi50 rotor, then using a syringe and metal cannula underlayer with 12.5 ml of 20% Nycodenz® and 4 ml of 45% Nycodenz® solution. ⑧ ⑨

5. Layer the 2000g supernatant on top, to fill the tube, as specified by the manufacturer.

6. Centrifuge at 200 000g for 1 h, using a slow acceleration and deceleration programs up to and below 2000 rpm (alternatively turn off the brake below 2000 rpm).

7. Harvest the gradient in a series of equal volume fractions prior to analysis. Lysosomes band at the 45/20%

Nycodenz® interface, very dense endosomes at the 20% polysucrose/20% Nycodenz®; all other endosomes band at the sample/20% polysucrose interface. ⑩

Notes

① Polysucrose is available commercially under this name (Axis-Shield and its distributors) or under the trade name Ficoll® (Amersham Pharmacia Biotech and its distributors). When making up solutions of these high molecular weight sucrose polymers, it is better to add small aliquots (2–3 ml) of the solvent to the weighed-out powder, using a glass rod to mix well after each addition.

② Ellis *et al.* [104] made up a more concentrated Ficoll solution in water (1 ml per g) before dialysing it for 2 h against a large volume of water and adjusting it to the appropriate concentration in 0.25 M sucrose, 1 mM EDTA, 10 mM TES-NaOH, pH 7.4. It is probably easier to make up a 25% (w/v) polysucrose in 0.25 M sucrose, 1 mM EDTA, 10 mM TES-NaOH, pH 7.4 and to dialyse it against the same medium, before checking and adjusting the volume to make it 20% with respect to polysucrose.

③ Either the Beckman Fraction Recovery System or the Labconco Auto Densi-flow is suitable. For more information see ref. 26.

④ Metal 'filling' cannulas (i.d. 0.8–1.0 mm) can be obtained from any surgical equipment supplies company.

⑤ If a vertical rotor is unavailable, either a fixed-angle or a swinging-bucket rotor may be used, but the longer sedimentation path length of these

rotors will require longer centrifugation times.

⑥ This methodology applies to rat liver and although in principle it may be applied to other tissues and cultured cells, in view of the functional uniqueness of liver, it is likely that some optimization of the gradient and centrifugation conditions may be necessary.

⑦ See ref. 104 for more information about liver perfusion.

⑧ A simplified method for separating lysosomes and endosomes in which the 20% polysucrose layer was omitted has also been used.

⑨ More detailed analysis of the light and dense endosomes may be carried out on continuous 1–22% polysucrose gradients (with a 45% Nycodenz® cushion) using approx. the same gradient volume, rotor and centrifugation conditions. The material containing lysosomes and dense endosomes at the cushion interface may be reanalysed in 0–35% or 0–45% Nycodenz® gradients; see refs 104–108 for more details. Analysis of *in vitro* systems to study the transfer of molecules between endosomes and lysosomes has also been carried out in 0–35% Nycodenz® gradients using a swinging-bucket rotor with a tube volume of approx. 14 ml, again at approx 200 000g for 1 h [109–111].

⑩ For gradient collection low-density end first, use either the tube puncture device of the Beckman Fraction Recovery System to deliver a dense liquid such as Fluorinert™ (Sigma Aldrich) or Flutec-Blue™ (F2 chemicals Ltd, Preston UK) to the bottom of the tube or the Labconco Auto Densi-flow device to aspirate from the meniscus. For collection dense-end first, use tube puncture. More information about gradient collection can be found in ref. 26.

References

1. Graham, J., Ford, T. and Rickwood, D. (1994) *Anal. Biochem.*, **220**, 367–373.
2. Hubbard, A. L., Wall, D. A. and Ma, A. (1983) *J. Cell Biol.*, **96**, 217–229.
3. Scott, L., Schell, M. J. and Hubbard, A. L. (1993) In: *Methods in Molecular Biology*, Vol. 19 (J. M. Graham and J. A. Higgins, eds), pp. 59–69. Humana Press, Totowa, NJ, USA.
4. Wisher, M. H. and Evans, W. H. (1975) *Biochem J.*, **146**, 375–388.
5. Meier, P. J., Sztul, E. S., Reuben, A. and Boyer, J. L. (1984) *J. Cell Biol.*, **98**, 991–1000.
6. Graham, J. M. and Northfield, T. C. (1987) *Biochem. J.*, **242**, 825–834.
7. Ali, N., Aligue, R. and Evans, W. H. (1990) *Biochem. J.*, **271**, 185–192.
8. Ekblad, L. and Jergil, B. (2001) *Biochim. Biophys. Acta*, **1531**, 209–221.
9. Sztul, E. S., Howell, K. E. and Palade, G. E. (1985) *J. Cell Biol.*, **100**, 1255–1261.
10. Howell, K. F., Gruenberg, J., Ito, K. and Palade, G. E. (1988) In: *Cell-free Analysis of Membrane Traffic* (D. J. Morré, K. E. Howell, G. M. C. Cook and W. H.Evans, eds), pp. 77–90. Alan R Liss Inc., New York.
11. Luzio, J. P., Mullock, B. M., Branch, W. J. and Richardson, P. J. (1988) in *Cell-free Analysis of Membrane Traffic* (D. J. Morré, K. E.Howell, G. M. C. Cook and W. H. Evans, eds), pp. 91–100. Alan R Liss Inc., New York.
12. Ellis, J. A., Jackman, M. R. and Luzio, J. P. (1992) *Biochem. J.*, **283**, 553–560.
13. Yeaman, C. (2003) *Methods*, **30**, 198–206.
14. Amieva, M. R., Vogelman, R., Covacci, A., Tompkins, L. S., Nelson, W. J. and Falkow, S. (2003) *Science*, **300**, 1430–1434.
15. Oliferenko, S., Paiha, K., Harder, T., Gerke, V., Schwarzler, C., Schwarz, H., Beug, H., Gunthert, U. and Huber, L. A. (1999) *J. Cell Biol.* **146**, 843–854.
16. Lafont, F., Verkade, P., Galli, T., Wimmer, C., Louvard, D. and Simons, K. (1999) *Proc. Natl. Acad. Sci., USA*, **96**, 3734–3738.
17. Smart E. J., Ying, Y-S., Mineo, C. and Anderson, R. G. W. (1995) *Proc. Natl. Acad. Sci., USA*, **92**, 10104–10108.
18. Uittenbogaard, A., Ying, Y. and Smart, E. J. (1998) *J. Biol. Chem.*, **273**, 6525–6532.
19. Lodish, H. F., Kong, N., Hirani, S. and Rasmussen, J. (1987) *J. Biol. Chem.*, **104**, 221–230.
20. Miller, S. G., Carnell, L. and Moore, H. H. (1992) *J. Cell Biol.*, **118**, 267–283.
21. Moore, S. E. and Spiro, R. G. (1992) *J. Biol. Chem.*, **267**, 8443–8451.
22. Yang, M., Ellenberg, J., Bonifacino, J. S. and Weissman, A. M. (1997) *J. Biol. Chem.*, **272**, 1970–1975.
23. Zhang, J., Kang, D. E., Xia, W., Okochi, M., Mori, H., Selkoe, D. J. and Koo, E. H. (1998) *J. Biol. Chem.*, **273**, 12436–12442.
24. Tekirian, T. L., Merriam, D. E., Marshansky, V., Miller, J., Crowley, A. C., Chan, H., Ausiello, D., Brown, D., Buxbaum,J. D., Xia, W. and Wasco, W. (2001) *Mol. Brain. Res.*, **96**, 14–20.
25. Woods, A. J., Roberts, M. S., Choudhary, J., Barry, S. T., Mazaki, Y., Sabe, H., Morley, S. J., Critchley, D. R. and Norman,J. C. (2002) *J. Biol. Chem.*, **277**, 6428–6437.
26. Graham, J. M. (2001) In: *Biological Centrifugation*, pp. 61–84. Taylor and Francis Books Ltd, Oxford, UK.
27. Petanceska, S. S., Seeger, M., Checler, F. and Gandy, S. (2000) *J. Neurochem.*, **74**, 1878–1884.
28. Dunn, M. J. and Bradd, S. J. (1993) In: *Methods in Molecular Biology*, Vol. 19 (J. M. Graham and J. A. Higgins, eds), pp. 203–210. Humana Press, Totowa, NJ.
29. Bradd, S. J. and Dunn, M. J. (1993) In: *Methods in Molecular Biology*, Vol. 19 (J. M. Graham and J. A. Higgins, eds), pp. 211–218. Humana Press, Totowa, NJ.
30. Dunn, M. J. (1993) *Gel Electrophoresis: Proteins*. Taylor and Francis Books Ltd, Oxford, UK.
31. Gjoen, T., Berg, T. O. and Berg T. (1997) In: *Subcellular Fractionation – a Practical Approach* (J. M. Graham and D. Rickwood, eds), pp. 169–203. Oxford University Press, Oxford, UK.

32. Courtoy, P. J. (1993) In: *Endocytic Compartments: Identification and Characterization* (J. J. M. Bergeron and J. R. Harris, eds), pp. 29–68. Plenum Press, New York.

33. Gupta, D. and Tartakoff, A. M. (1989) *Meth. Cell Biol.*, **31**, 247–272.

34. Courtoy, P. J., Quintart, J. and Baudhuin, P. (1984) *J. Cell Biol.*, **98**, 870–876.

35. Graham J. M. (1997) In: *Subcellular Fractionation – a Practical Approach* (J. M. Graham and D. Rickwood, eds), pp. 1–29. Oxford University Press, Oxford, UK.

36. Lafont, F., Lecat, S., Verkade, P. and Simons, K. (1998) *J. Cell Biol.*, **142**, 1413–1427.

37. Benting, J. H., Rietveld, A. G. and Simons, K. (1999) *J. Cell Biol.*, **146**, 313–320.

38. Plant, P. J., Lafont, F., Lecat, S., Verkade, P., Simons, K. and Rotin, D. (2000) *J. Cell Biol.*, **149**, 1473–1483.

39. Lecat, S., Verkade, P., Thiele, C., Fiedler, K., Simons, K. and Lafont, F. (2000) *J. Cell Sci.*, **113**, 2607–2618.

40. Hanwell, D., Ishikawa, T., Saleki, R. and Rotin, D. (2002) *J. Biol. Chem.*, **277**, 9772–9779.

41. Rietveld, A., Neutz, S., Simons, K. and Eaton, S. (1999) *J. Biol. Chem.*, **274**, 12049–12054.

42. Manes, S., Mira, E., Gomez-Mouton, C., Lacalle, R. A., Keller, P., Labrador, J. P. and Martinez-A, C. (1999) *EMBO J.*, **18**, 6211–6220.

43. Bruckner, K., Labrador, J. P., Scheiffele, P., Herb, A., Seeburg, P. H. and Klein, R. (1999) *Neuron*, **22**, 511–524.

44. Gimpl, G. and Fahrenholz, F. (2000) *Eur. J. Biochem.*, **267**, 2483–2497.

45. Bagnat, M., Keranen, S., Shevchenko, A., Shevchenko, A. and Simons, K. (2000) *Proc. Nat. Acad. Sci. USA*, **97**, 3254–3259.

46. Lee, M. C. S., Hamamoto, S. and Schekman, R. (2002) *J. Biol. Chem.*, **277**, 22395–22401.

47. Harder, T. and Kuhn, M. (2000) *J. Cell Biol.*, **151**, 199–207.

48. Simons, M., Kramer, E-M., Thiele, C., Stoffel, W. and Trotter, J. (2000) *J. Cell Biol.*, **151**, 143–153.

49. Ledesma, M. D., Da Silva, J. S., Crassaerts, K., Delacourte, A., De Stropper, B. and Dotti, C. G. (2000) *EMBO Rep.*, **1**, 530–535.

50. Tansey, M. G., Baloh, R. H., Milbrandt, J. and Johnson, E. M. (2000) *Neuron*, **25**, 611–623.

51. Bruses, J. L., Chauvet, N. and Rutishauser, U. (2001) *J. Neurosci.*, **21**, 504–512.

52. Lindwasser, O. W. and Resh, M. D. (2001) *J. Virol.*, **75**, 7913–7924.

53. Ehehalt, R., Keller, P., Haass, C., Thiele, C. and Simons, K. (2003) *J. Cell Biol.*, **160**, 113–123.

54. Liu, P., Ying, Y., Ko, Y. G. and Anderson, R. G. W. (1996) *J. Biol. Chem.*, **271**, 10299–10303.

55. Liu, P., Ying, Y-S. and Anderson, R. G. W. (1997) *Proc. Natl. Acad. Sci., USA*, **94**, 13666–13670.

56. Ikezu, T., Trapp, B. D., Song, K. S., Schegel, A., Lisanti, M. P. and Okamato, T. (1998) *J. Biol. Chem.*, **273**, 10485–10495.

57. Brouillett, E., Trembleau, A., Galanaud, D., Volovitch, M., Bouillot, C., Valenza, C., Prochiantz, A. and Allinquant, B. (1999) *J. Neurosci.*, **19**, 1717–1727.

58. Graf, G. G., Connell, P. M., van der Westhuyzen, D. R. and Smart, E. J. (1999) *J. Biol. Chem.*, **274**, 12043–12048.

59. Uittenbogaard, A., Everson, W. V., Matveev, S. V. and Smart, E. J. (2002) *J. Biol. Chem.*, **277**, 4925–4931.

60. Blair, A., Shaul, P. W., Yuhanna, I. S., Conrad, P. A. and Smart, E. J. (1999) *J. Biol. Chem.*, **274**, 32512–32519.

61. Czarny, M., Liu, J., Oh, P. and Schnitzer, J. E. (2003) *J. Biol. Chem.*, **278**, 4424–4430.

62. Cameron, P. L., Ruffin, J. W., Bollag, R., Rasmussen, H. and Cameron, R. S. (1997) *J. Neurosci.*, **17**, 9520–9535.

63. Frank, S., Upender, S., Hansen, S. H. and Casanova, J. E. (1998) *J. Biol. Chem.*, **273**, 23–27.

64. Pedrazzini, E., Villa, A., Longhi, R., Bulbarelli, A. and Borgese, N. (2000) *J. Cell Biol.*, **148**, 899–913.

65. Roy, M-O., Leventis, R. and Silvius, J. R. (2000) *Biochemistry*, **39**, 8298–8307.

66. Cai, Y., Maeda, Y., Cedzich, A., Torres, V. E., Wu, G., Hayashi, T., Mochizuki, T., Park, J. H., Witzgall, R. and Somlo, S. (1999) *J. Biol. Chem.*, **274**, 28557–28565.

67. Drummer, H. E., Maerz, A., Poumbourios, P. (2003) *FEBS Lett.*, **546**, 3385–3390.

68. Choukroun, G. J., Marshansky, V., Gustafson, C. E., McKee, M., Hajjar,R. J., Rosenzweig, A., Brown, D. and Bonventre, J. V. (2000) *J. Clin. Invest.*, **106**, 983–993.

69. Van't Hoff, W. and Resh, M. D. (1997) *J. Cell Biol.*, **136**, 1023–1035.

70. Newby, L. J., Streets, A. J., Zhao, Y., Harris, P. C., Ward, C. J. and Ong, A. C. M. (2002) *J. Biol. Chem.*, **277**, 20763–20773.

71. Kim,S-H., Lah, J. J., Thinakaran, G., Levey, A. and Sisodia, S. S. (2000) *Neurobiol. Disease*, **7**, 99–117.

72. Chahdi, A., Choi, W. S., Kim, Y. M. and Beaven, M. A. (2003) *J. Biol. Chem.*, **278**, 12039–12045.

73. Puglielli, L., Konopka, G., Pack-Chung, E., MacKenzie Ingano, L. A., Berezovska, O., Hyman,B. T., Chang,T. Y., Tanzi, R. E. and Kovacs, D. M. (2001) *Nature Cell Biol.*, **3**, 905–912.

74. Koulen, P., Cal, Y., Geng, L., Maeda, Y., Nishimura, S., Witzgall, R., Ehrlich, B. E. and Somlo, S. (2002) *Nature Cell Biol.*, **4**, 191–197.

75. Wu, C., Lai, C-F. and Mobley,W. C. (2001) *J. Neurosci.*, **21**, 5406–5416.

76. Xia, W., Zhang, J., Ostrazewski,B. L., Kimberly, W. T., Seubert, P., Koo, E. H., Shen, J. and Selkoe, D. J. (1998) *Biochemistry*, **37**, 16465–16471.

77. Campbell, W. A., Reed,M. L. O., Strahle,J., Wolfe, M. S. and Xia, W. (2003) *J. Neurochem.*, **85**, 1563–1574.

78. Kimberly, W. T., LaVoie, M. J., Ostrazewski, B. L., Ye, W., Wolfe, M. S. and Selkoe,D. J. (2002) *J. Biol. Chem.*, **277**, 35113–35117.

79. Iwata, H., Tomita, T., Maruyama, K. and Iwatsubo, T. (2001) *J. Biol. Chem.*, **276**, 21678–21685.

80. Tomita, T., Wayabiki, T., Takikawa, R., Morohashi, Y., Takasugi, N., Kopan, R., De Stropper, B. and Iwatsubo, T. (2001) *J. Biol. Chem.*, **276**, 33273–33281.

81. Chen,F., Yang,D-S., Petanceska,S., Yang,A., Tandon, A., Yu, G., Rozmahel, R., Ghiso,J., Nishimura, M., Zhang, D. M., Kawarai, T., Levesque, G., Mills, J., Levesque, L., Song, Y-Q., Rogaeva, E., Westaway, D., Mount, H., Gandy, S., St. George-Hyslop,P. and Fraser, P. E. (2000) *J. Biol. Chem.*, **275**, 36794–36802.

82. Schroder, M., Schafer, R. and Friedl, P. (2002) *Biotech. Bioeng.*, **78**, 131–140.

83. Yu,G., Chen,F., Nishimura,M., Steiner,H., Tandon,A., Kawarai,T., Arawaka,S., Supala,A., Song, Y-Q., Rogaeva, E., Holmes, E., Zhang, D. M., Milman, P., Fraser, P. E., Haass, C. and St George-Hyslop, P. (2000) *J. Biol. Chem.*, **275**, 27348–27353.

84. Yang, D-S., Tandon, A., Chen, F., Yu,G., Yu,H., Arawaki,S., Hasegawa,H.,Duthie,M., Schmidt, S. D., Ramabhadran,T. V., Nixon, R. A., Mathews, P. M., Gandy, S. E., Mount, H. T. J., St George-Hyslop, P. and Fraser, P. E. (2002) *J. Biol. Chem.*, **277**, 28135–28142.

85. Gu, Y-J., Chen, F., Sanjo, N., Kawarai, T., Hasegawa, H., Duthie,M., Li,W., Ruan, X., Luthra, A., Mount, H. T. J., Tandon, A., Fraser, P. E. and St George Hyslop, P. (2003) *J. Biol. Chem.*, **278**, 7374–7380.

86. Sabo, S. L., Lanier, L. M., Ikin, A. F., Khorkova, O., Sahasrabudhe, S., Greengard,P. and Buxbaum, J. D. (1999) *J. Biol. Chem.*, **274**, 7952–7957.

87. Chen, F., Tandon, A., Sanjo, N., Gu, Y-J., Hasegawa, H., Arawaka, S., Lee, F. J. S., Ruan, X., Mastrangelo, P., Erdebil, S., Wang, L., Westaway, D., Mount, H. T. J., Yankner, B., Fraser, P. E. and St George Hyslop, P. (2003) *J. Biol. Chem.*, **278**, 19974–19979.

88. Majoul,I.V., Bastiaens,P. I.H. and Soling,H-D. (1996) *J. Cell Biol.*, **133**, 777–789.

89. Majoul, I., Ferrari, D. and Soling, H-D. (1997) *FEBS Lett.*, **401**, 104–108.

90. Kuiper, R. P., Bouw, G., Janssen, K. P. C., Rotter, J., van Herp, F. and Martens, G. J. M. (2001) *Biochem. J.,* **360**, 421–429.

91. Scheiffele, P., Verkade, P., Fra, A. M., Virta, H., Simons, K. and Ikonen, E. (1998) *J. Cell. Biol.*, **140**, 795–806.

92. Love, H. D., Lin, C-C., Short, C. S. and Ostermann, J. (1998) *J. Cell Biol.,* **140**, 541–551.

93. De Leeuw, E., Poland, D., Mol, O., Sinning, I., ten Hagen-Jongman, C. M., Oudega, B. and Luirink, J. (1997) *FEBS Lett.*, **416**, 225–229.

94. Joglekar, A. P., Xu, D., Rigotti, D. J., Fairman, R. and Hay, J. C. (2003) *J. Biol. Chem.*, **278**, 14121–14133.

95. Satyanarayana, C., Schroder-Kohne, S., Craig, E. A., Schu, P. V. and Horst, M. (2000) *FEBS Lett.*, **470**, 232–238.

96. Sheff, D. R., Daro,E. A., Hull,M. and Mellman, I. (1999) *J. Cell Biol.*, **145**, 123–139.

97. Sugii, S., Reid, P. C., Ohgami, N., Du, H. and Chang, T-Y. (2003) *J. Biol. Chem.*, **278**, 27180–27189.

98. Chin, L-S., Raynor, M. C., Wei, X., Chen, H-Q. and Li, L. (2001) *J. Biol. Chem.*, **276**, 7069–7078.

99. Meyers, J. M. and Prekeris, R. (2002) *J. Biol. Chem.*, **277**, 49003–49010.

100. Pol, A., Lu, A., Pons, M., Peiro, S. and Enrich, C. (2000) *J. Biol. Chem.*, **275**, 30566–30572.

101. Biemesderfer, D., Mentone, S. A., Mooseker, M. and Hasson, T. (2002) *Am. J. Physiol., Ren. Physiol.*, **282**, F785–F794.

102. Billington, D., Maltby, P. J., Jackson, A. P. and Graham, J. M. (1998) *Anal. Biochem.*, **258**, 251–258.

103. Molinari, M., Galli, C., Norais, N., Telford, J. L., Rappuoli, R., Luzio, J. P. and Montecucco, C. (1997) *J. Biol. Chem.*, **272**, 25339–25344.

104. Ellis, J. A., Jackman, M. R., Perez, J. H., Mullock, B. M. and Luzio, J. P. (1992) In: *Protein Targeting: a Practical Approach* (A. I. Magee, and T. Wileman, eds), pp. 25–57. IRL Press at Oxford University Press, Oxford, UK.

105. Mullock, B. M., Perez, J. H., Kuwana, T., Gray, S. R. and Luzio, J. P. (1994) *J. Cell Biol.*, **126**, 1173–1182.

106. Mullock, B. M., Bright, N. A., Fearon, C. W., Gray, S. R. and Luzio, J. P. (1998) *J. Cell. Biol.*, **140**, 591–601.

107. Mullock, B. M., Smith, C. W., Ihrke, G., Bright, N. A., Lindsay,M., Parkinson,E. J., Brooks, D. A. Parton, R. G., James, D. E., Luzio, J. P. and Piper, R. C. (2000) *Mol. Biol. Cell*, **11**, 3137–3153.

108. Pryor, P. R., Mullock, B. M., Bright, N. A., Gray, SD. R. and Luzio, J. P. (2000) *J. Cell Biol.*, **149**, 1053–1062.

109. Authier, F. and Chauvet, G. (1999) *FEBS Lett.*, **461**, 25–31.

110. Authier, F., Danielsen, G. M., Kouach, M., Briand, G. and Chauvet, G. (2001) *Endocrinology*, **142**, 276–289.

111. Desbuquois, B., Chauvet, G., Kouach, M. and Authier, F. (2003) *Endocrinology*, **144**, 5308–5321.

6

In Vitro Techniques

Edited by J. Robin Harris

Nuclear components

Protocol 6.1	Nucleosome assembly coupled to DNA repair synthesis using a human cell free system	204
Protocol 6.2	Single labelling of nascent DNA with halogenated thymidine analogues	210
Protocol 6.3	Double labelling of DNA with different halogenated thymidine analogues	214
Protocol 6.4	Simultaneous immunostaining of proteins and halogen-dU-substituted DNA	217
Protocol 6.5	Uncovering the nuclear matrix in cultured cells	220
Protocol 6.6	Nuclear matrix–lamin interactions: *in vitro* blot overlay assay	228
Protocol 6.7	Nuclear matrix–lamin interactions: *in vitro* nuclear reassembly assay	230
Protocol 6.8	Preparation of *Xenopus laevis* egg extracts and immunodepletion	234
Protocol 6.9	Nuclear assembly *in vitro* and immunofluorescence	237
Protocol 6.10	Nucleocytoplasmic transport measurements using isolated *Xenopus* oocyte nuclei	240
Protocol 6.11	Transport measurements in microarrays of nuclear envelope patches by optical single transporter recording	244

Cells and membrane systems

Protocol 6.12	Cell permeabilization with Streptolysin O	248
Protocol 6.13	Nanocapsules: a new vehicle for intracellular delivery of drugs	250
Protocol 6.14	A rapid screen for determination of the protective role of antioxidant proteins in yeast	255
Protocol 6.15	*In vitro* assessment of neuronal apoptosis	259

Cell Biology Protocols. Edited by J. Robin Harris, John Graham, David Rickwood
© 2006 John Wiley & Sons, Ltd

Protocol 6.16   The mitochondrial permeability transition: PT and
                    $\Delta\Psi$m loss determined in cells or isolated mitochondria
                    with confocal laser imaging   265

Protocol 6.17   The mitochondrial permeability transition: measuring
                    PT and $\Delta\Psi$m loss in isolated mitochondria with Rh123
                    in a fluorometer   268

Protocol 6.18   The mitochondrial permeability transition: measuring
                    PT and $\Delta\Psi$m loss in cells and isolated mitochondria on
                    the FACS   270

Protocol 6.19   Measuring cytochrome *c* release in isolated
                    mitochondria by Western blot analysis   271

Protocol 6.20   Protein import into isolated mitochondria   272

Protocol 6.21   Formation of ternary SNARE complexes *in vitro*   274

Protocol 6.22   *In vitro* reconstitution of liver
                    endoplasmic reticulum   277

Protocol 6.23   Asymmetric incorporation of glycolipids into
                    membranes and detection of lipid flip-flop movement   280

Protocol 6.24   Purification of clathrin-coated vesicles from rat brains   286

Protocol 6.25   Reconstitution of endocytic intermediates on a lipid
                    monolayer   288

Protocol 6.26   Golgi membrane tubule formation   293

Protocol 6.27   Tight junction assembly   296

Protocol 6.28   Reconstitution of the major light-harvesting chlorophyll
                    a/*b* complex into liposomes   300

Protocol 6.29   Reconstitution of photosystem 2 into liposomes   305

Protocol 6.30   Golgi–vimentin interaction *in vitro* and *in vivo*   307

Cytoskeletal and fibrillar systems

Protocol 6.31   Microtubule peroxisome interaction   313

Protocol 6.32   Detection of cytomatrix proteins by immunogold
                    embedment-free electron microscopy   317

Protocol 6.33   Tubulin assembly induced by taxol and other
                    microtubule assembly promoters   326

Protocol 6.34   Vimentin production, purification, assembly and study
                    by EPR   331

Protocol 6.35   Neurofilament assembly   337

Protocol 6.36   α-Synuclein fibril formation induced by tubulin   342

Protocol 6.37   Amyloid-β fibril formation *in vitro*   345

Protocol 6.38   Soluble $A\beta_{1-42}$ peptide induces tau
                    hyperphosphorylation *in vitro*   348

Protocol 6.39   Anti-sense peptides   353

Protocol 6.40   Interactions between amyloid-ß and enzymes   359

Protocol 6.41   Amyloid-ß phosphorylation   364

Protocol 6.42   Smitin–myosin II coassembly arrays *in vitro*   369

Protocol 6.43   Assembly/disassembly of myosin filaments in the
                    presence of EF-hand calcium-binding protein S100A4
                    in vitro   372

Protocol 6.44   Collagen fibril assembly *in vitro*   375

Introduction

Modern cell biology is increasing moving onwards from the intact cell or component isolated directly from the cell, to a consideration of the experimental manipulation and reassembly of such components.

This chapter dealing with *in vitro* techniques in cell biology includes a range of recent methods that loosely fall within the sphere of experimental reconstitution/assembly procedures. The topics included integrate well with those within the other sections of the book and often rely upon both EM and advanced light microscopical techniques for assessment of what is achieved or produced experimentally.

Several assays for nuclear components, cellular and membrane systems are presented, together with the assembly of cytoskeletal proteins. In addition, protocols relating to *in vitro* collagen fibrillogenesis, amyloid-β fibrillogenesis, amyloid-β-enzyme interaction and amyloid-β phosphorylation are also included.

Because of the diversity of this material, most protocols (or a small group of protocols) includes a brief introduction, its own list of references and sometimes examples of typical data. It is hoped that these protocols will enable the book to indicate a way ahead for cell biology, as an addition to the well-used classical procedures dealing with cells and subcellular components presented in the other chapters of the book.

Nucleosome assembly coupled to DNA repair synthesis using a human cell free system

Geneviève Almouzni and Doris Kirschner

Introduction

The assembly of nucleosomes onto DNA, in two steps, involves the loading of a H3–H4 histone tetramer followed by the subsequent addition of two H2A–H2B histone dimers. *In vivo*, histone deposition is promoted by histone chaperones, among which the evolutionary conserved three-subunit complex called the chromatin assembly factor 1 (CAF-1) is the only one that links nucleosome formation to DNA synthesis.

Here we describe an assay to monitor nucleosome assembly coupled to DNA repair in a human cell free system. The nucleotide excision repair (NER) pathway used in this assay repairs most of the UV-photoproducts (mainly cyclobutane pyrimidine dimers and 6–4 photoproducts) can be reproduced *in vitro* [4, 6].

A cell free extract (cytosolic extract) derived from Hela cells is thus used for its ability to support NER reactions on a damaged DNA template. This extract, however, cannot ensure the assembly into chromatin of the repaired DNA molecules. It can be complemented using either a nuclear extract as a crude source of assembly factors or using purified recombinant assembly factors such as CAF-1 [1, 3].

Reagents

Agarose (Ultra Pure, Sigma)

Ammonium acetate (5 M)

100 mM ATP (Pharmacia) stored at $-80\,^{\circ}$C (each aliquot is used only once)

Bacculovirus produced recombinant CAF-1 4 ng/μl [5]

Creatine phosphokinase (Boehringer) 2.5 mg/ml in H_2O ①

Cytosolic extracts derived from HeLa cells [2] ② ③

Ethidium bromide, 10 mg/ml (GIBCO BRL)

Glycogen, 20 mg/ml (Roche), store at $-20\,^{\circ}$C

Loading buffer 5×: 0.42% bromophenol blue, 50% glycerol

Nuclear extracts from HeLa (4C Biotech/ Belgium or [2])

αP^{32} dCTP, 3000 Ci/mmol (ICN)

Phenol : chloroform : isoamyl alcohol (25 : 24 : 1) (GIBCO BRL)

Proteinase K, 20 mg/ml (Roche), aliquots stored at $-20\,^{\circ}$C

5× Reaction buffer: 25 mM $MgCl_2$, 200 mM Hepes-KOH (pH 7.8), 2.5 mM dithiothreitol (DTT; Sigma), 200 mM phosphocreatine (di-Tris salt, Sigma) ①

RNase A (Roche) 10 mg/ml in 0.01 M sodium acetate (pH 5.2), heat for 15 min at $100\,^{\circ}$C and adjust the pH with 1 vol of Tris-HCL (pH 7.4), aliquots stored at $-20\,^{\circ}$C

Stop-mix: 30 mM EDTA, 0.7% SDS

TAE 50×: 242 g Tris base, 57.1 ml glacial acetic acid, 37.2 g Na$_2$EDTA, (pH 8.0), adjust to a final volume of 1 l with deionized water (or 50× TAE from Bio Media)

TE: 10 mM Tris-HCl (pH 8.0), 1 mM EDTA (pH 8.0)

UV treated Bluescript plasmid (pBS) DNA (50 ng/μl) ④ [1]

Equipment

Eppendorf thermomixer 5436

Eppendorf tubes (1.5 ml, siliconized) (pre-cooled on ice)

Gelbox (35 × 20 cm, Biorad)

Germicidal lamp with a 254 nm peak (Philips)

Microfuge (e.g. Eppendorf centrifuge 5451)

3 MM Paper (Whatman)

Phosphoimager (Storm 860–Molecular Dynamics), and Image Quant 5.2 software (or films and standard film developer)

Power supply (Pharmacia Biotech, EPS 3500)

UV transilluminator (BIORAD Gel Doc 2000) equipped with a camera

Vacuum gel dryer (slab gel Dryer, SGD 2000, SAVANT)

Waterbath (e.g. polystat, Bioblock Scientific)

Procedure

A standard reaction contains 150 ng of plasmid DNA, ④ 200–400 μg of cytosolic extract, 5 mM MgCl$_2$, 40 mM Hepes-KOH (pH 7.8), 0.5 mM DTT, 40 mM phosphocreatine, 4 mM ATP, 2.5 μg creatine phosphokinase and 5 μCi α^{32}P dCTP in 25 μl.

1. First, to prepare the DNA-mix, in a precooled Eppendorf tube, assemble on ice the following reagents (prepare a proportionally larger mix for more than one reaction). ①
DNA-mix (amounts for one reaction): 3 μl DNA (50 ng/μl), 5 μl 5× reaction buffer, 1 μl ATP, 1 μl creatine kinase, 2.5 μl dilution buffer ③

2. For each reaction, distribute 1 μl of HeLa nuclear extract to precooled Eppendorf tubes on ice (or a protein complex to test for nucleosome assembly or buffer controls). In an assay designed for an antibody inhibition strategy, you can add the antibodies or the appropriate control serum at this step. ⑤

3. Add 10 μl of the cytosolic extract (as source of histones and repair factors) to each tube except to the DNA only control in which buffer is added instead. To preserve optimal activity of cytosolic extracts, defrost the extract just prior to use, work quickly and discard any remaining material. ②

4. Add α^{32}P dCTP to the prepared DNA-mix (0.5 μl for 150 ng DNA), homogenize gently by pipetting up and down.

5. Start the reaction by adding 13 μl of DNA-mix containing the labelled precursor to each reaction tube, and immediately transfer to a preheated water bath, or to a thermo-mixer at 37°C for a 3 h incubation, see note 6 and ref. 1.

6. Stop the reaction with 50 μl of stop-mix and 25 μl of H$_2$O.

7. Add 5 μl of RNase A (at 2 mg/ml) and incubate 30 min at 37°C.

8. Add 2 μl of proteinase K (at 20 mg/ml) and incubate 30 min at 37°C. ⑦

9. Add 110 μl of phenol : chloroform : isoamyl alcohol (25 : 24 : 1) and vortex

each tube at least 10 s. Centrifuge 10 min at $14\,000g$ in the Eppendorf centrifuge at room temperature and recover 90 µl of the aqueous phase in a fresh Eppendorf tube; do not take any of the interphase.

10. Precipitate the DNA by adding of 90 µl ammonium acetate (5 M), 1 µl of glycogen (20 mg/ml) and 300 µl (2 vol) of ice-cold ethanol, leave 30 min at $-20\,^{\circ}$C.

11. Centrifuge 30–45 min at $14\,000g$ at $4\,^{\circ}$C, recover the DNA pellet, wash with 500 µl of cold 70% ethanol, centrifuge for 5–10 min at $14\,000\,g$ at $4\,^{\circ}$C. Remove the supernatant carefully and air-dry the pellet.

12. Dissolve the pellet in 8 µl of TE buffer and add 2 µl of 5× loading dye.

13. Prepare a 1% agarose gel in 1× TAE without ethidium bromide (EtBr). ⑧ ⑨

14. Load the samples and migrate at 1.5 V/cm for 20 h at $4\,^{\circ}$C overnight, for optimal resolution of the topoisomers. ⑦

15. Soak the gel in EtBr staining solution (1µg/ml), rinse for 30 min in water at room temperature, and place the gel on a UV transilluminator system equipped with a camera to visualize the migration pattern of the total DNA and make your image acquisition.

16. Transfer the gel onto a 3 MM Whatman paper of equal size, cover with saran wrap and dry for 2 h at $80\,^{\circ}$C in a vacuum gel dryer. Expose the covered and dried gel to a Phosphoimager screen to visualize the labelled DNA (repaired), if not available use an X-ray film. Quantification of the topoisomer distribution is carried out on both the labelled DNA (repaired) and the total DNA by densitometric scanning of the images. ⑩

Notes

① Thawing and freezing of the 5× reaction buffer alter the concentration of DTT, phosphocreatine and dNTPs. The buffer is prepared in advance, aliquoted and stored at $-80\,^{\circ}$C, use only once.

② Cytosolic and nuclear extracts are prepared from unsynchronized adherent HeLa cells, under cooling conditions ($4\,^{\circ}$C), for a detailed protocol see [2]. The protein concentration is determined by Bradford assay and usually ranges between 10 and 20 mg/ml [2]. Small aliquots of 50–100 µl frozen in liquid nitrogen are stored at $-80\,^{\circ}$C to avoid waste.

③ The cytosolic extract used in this assay, which contains only trace amounts of p60 and p150, needs to be complemented with the nuclear extract or baculovirus recombinant CAF-1 to ensure an efficient chromatin assembly coupled to DNA repair synthesis [1]. For each extract, optimal conditions for both chromatin assembly and DNA repair synthesis have to be adjusted (protein/DNA ratio and ionic conditions). For example, the composition of the dilution buffer for commercial (4C) HeLa nuclear extract is: 20 mM Hepes (pH 7.8), 100 mM KCl, 0.2 mM EDTA, 0.2 mM PMSF, 0.5 mM DTT and 20% glycerol. This has to be taken into account.

④ Plasmid DNA is treated with germicidal UV light (Philips) with 254 nm peak and a dose of 500 J/m^2. The general estimation of DNA damage produced by 100 J/m^2 is roughly 1 pyrimidine dimer photoproduct per 1000 bp (the ratio is about 0.75 cyclobutane pyrimidine and 0.25 of 6-4 photo product) [1, 5].

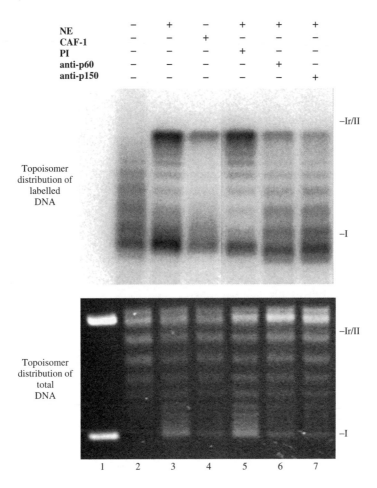

NE	−	+	−	+	+	+
CAF-1	−	−	+	−	−	−
PI	−	−	−	+	−	−
anti-p60	−	−	−	−	+	−
anti-p150	−	−	−	−	−	+

Figure 6.1 Chromatin assembly coupled to nucleotide excision repair synthesis mediated by CAF-1. UV-damaged DNA substrate was incubated in cytosolic extract (CE) competent for NER and complemented with nuclear extracts (NE) or recombinant CAF-1 to ensure the nucleosome assembly reaction. Lane 1 shows a DNA-only control (supercoiled DNA molecules) and lane 2 the DNA after incubation with cytosolic extract, which promotes the relaxation of the DNA and NER, which is visualized by the incorporation of radiolabelled nucleotides. Subsequent supercoiling of the repaired DNA is dependent upon addition of CAF-1, either as nuclear extract (lane 3) or as recombinant complex (lane 4). In this assay the CAF-1 dependent nucleosome formation was largely impaired by the addition of 1 μl of polyclonal antibodies directed respectively against the p60 subunit (lane 6) or the p150 subunit (lane 7) of CAF-1 and compared with a pre-immune (PI) serum (lane 5). Total DNA is visualized after ethidium bromide staining (lower panel) and radiolabel incorporation due to repair synthesis (labelled DNA) is visualized on a phosphoimager screen of the dried gel (upper panel). Relaxed/nicked circular DNA (Ir/II) and supercoiled DNA (I) are indicated

⑤ When antibodies are used as in Figure 6.1, the quantity added has to be adjusted for each serum or clone.

⑥ Nucleotide excision repair as followed by the DNA synthesis and chromatin assembly reactions is usually completed after 3 h. Although specific UV-dependent nucleotide incorporation can be detected already after 15 min, the chromatin assembly reaction is a slower process [1].

⑦ To monitor chromatin assembly, the supercoiling assay makes use of the topological properties of closed circular molecules. In the extracts, which contain topoisomerase activity, the deposition of nucleosomes induces topological stress that is relieved progressively. Therefore, after deproteinization, accumulation of topoisomers with an increasing number of negative supercoils can be used as an indication of the effectiveness of the assembly reaction. The repaired molecules (labelled DNA) are compared to the bulk DNA (total DNA) to evaluate a differential effect.

⑧ The concentration of the agarose gel is adjusted according to the size of the plasmid used. As a guideline, use 1% agarose gels for plasmids of 3–4 kb and 0.8% gels for plasmids of 6–7 kb in size.

⑨ It is crucial to run the supercoiling gel in the absence of EtBr. This intercalating agent would interfere with the analysis of the topological state of the molecules.

⑩ The radiolabelled DNA corresponds to incorporation of nucleotides during repair synthesis (one of the steps in NER) and can be analysed quantitatively and qualitatively using the Image Quant 5.2 software. Adapt the exposure time depending on the repair efficiency of the extract used. A higher proportion of supercoiled molecules in the radiolabelled DNA as compared to total DNA is indicative of a preferential assembly of repaired molecules, reflecting a link between repair synthesis and nucleosome assembly.

References

1. Gaillard, P. H., Roche, D. M. and Almouzni, G. A. (1999) *Meth. Mol. Biol.*, **119**, 231–243.
2. Martini, E., Roche, D. M., Marheineke, K., Verreault, A., and Almouzni, G. A. (1998) Recruitment of phosphorylated chromatin assembly factor 1 to chromatin after UV irradiation of human cells. *J. Cell Biol.*, **143**, 563–575.
3. Mello, J. A., Sillje, H. H., Roche, D. M., Kirschner, D. B., Nigg, E. A. and Almouzni, G. A. (2002) Human Asf1 and CAF-1 interact and synergize in a repair-coupled nucleosome assembly pathway. *EMBO Rep.* **3**(4), 329–334.
4. Moggs, J. G. and Almouzni, G. A. (1999) Assays for chromatin remodeling during DNA repair. *Meth. Enzymol.*, **304**, 333–351.
5. Verreault, A., Kaufman, P. D., Kobayashi, R. and Stillman, B. (1996) Nucleosome assembly by a complex of CAF-1 and acetylated histones H3/H4. *Cell*, **87**, 95–104.
6. Wood, R. D., Biggerstaff, M. and Shivji, M. K. K. (1995) Detection and measurement of nucleotide excision repair synthesis by mammalian cell extracts *in vitro*. *Methods: A Companion to Methods Enzymol.*, **7**, 163–175.

Acknowledgements

We thank Alain Verreault for the generous gift of CAF-1, Jill Mello for advice and Catherine Green for reading and critical comments. D.B.K. was supported by the EEC-RTN (to G.A.), the team of G.A. is supported by the Ligue Contre le Cancer, Euratom, EU-RTN and LCR of CEA.

PROTOCOLS 6.2–6.4

Immunocytochemical studies of DNA replication in mammalian nuclei

Daniela Dimitrova

Introduction

Indirect immunofluorescent staining of newly synthesized DNA and/or proteins of the replication machinery has emerged as a powerful technique for identification of proliferating cells [1] and for cancer diagnosis or prognosis [2, 3], as well as for studies of cell cycle kinetics [1, 4], intra-S-phase checkpoint control [5], organization of nuclear DNA replication sites [6, 7], establishment and execution of the temporal program for chromosomal DNA replication during the S-phase [8].

Single labelling of nascent DNA with halogenated thymidine analogues

Reagents

Phosphate-buffered saline (PBS): 8 g NaCl, 0.2 g KCl, 1.44 g Na_2HPO_4 and 0.24 g KH_2PO_4 per litre at pH 7.2

PBS-T: PBS containing 0.5% (v/v) Tween 20 (Sigma)

Halogenated nucleosides: stock solutions of 10 mg/ml of 5-bromo-2′-deoxyuridine (BrdU, Sigma), 5-chloro-2′-deoxyuridine (CldU, Sigma), or 5-iodo-2′-deoxyuridine (IdU, Sigma) are prepared in cell culture medium (without serum) at pH 7.2. Store frozen at $-20\,°C$ in small aliquots

Antibodies: (A) Mouse anti-BrdU (No. 347580; Becton Dickinson) is used for detection of BrdU, CldU and IdU [1]. (B) Rat anti-BrdU (MAS250b; Harlan-Sera lab) is used for detection of BrdU and CldU [1]. (C) Secondary antibodies: FITC-conjugated donkey anti-rat IgG (No. 712-095-153; Jackson ImmunoResearch Laboratories) or FITC-conjugated donkey anti-mouse IgG (No. 715-095-151; Jackson ImmunoResearch Laboratories) are used to detect the rat or mouse anti-BrdU antibodies, respectively

Blocking buffer: to suppress non-specific binding of antibodies to coverslips or to cellular material, antibody dilutions are made in PBS containing 2% bovine serum albumin and either 10% normal donkey serum, or 10% fetal bovine serum. Supplement with NaN_3 at 1 mM final concentration and store at $4\,°C$

DNA staining dye: a stock solution of 1 mg/ml of 4′,6-diamidino-2-phenylindole (DAPI; Sigma) is prepared in ddH_2O and stored frozen in small aliquots at $-20\,°C$. A working solution of 10 µg/ml can be kept at $4\,°C$

Materials and equipment

Sterile glass coverslips

Cell culture dishes and medium

A pair of fine sharp forceps

Dark humid incubation chamber

Plain glass slides (3 × 1″)

Mounting medium for fluorescent microscopy, e.g. Vectashield (Vector Laboratories)

Epifluorescent microscope

Procedure

1. Grow cells directly on glass coverslips ① submerged in medium inside cell culture dishes. ② Multi-well dishes can be used if cell cultures need to be grown in separate chambers. Alternatively, several coverslips can be placed within a single dish. Make sure that the coverslips do not overlap.

2. Add the nucleotide precursor of choice (i.e. BrdU, CldU or IdU) to the culture medium to a final concentration of

10 μM. Mix gently by swirling the dish, taking care not to disturb the coverslips, and quickly return the dish to the cell culture incubator.

3. Incubate at 37 °C for several minutes to allow incorporation of the thymidine analogue into nascent DNA within the S-phase cells. ③

4. Aspirate the medium and wash the cells with ice-cold PBS to stop the labelling and to remove traces of medium and unincorporated nucleosides.

5. Fix the cells with cold 70% ethanol for at least 30 min at 4 °C. ④

6. It is convenient to perform the subsequent immunostaining steps by transferring the desired number of coverslips into multi-well dishes. In this way many coverslips can be processed in batches, while ensuring easy and precise tracking of the individual coverslips. ⑤

7. Remove the ethanol and rinse the cells once with PBS.

8. Incubate the cells in 1.5 N HCl for 30 min at room temperature. ⑥

9. Remove the HCl and rinse several times with ample volumes of PBS. Make sure that the pH is neutral at the end of the washing.

10. Wash once with PBS-T. ⑦

11. Remove the coverslips from the wells of the dish and blot the excess liquid by placing them on a piece of soft tissue (e.g. Kleenex or similar). Do not handle more than five to six coverslips at a time to avoid drying of the cells.

12. Invert each coverslip (cell side down) ⑧ over a drop of anti-BrdU antibody diluted in blocking solution. Drops of 3–4 μl, or 7–8 μl, work

well with 12 mm, or 18 mm coverslips, respectively. Dilutions (usually in the range of 1 : 20–1 : 500) must be optimized empirically for each antibody. The drops can be placed on any type of clean hydrophobic surface, e.g. a piece of Parafilm. In cases when numerous coverslips must be processed simultaneously, it is best to invert the coverslips onto microscope glass slides, which can then be stacked into tight-closing slide storage boxes.

13. Incubate for 1 h at room temperature ⑨ in a humid chamber to prevent the cells from drying. A humid chamber can be improvised by placing a layer of wet filter paper inside a covered plastic dish or along the inner side of the cover of a microscope slide storage box.

14. Lift the coverslips carefully (avoid dragging along the surface, which can cause cell damage and/or loss) and return to the washing dish. Wash twice with PBS-T.

15. Remove the coverslips from the wells as described in steps 11–12 and place over drops with an appropriate fluorochrome-conjugated secondary antibody diluted in blocking solution (1 : 100–1 : 1000).

16. Incubate for 1 h at room temperature in a dark humid chamber. ⑩

17. Return the coverslips to the washing dish. Wash twice with PBS-T.

18. Counterstain the nuclei by incubating in 0.2 μg/ml solution of DAPI in PBS for 10 min at room temperature. Alternatively, this step can be skipped and DAPI can be diluted directly in the mounting medium.

19. Mount the coverslips (cell side down) over drops of Vectashield placed on clean microscope slides and observe

using an epifluorescent microscope. Do not allow the coverslips to dry out. If long-term preservation is desired, blot excess mounting medium and seal the coverslips with colourless nail polish. Store inside microscope slide boxes at 4 °C or at −20 °C. ⑪

Notes

This procedure will take approximately 5 h.

① Use clean forceps when handling coverslips and always open the box with coverslips under a laminar hood to avoid contaminations. If these conditions are observed, additional sterilization of the coverslips is not usually required. However, if contamination occurs, the coverslips can be sterilized by immersing in ethanol and flaming.

② For best results, seed the cells ∼24 h before labelling to allow sufficient time for attachment and good spreading of the cells. Seed the cells sparsely, so that the culture is ∼30–50% confluent at the time of labelling. This ensures optimal cell growth conditions and also helps to reduce nonspecific background fluorescence.

③ The duration of the pulse would depend on the purpose of the experiment. Limit pulses to 2–5 min when labelling of short nascent DNA chains close to the replication forks is desired. Longer pulses (10–30 min), resulting in brighter staining, can be administered when the length of the labelled DNA segment is of no importance (e.g. for scoring the percentage of S-phase cells in the cell population).

④ Fixed cells can be stored indefinitely under these conditions, provided that care is taken not to allow the ethanol

to evaporate (e.g. seal the dishes with Parafilm). After this step, it is critical not to allow the cells to dry out, since this will result in destruction of cell/nuclear morphology.

⑤ The fragile coverslips must be handled carefully during transfers to avoid breaking. Use fine sharp point forceps to lift the coverslips from inside the wells of the dish. Keeping the wells full facilitates the process, since the liquid momentarily supports the lifted coverslip in an upright position.

⑥ This step is critical for successful immunostaining. The brief HCl treatment causes mild depurination of DNA, thus exposing the BrdU bases buried inside the double helix for binding by the anti-BrdU antibodies. If staining is weak or absent, troubleshoot by increasing the HCl concentration (e.g. incubate in 3 N HCl for 15 min at room temperature).

⑦ Typically, each washing step is performed for 5–10 min at room temperature. After incubations with fluorochrome-conjugated antibodies, the washing steps must be performed in the dark (e.g. cover the dishes with aluminum foil).

⑧ Since the small fragile coverslips are hard to mark, special attention must be paid to track the side on which the cells are attached and not to flip the coverslips.

⑨ The incubation times are flexible. To accelerate the procedure, incubations can be carried out for 30 min. at 37 °C. Alternatively, longer incubations (e.g. several hours, overnight, or up to few days) can be performed at 4 °C.

⑩ If a cell culture dish is used as a humid chamber, place a piece of aluminum foil over it to prevent exposure to light. Alternatively, place the dish inside a dark cabinet or drawer.

(11) Two types of control experiments must be performed to eliminate the possibility of non-specific immunostaining:

(A) Perform the procedure with cells that have not been labelled with a halogenated thymidine analogue, i.e. skip steps 1–3. No immunofluorescent signal should be visible and all cells must appear unlabelled in the FITC (green) channel. The DAPI-stained nuclei can be observed and/or counted in the UV (blue) channel.

(B) Perform the procedure with cells that have been labelled with a halogenated thymidine analogue, but omit the primary anti-BrdU antibody, i.e. skip steps 12–14. No immunofluorescent signal should be visible and all cells must appear unlabelled in the FITC (green) channel. The DAPI-stained nuclei can be observed and/or counted in the UV (blue) channel.

If non-specific fluorescent background is present:

(i) increase the number and/or duration of the washing steps;

(ii) decrease the concentration of the primary or secondary antibodies;

(iii) test different batches of antibodies.

Safety precautions

Some chemicals used in the procedure are hazardous (e.g. HCl, Tween 20, concentrated BrdU, IdU or CldU) and contact with skin must be avoided. Wear gloves when performing the procedure.

References

1. Aten, J. A., Bakker, P. J. M., Stap, J., Boschman, G. A. and Veenhof, C. H. N. (1992) DNA double labelling with IdUrd and CldUrd for spatial and temporal analysis of cell proliferation and DNA replication. *Histochem. J.*, **24**, 251–259.

2. Freeman, A., Morris, L. S., Mills, A. D., Stoeber, K., Laskey, R. A., Williams, G. H. and Coleman, N. (1999) Minichromosome maintenance proteins as biological markers of dysplasia and malignancy. *Clin. Cancer Res.*, **5**, 2121–2132.

3. Hunt, D. P., Freeman, A., Morris, L. S., Burnet, N. G., Bird, K., Davies, T. W., Laskey, R. A. and Coleman, N. (2002) Early recurrence of benign meningioma correlates with expression of mini-chromosome maintenance-2 protein. *Br. J. Neurosurg.*, **16**, 10–15.

4. Shibui, S., Hoshino, T., Vanderlaan, M. and Gray, J. W. (1989) Double labeling with iodo- and bromodeoxyuridine for cell kinetics studies. *J. Histochem. Cytochem.*, **37**, 1007–1011.

5. Dimitrova, D. S. and Gilbert, D. M. (2000) Temporally coordinated assembly and disassembly of replication factories in the absence of DNA synthesis. *Nat. Cell Biol.*, **2**, 686–694.

6. Dimitrova, D. S. and Berezney, R. (2002) The spatio-temporal organization of DNA replication sites is identical in primary, immortalized and transformed mammalian cells. *J. Cell Sci.*, **115**, 4037–4051.

7. Manders, E. M., Stap, J., Strackee, J., van Driel, R. and Aten, J. A. (1996) Dynamic behavior of DNA replication domains. *Exp. Cell Res.*, **226**, 328–335.

8. Dimitrova, D. S. and Gilbert, D. M. (1999) The spatial position and replication timing of chromosomal domains are both established in early G1 phase. *Mol. Cell*, **4**, 983–993.

PROTOCOL 6.3

Double labelling of DNA with different halogenated thymidine analogues

Reagents, materials and equipment

Most reagents and equipment are the same as in *Protocol 6.2*.

Antibodies: in addition to the antibodies listed in *Protocol 6.2*, Texas Red-conjugated donkey anti-rat IgG (No. 712-075-153; Jackson ImmunoResearch Laboratories) or Texas Red-conjugated donkey anti-mouse IgG (No. 715-075-151; Jackson ImmunoResearch Laboratories) are used to detect the rat or mouse anti-BrdU antibodies, respectively

High-salt washing buffer (TNT): 50 mM Tris-HCl, 0.5 M NaCl and 0.5% (v/v) Tween 20, pH 8

Procedure

1. Grow cells on coverslips as described in step 1 of Protocol 6.2.

2. Add IdU to the culture medium to a final concentration of 10 μM. Mix gently by swirling the dish and quickly return the dish to the cell culture incubator.

3. Incubate at 37 °C for the desired time to allow IdU incorporation into nascent DNA within the S-phase cells. ①

4. Aspirate the medium and rinse the cells two to three times with ample volumes of warm (37 °C) PBS or medium to remove all traces of exogenous IdU. ②

5. Fill the dish with warm free medium and return to the cell culture incubator for the desired period of time. ①

6. Administer a second pulse-label by adding CldU ③ to the medium to a final concentration of 10–100 μM. ④ Mix gently by swirling the dish and quickly return the dish to the cell culture incubator.

7. Incubate at 37 °C for the desired time to allow CldU incorporation into nascent DNA. ①

8. Aspirate the medium and wash the cells with ice-cold PBS to stop the labelling and to remove traces of medium and unincorporated CldU.

9. Fix the cells with cold 70% ethanol and depurinate DNA following steps 5–11 of Protocol 6.2. The subsequent immunostaining procedure for double-labelled DNA generally follows the protocol developed by Aten *et al.* [1].

10. Incubate the cells for 1 h at room temperature ⑤ with the rat anti-BrdU antibody diluted in blocking solution. ⑥

11. Wash twice with PBS-T. ⑦

12. Incubate the cells for 1 h at room temperature with the FITC-conjugated

donkey anti-rat IgG diluted in blocking solution.

13. Wash twice with PBS-T.

14. Incubate the cells for 1 h at room temperature with the mouse anti-BrdU antibody ⑧ diluted in blocking solution.

15. Wash twice with TNT.

16. Wash once with PBS-T.

17. Incubate the cells for 1 h at room temperature with the Texas Red-conjugated ⑨ donkey anti-mouse IgG diluted in blocking solution.

18. Wash twice with PBS-T.

19. Counterstain the nuclei with DAPI and mount the coverslips as described in steps 18–19 of Protocol 6.2. Observe using an epifluorescent microscope. ⑩

Notes

Depending on the length of the chase period, this procedure may take from several hours to several days.

① The duration of the pulse and chase periods is determined by the purpose of the experiment. When cultured during long chase periods (e.g. several days), the cells must be passaged to ensure proper cell density and optimal growth conditions. In this case, the cells are initially grown directly in tissue culture dishes during the first label and most of the chase period, and are subsequently seeded on coverslips, preferably ∼24 h before the second label.

② When followed by long chase periods (>2 h), 200 μM thymidine can be added to the PBS during the first washing step to promote the immediate cease of IdU incorporation. However, use of concentrated thymidine is

not recommended when short chase periods (≤1 h) separate the two labels, because it prevents the efficient incorporation of CldU during the second pulse-label.

③ The order of addition of IdU and CldU is of no importance and can be reversed.

④ Following short chase periods (<1 h), it is recommended to use higher concentrations of CldU (e.g. 100 μM) during the second label to ensure the preferential incorporation of CldU over any residual IdU persisting inside the cells.

⑤ All incubations must be performed in a dark humid chamber to prevent the cells from drying and the fluorochromes from bleaching. See also note 9 in Protocol 6.2.

⑥ The final concentrations must be optimized for each batch of antibodies. Start by testing primary antibody dilutions in the range of 1:20–1:500, and secondary antibody dilutions in the range of 1:100–1:1000.

⑦ Typically, each washing step is performed for 5–10 min at room temperature. After incubations with fluorochrome-conjugated antibodies, the washing steps must be performed in the dark (e.g. cover the dishes with aluminum foil).

⑧ The order of incubation with the primary rat and mouse anti-BrdU antibodies is important [1] and the sequence described in the procedure must be followed strictly.

⑨ The exact combination of fluorochrome-conjugated secondary antibodies used to detect the rat and mouse anti-BrdU antibodies is of no importance and fluorochromes can be selected at freedom with consideration for the specific needs of each individual experiment (e.g. Texas Red-conjugated donkey anti-rat IgG in combination with FITC-conjugated donkey anti-mouse IgG can be used with equal success). When choosing secondary antibodies, however, it is essential to select the highly cross-adsorbed forms to avoid non-specific cross-species reactivity.

⑩ Three types of control experiments must be performed to eliminate the possibility of non-specific immunostaining:

(A) Perform the complete procedure with cells that have not been labelled, i.e. skip steps 1–8. No immunofluorescent signal should be visible and all cells must appear unlabelled in the FITC (green) and Texas Red (red) channels. The DAPI-stained nuclei can be observed and/or counted in the UV (blue) channel. This experiment will control for non-specific reactivity by all antibodies.

(B) Perform the procedure with cells that have been labelled with a single halogenated thymidine analogue (e.g. either IdU or CldU). Incubate the cells with the primary anti-BrdU antibody, which recognizes the respective halogenated nucleoside (e.g. the mouse anti-BrdU antibody in the case of IdU labelling, or the rat anti-BrdU antibody in the case of CldU labelling). Then use 'the wrong secondary antibody', i.e. the FITC-conjugated donkey anti-rat IgG to detect the mouse anti-BrdU antibody within IdU-labelled cells, or the Texas Red-conjugated donkey anti-mouse IgG to detect the rat anti-BrdU

antibody within CldU-labelled cells. No immunofluorescent signal should be visible and all cells must appear unlabelled in the FITC (green) and Texas Red (red) channels. This experiment will provide control for the specificity of the secondary antibodies.

(C) Perform the complete procedure with cells that have been labelled with a single halogenated thymidine analogue (e.g. either IdU or CldU). The positive (S-phase) nuclei must be stained in one colour only, i.e. nuclei must be stained only in red within IdU-labelled cells, or only in green within CldU-labelled cells. This experiment will provide control for the specificity of the primary antibodies.

If non-specific fluorescent background is present:

(i) increase the number and/or duration of the washing steps;

(ii) decrease the concentration of the primary or secondary antibodies;

(iii) test different batches of antibodies.

Safety precautions

Some chemicals used in the procedure are hazardous (e.g. HCl, Tween 20, concentrated BrdU, IdU or CldU) and contact with skin must be avoided. Wear gloves when performing the procedure.

Reference

1. Aten, J. A., Bakker, P. J. M., Stap, J., Boschman, G. A. and Veenhof, C. H. N. (1992) DNA double labelling with IdUrd and CldUrd for spatial and temporal analysis of cell proliferation and DNA replication. *Histochem. J.*, **24**, 251–259.

Simultaneous immunostaining of proteins and halogen-dU-substituted DNA

Reagents, materials and equipment

Most reagents and equipment are the same as in *Protocols 6.2* and *6.3*.

1. Antibodies: in addition to the antibodies listed in *Protocols 6.2* and *6.3*, primary antibodies recognizing specific cellular proteins must be available. Appropriate highly cross-adsorbed fluorochrome-conjugated secondary antibodies must also be obtained to detect the respective primary antibodies.

2. Fixative: use fresh formaldehyde solution prepared by depolymerization of paraformaldehyde (Sigma). Formaldehyde is toxic and the fixative must be prepared in a chemical fume hood. To prepare 100 ml 4% formaldehyde in PBS, dissolve 4 g paraformaldehyde in ~40 ml ddH$_2$O by heating to 60–65 °C and vigorous mixing. Add 10 N NaOH drop by drop until the white powder is completely dissolved (around neutral pH). Remove from the heat and add 50 ml of 2× concentrated PBS. Adjust the pH to 7.2, then bring to 100 ml by adding ddH$_2$O. The fixative can be stored at 4 °C for 1–2 weeks.

3. Permeabilizing solution: 0.5% (v/v) Triton X-100 (Sigma) in PBS.

Procedure

1. Grow cells on coverslips as described in step 1 of Protocol 6.2.

2. Depending on the purpose of the experiment, single- or double-label the cells with CldU and/or IdU following the procedures described in Protocols 6.2 and 6.3.

3. Aspirate the cell culture medium and wash the cells with ice-cold PBS.

4. Remove the PBS and add appropriate fixative. ① Three frequently used fixatives are described below, but other reagents or combinations of fixatives can be tested.
   Incubate the cells in:

   (A) 4% formaldehyde in PBS for 10 min, ② or

   (B) cold methanol for 30 min. at −20 °C, or

   (C) cold ethanol : acetic acid (19 : 1) for 30 min at −20 °C.

5. Remove the fixative and wash several times with ample volumes of PBS.

6. Incubate the cells for 10 min at room temperature with permeabilizing solution ③ to ensure access of the antibodies to their targets. ④

7. Remove the permeabilizing solution and wash several times with ample volumes of PBS.

8. Incubate the cells for 1 h at room temperature ⑤ with a primary antibody (diluted in blocking solution ⑥) that recognizes the protein of interest.

9. Wash twice with PBS-T. ⑦

10. Incubate the cells for 1 h at room temperature with appropriate fluorochrome-conjugated secondary antibody to detect the protein target-bound primary antibody.

11. Wash twice with PBS-T.

12. Fix the cells again with 4% formaldehyde in PBS for 10 min. ⑧

13. Remove the fixative and wash several times with ample volumes of PBS.

14. Incubate the cells in 3 N HCl for 15 min at room temperature to depurinate DNA.

15. Remove the HCl and rinse several times with ample volumes of PBS. Make sure that the pH is neutral at the end of the washing.

16. Wash once with PBS-T.

17. Depending on the type of DNA labelling, follow steps 11–17 of Protocol 6.2, or steps 10–18 of Protocol 6.3 to immunostain the halogen-dU-substituted DNA.

18. Counterstain the nuclei with DAPI and mount the coverslips as described in steps 18–19 of Protocol 6.2. Observe using an epifluorescent microscope. ⑨

Notes

Depending on the length of the pulse/chase period(s), this procedure may take from several hours to several days.

① Since the immunostaining of halogen-dU-substituted DNA is not dependent on the fixation method, the choice of fixative is dictated by the properties and sensitivity of the protein target. Try several different fixation procedures when testing new primary antibodies.

② The type, concentration and temperature of fixative and the length of treatment need to be determined empirically for each protein.

③ Treatment with acetone for 1–2 min can be used as an alternative to Triton-PBS.

④ Since methanol and ethanol both fix and simultaneously permeabilize the cells, additional permeabilization is usually not required.

⑤ All incubations must be performed in a dark humid chamber to prevent the cells from drying and the fluorochromes from bleaching. See also note 9 in Protocol 6.2.

⑥ The final concentrations must be optimized for each antibody. Start by testing primary antibody dilutions in the range of 1 : 10–1 : 500, and secondary antibody dilutions in the range of 1 : 100–1 : 1000.

⑦ Typically, each washing step is performed for 5–10 min at room temperature. After incubations with fluorochrome-conjugated antibodies, the washing steps must be performed in the dark (e.g. cover the dishes with aluminum foil).

⑧ During this step, the antibodies used to detect proteins are covalently fixed in their positions. It is essential to perform the immunostaining in this order, since the subsequent depurination of DNA with HCl can be destructive for many cellular components and, if performed first, can cause the loss or redistribution of protein antigen epitopes [1, 2].

⑨ Control experiments similar to those described in note 11 of Protocol 6.2 and note 10 of Protocol 6.3 must be performed to test the specificity of all antibodies.

Safety precautions

Some chemicals used in the procedure are hazardous (e.g. HCl, Tween 20, concentrated BrdU, IdU or CldU) and contact with skin must be avoided. Wear gloves when performing the procedure. Formaldehyde solutions must be prepared and handled in a chemical fume hood.

References

1. Dimitrova, D. S. and Berezney, R. (2002) The spatio-temporal organization of DNA replication sites is identical in primary, immortalized and transformed mammalian cells. *J. Cell Sci.*, **115**, 4037–4051.
2. Humbert, C. and Usson, Y. (1992) Eukaryotic DNA replication is a topographically ordered process. *Cytometry*, **13**, 603–614.

Uncovering the nuclear matrix in cultured cells

Jeffrey A. Nickerson, Jean Underwood and Stefan Wagner

Introduction

Nucleic acid metabolism is temporally and architecturally organized in the cell nucleus on two nucleic acid containing structures: the DNA containing chromatin and an RNA containing fibrogranular ribonucleoprotein (RNP) network [1]. This RNP network seen in unextracted nuclei by selective staining for RNA is the nuclear matrix (reviewed in ref. 2). Electron microscopy shows that the RNP network consists of interconnected structures including the splicing factor-rich interchromatin granule clusters and the perichromatin fibrils which are enriched in new transcripts [1, 3].

The spatial distribution of nuclear domains where transcription, RNA splicing, DNA replication and other nuclear functions occur, remains unchanged after the removal of chromatin, suggesting that the nuclear matrix and not chromatin is the fundamental structure organizing the nucleus (reviewed by Nickerson [2]). Over the years numerous protocols have been developed to separate the nuclear matrix from chromatin, based on the first procedure of Berezney and Coffey [4]. A conservative nuclear matrix preparation should preserve the ultrastructure of the RNP network. In this protocol, we present two methods for uncovering the nuclear matrix in cultured cells. The first uncovers a structure with excellent preservation of RNP network ultrastructure [5].

This is achieved by cross-linking proteins before the removal of chromatin. The second method, performed without cross-linking, is less conservative and removes more proteins but reveals that the nuclear matrix is constructed on an underlying network of branched 10 nm filaments [6] (see Figure 6.2).

Reagents

Electron microscopy grade formaldehyde (16% w/v solution stored under inert gas) Ted Pella (cat. no. 18505)

RNase-free DNase I, e.g. Roche (cat. no. 776 785)

Serine protease inhibitor, 4-(2-aminoethyl) benzenesulfonyl fluoride (AEBSF), Roche under the trade name Pefabloc SC, or from Sigma-Aldrich (cat. no. 76307)

Triton X-100, protein grade, as a 10% solution from Roche

Other extraction chemicals (Roche)

Vanadyl riboside complex (VRC), Sigma-Aldrich Co. (cat. no. 94740) or New England Biolabs (cat. no. S1402S)

Solutions

Cytoskeletal buffer: (10 mM PIPES, pH 6.8, 300 mM sucrose, 100 mM NaCl, 3 mM MgCl$_2$ and 1 mM EGTA)

To make 1 l of the stock solution use 3.024 g of PIPES, 102.69 g of sucrose,

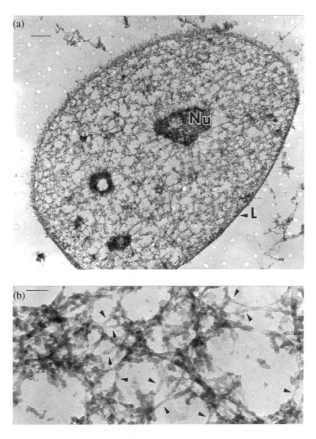

Figure 6.2 The ultrastructure of the cross-link stabilized nuclear matrix. The nuclear matrix of a CaSki cell was prepared by the cross-link stabilized nuclear matrix preparation procedure and visualized by resinless section electron microscopy. (a) The nuclear matrix consisted of two parts, the nuclear lamina (L) and a network of intricately structured fibres connected to the lamina and well distributed through the nuclear volume. The matrices of nucleoli (Nu) remained and were connected to the fibres of the internal nuclear matrix. Three remnant nucleoli may be seen in this section. Few intermediate filaments were connected to the outside of the lamina. (b) Seen at higher magnification the highly structured fibres of the internal nuclear matrix seemed to be built on an underlying structure of 10 nm filaments which are occasionally branched. These were seen most clearly when, for short stretches, they were free of covering material (arrowheads). The classical nuclear matrix procedure, when used with the 2 M NaCl step, uncovers this network of core filaments. The bar shown in panel (a) represents 1 μM and in panel (b) it is 100 nm. This figure is reproduced with permission from ref. 6

5.844 g of NaCl, 0.6099 g of $MgCl_2 \cdot 6H_2O$ and 0.3804 g of EGTA. Titrate to pH 6.8 with 1 M NaOH. Freeze in aliquots at −20 °C.

Before use VRC is added to a final concentration of 2 mM from the 100× stock solution, and AEBSF is added to a final concentration of 1 mM from the 100× stock solution. Additionally, for some experiments Triton X-100 is added to a final concentration of 0.5% from the 20× stock solution.

Extraction buffer: (10 mM PIPES, pH 6.8, 250 mM ammonium sulfate, 300 mM sucrose, 3 mM $MgCl_2$, 1 mM EGTA)

To make 1 l of this stock solution use 3.024 g of PIPES, 102.69 g of sucrose,

33.035 g of ammonium sulfate, 0.6099 g of $MgCl_2 \cdot 6H_2O$, and 0.3804 g of EGTA. Titrate to pH 6.8 with 1 M NaOH. Freeze in aliquots at $-20\,^{\circ}C$.

Prior to use, Triton X-100 is added to a final concentration of 0.5% from the 20× stock solution, VRC is added to a final concentration of 2 mM from the 100× stock solution, and AEBSF is added to a final concentration of 1 mM from the 100× stock solution.

Digestion buffer: (10 mM PIPES, pH 6.8, 300 mM sucrose, 50 mM NaCl, 3 mM MgCl₂, 1 mM EGTA)

To make 1 l of this stock solution use 3.024 g of PIPES, 102.69 g of sucrose, 2.922 g of NaCl, 0.6099 g of $MgCl_2 \cdot 6H_2O$ and 0.3804 g of EGTA. Titrate to pH 6.8 with 1 M NaOH. Freeze in aliquots at $-20\,^{\circ}C$.

Before use, Triton X-100 is added to a final concentration of 0.5% from the 20× stock solution, VRC is added to a final concentration of 2 mM from the 100× stock solution, and AEBSF is added to a final concentration of 1 mM from the 100× stock solution.

2M NaCl buffer: (10 mM PIPES, pH 6.8, 300 mM sucrose, 2 M NaCl, 3 mM MgCl₂, 1 mM EGTA)

To make 1 l of this stock solution use 3.024 g of PIPES, 102.69 g of sucrose, 116.88 g of NaCl, 0.6099 g of $MgCl_2 \cdot 6H_2O$ and 0.3804 g of EGTA. Titrate to pH 6.8 with 1 M NaOH. Freeze in aliquots at $-20\,^{\circ}C$.

Before using, VRC is added to a final concentration of 2 mM from the 100× stock solution and AEBSF is added to a final concentration of 1 mM from the 100× stock solution.

Triton stock: (10% (w/v) Triton X-100)

This is a 20× stock solution frozen in aliquots at $-20\,^{\circ}C$.

VRC stock: (200 mM Vanadyl riboside complex)

This is a 100× stock solution frozen in aliquots at $-20\,^{\circ}C$.

AEBSF Stock: (100 mM 4-(2-Aminoethyl)-benzenesulfonyl fluoride, hydrochloride)

This is a 100× stock solution frozen in aliquots at $-20\,^{\circ}C$.

To prepare the 100× stock solution, dissolve 100 mg in 4 ml water. Other protease inhibitors can be added if proteolysis is suspected. Do not use EDTA since divalent ions are necessary for structural integrity.

Phosphate buffered saline: (10 mM Na₂HPO₄, 1 mM KH₂PO₄, 137 mM NaCl and 2.7 mM KCl)

To make 1 l use 1.420 g of Na_2HPO_4, 0.136 g of KH_2PO_4, 8.006 g of NaCl and 0.2013 g of KCl. Autoclave and store in aliquots at room temperature.

Formaldehyde fixative: (4% solution)

The 4% formaldehyde fixative is prepared fresh, just before use in cytoskeletal buffer from a stock solution of 16% formaldehyde (EM-grade) stored under an inert gas. Alternatively, fresh formaldehyde can be prepared from paraformaldehyde powder.

Equipment

Tissue culture incubator

Low-speed centrifuge (for suspension cells)

Procedure

Cells grown in monolayers, either on coverslips or in dishes, can be extracted by exchanging solutions. Suspension cells or cells removed from a growth surface can be sequentially processed by centrifugation, removal of the supernatant and resuspension in the next solution. ① These procedures, as presented, uncover the nuclear matrix in whole cells without prior nuclear isolation. This allows the best preservation of ultrastructure and reveals the nuclear matrix–intermediate filament scaffold [7]. This consists of an internal nuclear matrix and the intermediate filaments of the cytoskeleton integrated into a single cell-wide structure by their attachments to the nuclear lamina. The same procedures can, however, be used to extract isolated nuclei in suspension and this is sometimes preferable for biochemical fractionation. One compatible method for nuclear isolation is that of Penman [8, 9].

A. Cross-link stabilized nuclear matrix

This method affords the best conservation of the nuclear matrix, as judged by the ultrastructural preservation of the nuclear RNP network [6]. This preparation is excellent for microscopy, but the cross-linking can make biochemical analysis difficult.

1. Wash cells in phosphate buffered saline at 4 °C. ①

2. Permeabilize cells in cytoskeletal buffer with 0.5% Triton X-100 at 4 °C for 2–5 min. This step will remove soluble proteins, both cytoplasmic and nucleoplasmic, and prevent their cross-linking.

3. Wash briefly in cytoskeletal buffer at 4 °C.

4. Cross-link structures using 4% formaldehyde in cytoskeletal buffer at 4 °C for 40 min.

5. Wash in cytoskeletal buffer at 4 °C three times for 2 min each to remove formaldehyde.

6. Cross-linked chromatin is removed by digestion with 400 units of RNase-free DNase I in digestion buffer at 32 °C for 50 min. Most DNA is removed from the structure at this step. Residual DNA can be removed by washing:

7. Wash 1: Wash cells with extraction buffer (which contains 0.25 M ammonium sulfate) at room temperature for 5 min.

8. Wash 2: Wash cells with 2 M NaCl buffer at room temperature for 5 min.

9. The structure can be processed for microscopy after a wash with cytoskeletal buffer.

B. Classical nuclear matrix

This method is more suitable for molecular analysis. The spatial distribution of nuclear components, for example those involved in transcription, RNA splicing and DNA replication, is well preserved as compared to the unextracted nucleus. The procedure can be stopped after the DNase I digestion and 0.25 M ammonium sulfate extraction, but further extraction in 2 M NaCl removes some proteins of the RNP fibres, uncovering a core structure of branched 10 nm filaments, the core filaments of the nuclear matrix [5].

1. Wash cells with phosphate buffered saline at 4 °C. ①

2. Permeabilize cells in cytoskeletal buffer with Triton X-100 at 4 °C for 2–5 min. This will remove soluble proteins, both cytoplasmic and nucleoplasmic.

3. Digest chromatin with 400 units of RNase-free DNase I in digestion buffer for 30–50 min at 32 °C. This step

will remove DNA and the nucleosomal histones. The nuclear structure at this point is the nuclear matrix. ② ③

4. Extract cells with extraction buffer at 4 °C for 3–5 min. This will remove histone H1 and will strip the cytoskeleton except for the intermediate filaments which remain tightly anchored to the outside of the nuclear lamina. ②

5. Optional: Extract the structure in 2 M NaCl buffer at 4 °C for 3–5 min. Better preservation is obtained by increasing the NaCl concentration slowly–or in steps. This step strips some proteins from the nuclear matrix uncovering a highly branched network of 10 nm filaments that form the core structure of the nuclear matrix. ④

6. For microscopy, fix immediately after fractionation. Incubate the nuclei in 4% formaldehyde in cytoskeletal buffer at 4 °C for 30–50 min.

Notes

① Cells in suspension are most conveniently processed for biochemical experiments following trypsinization or scraping. For suspended cells, we use about 1 ml for each 10^7 cells until the digestion step and then halve the volume. Suspension processing can be done by centrifuging at $1000 \times g$ for 3 min at 4 °C and sequentially resuspending cell pellets in the next wash or extraction solution between steps. The supernatant fractions can be saved for biochemical analysis. The extracted cell structure at each step is in the pellet.

② Steps 3 and 4 may be reversed, with equivalent results as judged by electron microscopy. The protein composition of the resulting nuclear matrix is also the same. An easy alternative is to perform the DNase I digestion first and then add ammonium sulfate slowly from a 1 M stock solution to a final concentration of 0.25 M.

③ DNA release can be evaluated microscopically by staining with a fluorescent DNA-binding dye such as 4′-6-diamidino-2-phenylindole (DAPI; 1–10 ug/ml in phosphate buffered saline for 5 min), or by pulse-labelling cells in ^{3}H-thymidine before fractionation and then measuring radioactivity.

④ Nuclear matrix proteins that are not part of this core filament network should be in the supernatant fraction.

References

1. Monneron, A. and Bernhard, W. (1969) Fine structural organization of the interphase nucleus in some mammalian cells. *J. Ultrastruct. Res.*, **27**, 266–288.
2. Nickerson, J. (2001) Experimental observations of a nuclear matrix. *J. Cell Sci.*, **114**, 463–474.
3. Nash, R. E., Puvion, E. and Bernhard, W. (1975) Perichromatin fibrils as components of rapidly labeled extranucleolar RNA. *J. Ultrastruct. Res.*, **53**, 395–405.
4. Berezney, R. and Coffey, D. S. (1974) Identification of a nuclear protein matrix. *Biochem. Biophys. Res. Commun.*, **60**, 1410–1417.
5. Nickerson, J. A., Krockmalnic, G., Wan, K. M. and Penman, S. (1997) The nuclear matrix revealed by eluting chromatin from a cross-linked nucleus, *Proc. Nat. Acad. Sci. USA*, **94**, 4446–4450.
6. He, D. C., Nickerson, J. A. and Penman, S. (1990) Core filaments of the nuclear matrix. *J. Cell. Biol.*, **110**, 569–580.
7. Fey, E. G., Wan, K. M. and Penman, S. (1984) Epithelial cytoskeletal framework and nuclear matrix–intermediate filament scaffold: three-dimensional organization and protein composition. *J. Cell Biol.*, **98**, 1973–1984.

8. Penman, S. (1966) RNA metabolism in the HeLa cell nucleus. *J. Mol. Biol.*, **17**, 117–130.

9. Capco, D. G., Krockmalnic, G. and Penman, S. (1984) A new method of preparing embedment-free sections for transmission electron microscopy: applications to the cytoskeletal framework and other three-dimensional networks. *J. Cell Biol.*, **98**, 1878–1885.

Nuclear matrix–lamin interactions

Barbara Korbei and **Roland Foisner**

Background

The nuclear envelope of the eukaryotic cell nucleus is composed of the outer and inner nuclear membranes, nuclear pore complexes, and a protein filament meshwork underlying the inner nuclear membrane, termed the nuclear lamina (see Figure 6.3). The lamina contains nucleus-specific (type V) intermediate filament proteins, the lamins, plus numerous integral and peripheral proteins of the inner membrane, which bind lamins [1, 2]. Two different types of lamins can be distinguished according to their expression patterns and biochemical properties. B-type lamins arc constitutively expressed throughout development and are essential for cell viability. A-type lamins, which are expressed only in differentiated cells, are not essential for development, but serve yet unknown functions in tissue homeostasis. More recent studies have also identified lamins in the nuclear interior, where they form complexes with nucleoplasmic lamin binding proteins, such as lamina-associated polypeptide 2α (LAP2α) [3].

The discovery of novel binding partners of lamins and of lamina-associated polypeptides has significantly changed our view of the functions of lamin complexes in recent years. Apart from the long-known function of the lamina as a scaffolding framework providing mechanical stability for the nucleus, recent studies suggested that lamin complexes are also involved in chromatin organization and chromosome segregation as well as in DNA replication and RNA processing [1, 4].

The hypothesis that many diverse nuclear functions are directly or indirectly linked to the nuclear lamina is supported by recent findings revealing that mutations in A-type lamins or in some of their binding partners give rise to an increasing scope of diverse genetic diseases known as laminopathies [5]. Mutations in the *LMNA* gene that encodes A-type lamins result in a broad range of disease phenotypes, affecting skeletal and heart muscle, adipose, nerve and bone tissue. The molecular mechanisms of these diseases are completely unknown. At least some of the pathological phenotypes may be the result of impaired interactions of mutant lamins with chromatin proteins and with proteins in transcriptional complexes. The identification of novel (tissue-specific) lamin binding partners is therefore an essential first step in unravelling lamin functions and has to be followed by a detailed molecular analysis of the interactions, including analysis of interaction domains, binding strengths, assembly dynamics and molecular structure.

Here we describe two *in vitro* techniques, which have been used to analyse the interaction of lamins with LAP2α

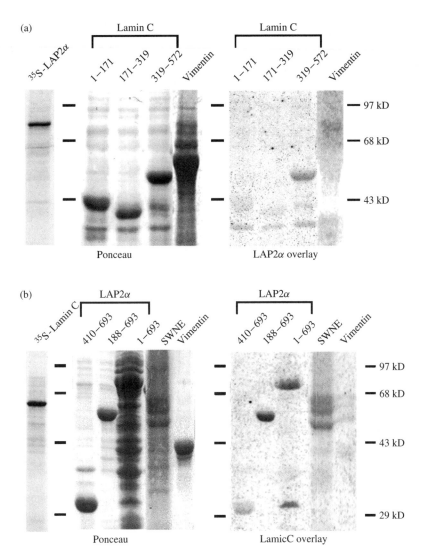

Figure 6.3 (a) Overlay of *in vitro* translated [^{35}S]-labelled LAP2α onto transblotted fragments of lamin C and the intermediate filament protein vimentin (negative control), showing that LAP2α interacts with the C-terminal domain of lamin C. (b) Overlay of *in vitro* translated [^{35}S]-labelled lamin C onto transblotted recombinant LAP2α fragments or salt-washed nuclear envelope fractions of rat liver (SWNE, containing both lamin A and C, as positive control) and vimentin (negative control), revealing binding of lamin C to the C-terminal 78 residues of LAP2α. Ponceau S stains of blots and autoradiograms of *in vitro* translated proteins (left) and of overlays (right) are shown

and with chromatin: the *in vitro* overlay binding assay, which allows identification of interaction domains, and the *in vitro* nuclear assembly assay, which reveals the dynamic properties of lamin–chromatin interactions during the cell cycle.

Nuclear matrix–lamin interactions: *in vitro* blot overlay assay

Introduction

Transblotted proteins immobilized on a nitrocellulose membrane are incubated with a radioactively labelled binding partner. If various fragments representing different domains of the protein are immobilized onto the membrane, this assay identifies interaction domains and may also reveal a rough determination of relative binding strengths of different domains. A major pitfall of this assay is the denaturation of immobilized proteins during SDS-PAGE and blotting, which may, despite partial renaturation of proteins on the membrane during incubation in a binding buffer, cause unspecific interactions. Therefore negative controls, applying non-interacting proteins, which have similar structure and/or properties as analysed proteins, should always be included.

Reagents

Blocking buffer: 2% BSA in overlay buffer

Dilution buffer: 1% BSA in overlay buffer with 1 mM PMSF (phenylmethylsulfonylfluoride)

Overlay buffer: 10 mM Hepes/KOH (pH 7.4), 100 mM NaCl, 5 mM $MgCl_2$, 2 mM EGTA, 0.1% Triton X-100

PBST (0.05% Tween 20 in PBS)

Phosphate-buffered saline (PBS): 2.6 mM KCl, 137 mM NaCl, 1.5 mM KH_2PO_4, 8 mM $Na_2HPO_4.7H_2O$, pH 7.4

Ponceau S: 0.2% (w/v) Ponceau S, 3% (w/v) trichloroaceticacid

Triton X-100, 1 mM dithiothreitol (DTT)

Procedure

1. Separate bacterially expressed recombinant lamina proteins (bacterial cell lysates are usually fine) by SDS-PAGE and transblot onto nitrocellulose membranes (0.2 μm). ①

2. Stain nitrocellulose membranes with immobilized proteins using Ponceau S to visualize proteins and estimate protein quantity, followed by thorough washing in PBST. ②

3. Incubate in overlay buffer for 1 h with three changes (to allow renaturation of proteins), then block the membranes for 30 min in blocking buffer.

4. In the meantime, the interaction partner to be tested is transcribed and translated *in vitro* and [35S]-labelled, using the TNT Quick coupled transcription/translation system (Promega, Madison, WI), according to the manufacturer's protocol.

5. Probe the membranes with 100 μl of the standard *in vitro* translation reaction mixture (diluted 1 : 50 in overlay buffer) containing the [35S]-labelled protein for 3 h at room temperature, while gently moving the incubation chamber.

6. Wash extensive in overlay buffer and air-dry the nitrocellulose.

7. Detect bound protein by autoradiography. Quantify signals on autoradiograms by densitometric scanning and normalize using the intensities detected on the Ponceau S-stained blot, thus providing a rough estimation of the relative binding strengths of the tested fragments.

Notes

① Various lamina fragments covering different regions of the protein can be applied to one gel in order to determine the binding domains.

② If renaturation of proteins is critical for interaction, avoid denaturing Ponceau S stain on the blot used for overlay, and use separate membranes prepared in parallel for estimation of protein amounts).

PROTOCOL 6.7

Nuclear matrix–lamin interactions: *in vitro* nuclear reassembly assay

Introduction

Metaphase cell lysates containing metaphase chromosomes and lamina proteins, or isolated chromosomes and chromosome-free metaphase cell lysates, are incubated at 37 °C for different time periods in order to investigate the dynamic interaction of different lamina proteins with chromatin during nuclear reassembly after mitosis. Furthermore, dominant negative mutants interfering with the assembly can be tested in this assay.

Reagents

Complete culture medium: Dulbecco's Modified Eagle's Medium (DMEM), 10% fetal calf serum (FCS), 50 U/ml penicillin, 50 µg/ml streptomycin

KHM buffer: 50 mM Hepes/KOH (pH 7.0), 78 mM KCl, 4 mM MgCl$_2$, 10 mM EGTA, 8.4 mM CaCl$_2$, 1 mM dithiothreitol (DTT), 20 µM cytochalasin B

3× SDS-PAGE sample buffer: 186 mM Tris/HCl (pH 6.8), 300 mM dithiothreitol (DTT), 6% sodium dodecyl sulfate (SDS), 0.1% bromphenol blue, 30% glycerol

Equipment

Heraeus Megafuge 1.0

Incubator (37 °C, 8.5% CO$_2$)

Metal ball cell cracker (EMBL, Heidelberg)

Plastic cell culture flasks 175 cm^2 (Nunc)

Sterile hood

Thermoblock for Eppendorf tubes

Procedure (adapted from ref. 6)

A. Synchronization of cells

1. Plate NRK (normal rat kidney) cells in ten 175 cm^2 cell culture flasks and allow to grow to approximately 60–70% confluency in complete medium at 37 °C and 8.5% CO$_2$.

2. Replace the medium with complete medium plus 2 mM thymidine in order to arrest cells at the G1-S phase boundary.

3. After 10 h incubation in thymidine medium, wash the monolayer of cells three times with sterile PBS and incubate in complete medium for 4 h.

4. Add 0.2 µg/ml nocodazole (from a 10 mg/ml stock solution in 100% DMSO) and incubate cells for a further 10 h.

5. Harvest the loosely attached prometaphase cells using a mechanical shake-off.

B. Preparation of mitotic lysates

1. Pellet mitotic cells from ten flasks at 1000 rpm (Heraeus Megafuge, 1.0R) for 3 min.

2. Wash twice in PBS and incubate in 30 ml complete medium containing

nocodazole (0.2 μg/ml) and cytochalasin B (20 μM, added from a 20 mM stock in 100% DMSO) for 45 min at 37 °C to destroy microtubules and actin filaments.

3. Sediment mitotic cells at 1000 rpm (Heraeus Megafuge, 1.0R) for 3 min, wash twice with ice-cold PBS containing 0.2 μg/ml nocodazole and resuspend in an equal volume of ice-cold KHM buffer supplemented with protease inhibitors (3.3 μg/ml leupeptin, 3.3 μg/ml aprotinin, 3.3 μg/ml pepstatin A, 1 mM PMSF (phenylmethylsulfonylfluoride)).

4. Homogenization is accomplished best by pressing the suspension 10–15 times through a metal ball cell cracker (EMBL, Heidelberg) equipped with a tight-fitting ball ($r = 8.008$ or 8.006 mm) on ice. Alternatively, cell suspension can be pressed several times through a bent needle (27G), or cells can be lysed in a potter using a tightly fitting glass pestle. Efficiency of cell breakage is controlled in the phase contrast microscope using a 40× objective.

C. In vitro reassembly

1. For control assembly reactions, dilute the mitotic NRK cell lysate, containing chromosomes and soluble phosphorylated lamina proteins, with half the lysis volume of ice-cold KHM buffer supplemented with protease inhibitors (see above).

2. Incubate for different time intervals between zero and 2 h at 37 °C, to allow dephosphorylation and assembly of lamin structures at the chromosomal surface.

3. At different time points, transfer aliquots of the incubation mixture to a precooled tube, mix with phosphatase

inhibitors (0.1 μM calyculin A; 0.1 μM okadaic acid (Invitrogen); 1 mM orthovanadate; 0.5 mM beta-glycero-phosphate (Sigma-Aldrich)) and place on ice to prevent further assembly.

4. Centrifuge the collected samples at 2000 rpm (Heraeus Megafuge, 1.0R) for 10 min at 4 °C to separate the soluble, cytoplasmic from the insoluble, cytoskeleton/chromatin fractions.

5. Resuspend the pellet in $1\frac{1}{2}$ times the original aliquot volume of 1× SDS-PAGE sample buffer.

6. Mix the supernatant with half its volume of 3× SDS sample buffer.

7. Analyse the samples by SDS-PAGE and immunoblotting using specific antibodies to the lamina proteins of interest (Figure 6.4A).

8. Samples can also be prepared for immunofluorescence microscopy (Figure 6.4B). For this, 70 μl aliquots of cell lysates are mixed with formaldehyde (37% w/v solution) to a final formaldehyde concentration of 3.7% (w/v).

9. Spin samples onto 18× 18 mm glass coverslips for 20–30 s at 500 rpm (Heraeus Megafuge, 1.0R) in a cytospin rotor.

10. Following centrifugation, coverslips should be immediately immersed in 1 ml 3.7% formaldehyde for 20 min, and then processed for immunolabelling.

D. Assembly with exogenous chromosomes

1. Remove endogenous chromosomes from mitotic cell lysates by centrifugation at 2000 rpm (Heraeus Megafuge, 1.0R) for 5 min at 4 °C.

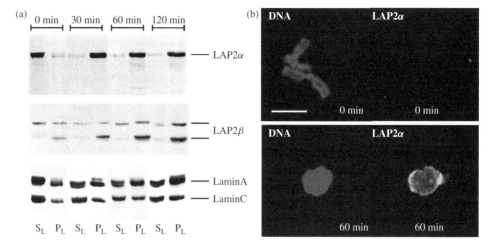

Figure 6.4 (a) *In vitro* nuclear assembly around endogenous chromosomes. Mitotic NRK cell lysates were incubated at $37\,^{\circ}$C and insoluble pellet fractions (P_L) and soluble supernatant fractions (S_L) were collected by centrifugation at $2000\ g$ at indicated time points. Samples were tested by immunoblot analysis using monoclonal antibodies to LAP2α, LAP2β (lamina associated polypeptide 2β) and to lamins A/C. Immunoblots are shown, revealing a redistribution of the proteins from a soluble chromosome-free fraction to a sedimentable chromosome-containing fraction within 30 min incubation. (b) Immunofluorescence analysis of *in vitro* assembled nuclear structures. Assembly mixtures containing exogenous chromosomes and chromosome-free mitotic cell lysates were spun on coverslips after 0 or 60 min incubation and processed for immunofluorescence microscopy using antibodies to the indicated proteins and Hoechst dye to stain DNA. Confocal images are shown. Bar, 5 μm

2. Mix chromosome-free supernatants with mitotic chromosomes ($OD_{260} = 3$) isolated from nocodazole arrested Chinese hamster ovary (CHO) cells by sucrose gradient centrifugation (for details see ref. 7).

3. Perform assembly reactions exactly as with endogenous chromosomes.

E. Assembly with recombinant proteins

1. To analyse the effects of proteins or protein fragments on the assembly, add bacterially expressed recombinant proteins, either purified or as whole bacterial cell lysates (if protein concentration is high) to the nuclear assembly mixtures, at concentrations ranging from twofold to eightfold the concentration of the endogenous proteins.

2. Perform the assembly reaction exactly as in the control sample.

3. Control assemblies should always be carried out in parallel, as assembly efficiencies may vary from experiment to experiment.

Acknowledgements

We thank Thomas Dechat, University of Vienna, for providing data shown in Figure 6.3, and Sylvia Vlcek, University of Vienna, for her helpful comments on the manuscript. Work in the authors' laboratory was supported by grants from the Austrian Science Research Fund (FWF P15312), the Austrian National Bank and the Österreichische Muskelforschung (to R.F.).

References

1. Goldman, R. D., Gruenbaum, Y., Moir, R. D., Shumaker, D. K. and Spann, T. P. (2002) *Genes Dev.*, **16**, 533–547.

2. Foisner, R. (2001) *J. Cell Sci.*, **114**, 3791–3792.

3. Dechat, T., Korbei, B., Vaughan, O. A., Vlcek, S., Hutchison, C. J. and Foisner, R. (2000) *J. Cell Sci.*, **113**, 3473–3484.

4. Cohen, M., Lee, K. K., Wilson, K. L. and Gruenbaum, Y. (2001) *Trends Biochem. Sci.*, **26**, 41–47.

5. Burke, B. and Stewart, C. L. (2002) *Nat. Rev. Mol. Cell Biol.*, **3**, 575–585.

6. Burke, B. and Gerace, L. (1986) *Cell*, **44**, 639–652.

7. Vlcek, S., Korbei, B. and Foisner, R. (2002) *J. Biol. Chem.*, **277**, 18898–18907.

Safety hazards

Working safely with radioactivity

$[^{35}S]$ emits β-rays with a maximum β energy of 0.167 MeV and a half-life of 87.4 days. A Geiger-Müller counter is suitable for the detection of β-emitters. β particles can affect superficial layers of tissues and represent an external hazard. They are also potential internal hazards, if radioactivity gets inside the body via inhaled gas particles or via the mouth or via skin cuts. Best protection from the external hazard can be attained by reducing exposure time to radioactive material. This can be achieved by planning and preparing the experiments carefully before starting bench work. Furthermore, intensity of electromagnetic radiation decreases with the square of the increasing distance. The use of shields between the body and the radioactive samples is also highly recommended, as β-rays may have a range of up to a few metres in air, while 1 cm of Plexiglas will stop any rays. This will, however, generate *Bremsstrahlung*, a form of X-rays, which are also potentially harmful. To protect from internal hazards, the most important rule is to prevent contamination of the working environment and the individual. This is achieved by restricting the working area for radioactive substances to a particular location in the lab. Furthermore, all equipment, materials and waste have to be labelled and the working area should be monitored regularly. Waste should be minimized and disposed of according to local rules and guidelines.

Other hazards

In principle, all substances which are cytotoxic or influence cellular functions and parameters, such as phosphorylation, cell cycle progression and proteolytic activities, are to be considered as potential hazards.

The phosphatase inhibitors are toxic if swallowed or inhaled, and upon prolonged or repeated exposure are also toxic if absorbed through the skin. Therefore, when handling such substances, one should wear gloves. They should be used only in a well-ventilated area and kept closed or covered when not in use.

Inhalation of protease inhibitor phenylmethylsulfonylfluoride may result in spasm, inflammation or oedema of the larynx and bronchi, chemical pneumonitis and pulmonary oedema. When handling, avoid dust formation.

Cytochalasin is a cell-permeable fungal toxin that disrupts contractile microfilaments by inhibiting actin polymerization and thus interferes with many cellular processes. It is a very powerful toxin and should be handled with extreme caution.

Nocodazole is an antimitotic agent that disrupts microtubules by binding to β-tubulin, thus affecting microtubule dynamics, spindle function and Golgi complex formation. It arrests the cell cycle at the G2/M phase boundary and induces apoptosis in several normal and tumour cell lines and is therefore also considered very toxic.

PROTOCOL 6.8

Preparation of *Xenopus laevis* egg extracts and immunodepletion

Tobias C. Walther

Introduction

Egg extracts from *Xenopus laevis* have been used as cell-free systems to study mitotic events and nuclear functions such as nuclear envelope (NE) formation and replication; see Lohka and Masui [1]. A major advantage of the system is that large volumes of *Xenopus* eggs can be obtained cheaply with relatively low effort. Importantly, in contrast to mammalian mitotic homogenates, egg extracts will efficiently package naked DNA into chromatin and NE precursors are not limiting in this system. In combination with immunodepletion of proteins, egg extracts provide a powerful system to study nuclear architecture, function and dynamics.

Reagents

10 × MMR: 1 M NaCl, 20 mM KCl, 10 mM $MgCl_2$, 20 mM $CaCl_2$, 1 mM EDTA, 50 mM HEPES/KOH, pH 8.0

D-buffer: 2% w/v cysteine in 0.25 × MMR, pH 7.8

Sucrose buffer 250 (S250): 250 mM sucrose, 50 mM KCl, 2.5 mM $MgCl_2$, 10 mM HEPES/KOH, pH 7.5

Sucrose buffer 250+ (S250+) take S250 and add to:

Final concentration	Stock	Per ml extract	Substance
1 mM	1 M	1 μl	DTT
44 μg/ml	20 mg/ml	2.5 μl	Cycloheximide (CHX)
5 μg/ml	10 mg/ml	0.5 μl	CytochalasinB (CytB)
1×	100×	10 μl	Trasylol
2 μg/ml	10 mg/ml	0.2 μl	Leupeptin
1 μg/ml	10 mg/ml	0.1 μl	Pepstatin

Blocking buffer: S250+ supplemented with 50 mg/ml BSA

Sucrose buffer 500 + (S500+): take S250+ and add 1.25 ml 2 M sucrose/10 ml

Ionophore A21387 2 mg/ml in DMSO

100 mM NaBO4 pH 9.0, 100 mM ethanolamine pH 8.0, 100 mM glycine pH 2.0, 10 mM Tris/HCL pH 7.4

Equipment

Beakers (800 ml)

Low-speed refrigerated centrifuge with swinging-bucket rotor

Microcolumns (800 μl)

Microfuge

Ultracentrifuge with swing-out rotor (5 ml tubes)

Procedure

Day 0

Inject 10 frogs with 500 U pregnant mare's serum gonadotropin (PMSG) into the dorsal lymph sac 4–14 days before the extract preparation.

Day 1

Inject 1000 U human chorionic gonadotropin (2000 U/ml in water) per frog, incubate at 16 °C for 16–18 h in 1 × MMR.

Day 2

A. Collect eggs

1. Pour off buffer with eggs (from frog containers). Avoid eggs that are in large clumps or 'ropes'!

2. Wash eggs with c.500 ml 1 × MMR.

3. Dejelly eggs in D-buffer up to 10 min, eggs become closely packed; swirl every 30 s.

4. Rinse eggs four times with c.500 ml 1 × MMR.

5. Activate eggs by adding 8 µl A23187 (2 mg/ml) per 100 ml 1 × MMR; animal cap contraction becomes visible after 3 min; leave up to 10 min (usually 7 min).

6. Wash three times with 1 × MMR, take care not to expose eggs to air, remove white eggs.

7. Incubate up to 25 min at 22 °C.

8. Rinse eggs three times with S250.

9. Wash eggs with S250(+).

B. Extract preparation

1. Transfer eggs to 5 ml tubes, spin 60 s at 2000 g to pack the eggs; remove excess buffer.

2. Spin 20 min at 20 000 g to crush eggs.

3. Take supernatant (= 'low-speed extracts') and add to:

f.c.	Stock	Per ml extract	Substance
1 mM	1 M	1 µl	DTT
44 µg/ml	20 mg/ml	2.5 µl	CHX
5 µg/ml	10 mg/ml	0.5 µl	CytB
Protease inhibitor cocktail			

4. Spin 35 min at 200 000 g.

5. Remove cytosolic (orange) phase, leaving lipids (on top) and pellet behind.

6. Dilute c.0.3 fold with S250+.

7. Spin 35 min at 200 000 g.

8. Remove cytosol, add glycerol to f.c. 3% (v/v) snap freeze and/or process for immunodepletion.

9. Wash membranes: resuspend membranes in c.20 × vol. in S250 (+) and lay carefully on top of 800 µl S500(+) buffer cushion.

10. Spin 15 min at 15 000 g.

11. Resuspend membranes in c.10th of cytosol vol. and snap freeze immediately.

C. Preparation of column for immunodepletion (should be done in advance)

1. Rotate 1–2 mg antibody with 0.5 ml protein A sepharose in a volume of 5 ml for 1 h at room temperature.

2. Wash beads with 2 × with 5 ml 0.2 M sodium borate buffer pH 9.0.

3. Resuspend beads in 5 ml 0.2 M sodium borate buffer pH 9.0 and add 13 mg of dimethylpimelimidate (DMP) from a freshly opened bottle (final concentration 10 mM).

4. Rotate for 30 minutes at room temperature.

5. Add another 13 mg of DMP, rotate 30 min at room temperature.

6. Stop reaction by washing beads 1× with 0.2 M ethanolamine pH 8.0 and incubating at room temperature for 2 h in this buffer. Check coupling efficiency by measuring protein concentration of antibody solution before and after cross-link.

7. Wash column with 5 ml PBS, 5 ml 10 mM Tris/HCl pH 7.4.

8. Pre-elute column with 5 ml 100 mM glycin pH 2.0.

9. Wash column with 5 ml PBS.

10. Transfer antibody (or mock) beads (250 µl each) to two 800 µl micro-columns.

11. Drain columns mounted on a 2 ml tube by spinning 10 s at 1000*g*.

12. Rotate 30 min with 400 µl blocking buffer.

D. Immunodepletion

1. Add 400 µl freshly prepared cytosolic extract to antibody or mock column, rotate 30 min to 1 h at 4 °C.

2. Collect first depletion as above (step 2), and add to second antibody column.

3. Rotate and collect as above, snap freeze.

Notes

Buffers should be around 16 °C before lysis of eggs and ice-cold for all subsequent steps.

When jelly coat has been removed wash eggs very carefully.

Reference

1. Lohka, M. J. and Masui, Y. (1983) Roles of cytosol and cytoplasmic particles in nuclear envelope assembly and sperm pronuclear formation in cell-free preparations from amphibian eggs. *J. Cell Biol.*, **98**, 1222–1230.

Nuclear assembly *in vitro* and immunofluorescence

Martin Hetzer

Introduction

The molecular mechanisms of nuclear assembly are not well understood. Many analyses have been carried out in a cell-free system based on *Xenopus* egg extracts that recapitulate pronuclear formation after fertilization. The first step in nuclear assembly is decondensation of sperm DNA. Subsequent events involve binding of membranes to chromatin, fusion of membranes to form the nuclear envelope (NE), insertion of nuclear pore complexes (NPCs) and nuclear growth accompanied by further chromatin swelling. For a detailed review see Gant and Wilson [1]. In egg extracts individual steps of nuclear assembly can be separated and analysed biochemically. The formation of a functional nucleus can be visualized by immunofluorescence using antibodies specific for NE proteins or NPC components.

Reagents

HSP: 250 mM sucrose, 15 mM Hepes/KOH, pH 7.4, 0.5 mM spermidine tetrachloride, 0.2 mM spermine

HSPP: HSP + 0.3 mM PMSF, 10 μg/μl leupeptin

PBS

20× Energy mix: Creatine phosphate (CP) 200 mM (51 mg/ml), ATP 10 mM, GTP 10 mM, CP-kinase 0.5 mg/ml First mix the CP, ATP and GTP. Measure pH and add 1 M Hepes/KOH, pH 7.3 for buffering if necessary. Add sucrose to a final concentration of 250 mM. Finally add kinase. Freeze in aliquots. Keep at −80 °C

20× glycogen: 150 mg/ml oyster glycogen in 10 mM Hepes/KOH, pH 7.5

Acetate buffer: 100 mM KAc, 3 mM MgAc, 5 mM EGTA, 20 mM Hepes/KOH, pH 7.4, 150 mM sucrose, 1 mM DTT

30%S: 30% sucrose (w/v) in acetate buffer

FIX: 8% formaldehyde in PBS

IF buffer: 0.1% Triton X-100, 0.02% SDS, 10 mg/ml BSA in PBS

Poly-L-lysine 0.1% w/v in water

Vectashield mounting solution

Equipment

Coverslips and slides

Low-speed refrigerated centrifuge with swinging-bucket rotor

Scissors, 1 ml syringe, glass plate, cheesecloth

Watchmaker's forceps

Water bath at 20 °C

Procedure

A. Preparation of sperm head chromatin from Xenopus laevis

(see Gurdon, J.B. [2]; for extract preparation, see *Protocol 6.8*)

1. Put male *Xenopus laevis* in ice water for ~40 min.

2. Take frog, cut away upper part of the head (use strong scissors), kill by sticking forceps into spinal chord.

3. Take testis, remove fat and connective tissue.

4. Resuspend in HSP buffer using forceps and chop testis into pieces.

5. Disperse further by pressing pieces through 1 ml syringe on glass plate.

6. Filter through cheesecloth, wash (final vol. from two testes ~6 ml).

7. Pellet by centrifugation 2000 rpm at $4\,°C$, wash once with HSPP 10 ml.

8. Resuspend in 1 ml HSPP, add 50 μl 10 mg/ml lysolecithin, incubate 5 min at room temperature.

9. Quench by adding 10 ml HSPP + 3% BSA, chill.

10. Pellet by centrifugation at 2000 rpm for 10 min.

11. Wash with 3 ml HSPP + 0.3% BSA, centrifuge as above.

12. Resuspend in 2.5 ml HSPP + 0.3% BSA + 30% glycerol.

13. Count 1/10 dilution: No. of sperm in 1 'grossquadrat' $\times 10^4$ = sperm/ml.

14. Dilute to 1000 sperm/μl.

15. Aliquot in 10 μl and freeze in liquid N_2.

16. Test: use 2 μl sperm heads + 2 μl Trypan Blue 0.4%; if there is no exclusion from sperm then permeabilization was successful.

B. Nuclear assembly

1. To assay NE formation 13–15 μl of cytosolic extract (see *Protocol 6.8*) is transferred into an Eppendorf tube. Demembrenated sperm heads are added to a concentration of 10^3/μl (stock is 10^3/μl). Gently mix with pipette and place reaction on ice for 5 min (to decondense sperm chromatin). Add 2 μl 10× membranes and 1 μl 20× glycogen.

2. Incubate at $20\,°C$ in a water bath for 60–90 min.

3. Reactions are stopped on ice and immediately fixed for 20 min by adding 100 μl AB-buffer and 100 μl 8% FIX (8% paraformaldehyde).

4. Layer sample on top of ice-cold 30% S in a tube containing a poly-L-lysine-coated coverslip (11 mm diameter).

5. Spin at 1000 *g* for 10 min.

6. Remove supernatant and recover the coverslip with watchmaker's forceps.

7. If the reactions contain fluorescently labelled components dip the coverslip in PBS and mount on a microscopy slide.

C. Immunofluorescence

1. Put coverslip on a grid or parafilm and permeabilize the nuclei for 30 min with IF buffer at room temperature.

2. Incubate for 1 h in IF buffer containing the primary antibody.

3. Wash 3× with 500 μl IF buffer.

4. Incubate for 1 h in IF buffer containing the secondary antibody coupled to a fluorophor.

5. Wash 3× with 500 μl IF buffer.

6. Wash the coverslip with IF buffer (+410 mM NaCl), incubate with Hoechst/DAPI for 1 min, mount the coverslip with Vectashield solution and analyse samples by confocal fluorescence microscopy.

Notes

Optimal salt concentration for the washing step has to be determined for each antibody. Incubate the coverslips in the dark.

References

1. Gant, T. M. and Wilson, K. L. (1997) Nuclear assembly. *Annu. Rev. Cell Dev. Biol.*, **13**, 669–695.

2. Gurdon, J. B. (1976) Injected nuclei in frog oocytes: fate, enlargement, and chromatin dispersal. *J. Embryol. Exp. Morphol.*, **36**, 523–540.

Nucleocytoplasmic transport measurements using isolated *Xenopus* oocyte nuclei

Reiner Peters

Introduction

The exchange of matter between cell nucleus and cytoplasm, involving some of the most significant cellular molecules and molecular assemblies such as ribonucleo-protein particles, ribosomal subunits and transcription factors, is mediated by the nuclear pore complex (NPC), a large transporter spanning the nuclear envelope. In the past, transport across the NPC has been studied almost exclusively in intact cells (e.g. by expression of GFP constructs, cell fusion and microinjection), in semi-intact cells (e.g. by selective permeabilization of the plasma membrane employing the detergent digitonin [1]), and in artificial nuclei reconstituted in cell extracts [2, 3]. Here, the protocol of a transport assay is given which employs manually isolated nuclei of *Xenopus* oocytes [4]. The assay is easy, fast and convenient, employs synchronized primary cells, completely avoids the use of detergents and yields quantitative kinetic data. The large size of *Xenopus* oocyte nuclei facilitates the combination of transport measurements with other functional as well as biochemical and structural studies. A complementary protocol pertaining to isolated nuclear envelopes is given as *Protocol 6.11*.

Reagents and materials

Tissue culture dishes, e.g. Falcon no. 353004

2.5 mm diameter drill

Amphibian Ringer's solution: 88 mM NaCl, 1 mM KCl, 0.8 mM $MgSO_4$, 1.4 mM $CaCl_2$, 5 mM Hepes (pH 7.4)

Mock 3 intracellular medium: ① 90 mM KCl, 10 mM NaCl, 2 mM $MgCl_2$, 0.1 mM $CaCl_2$, 1.0 mM *N*-(2-hydroxyethyl)ethylene-diaminetriacetic acid (HEDTA), 10 mM Hepes (pH 7.3)

Transport solution: mock 3 containing 0.5 µM of a fluorescent protein containing a nuclear localization sequence (NLS) such as GG-NLS [5] or Alexa488-P4K [4], 0.5 µM karyopherin α2, 0.5 µM karyopherin β1, 2 µm Texas-red-labelled 70 kDa dextran, energy mix (final concentrations: 2 mM ATP, 25 mM phosphocreatine, 30 units/ml creatine phosphokinase, 200 µM GTP, 20 g/l BSA ②

Piece of *Xenopus laevis* ovary

Equipment

Forceps, e.g. Dumont no. 5

GELoader Tips® (i.e. very fine pipette tips made by Eppendorf, Hamburg, Germany)

Stereomicroscope

Confocal laser scanning fluorescence microscope

Image processing program (e.g. ImageJ)

Procedure

1. Prepare a microchamber by drilling a hole of 2.5 mm diameter into the bottom of a tissue culture dish and gluing a cover slide to the bottom of the dish.

Fill the microchamber with ~5 μl of mock 3.

2. Manually isolate and purify a nucleus from a *Xenopus* oocyte. Remove a stage VI oocyte (~1.2 mm diameter, Figure 6.5(a)) from a piece of *Xenopus*

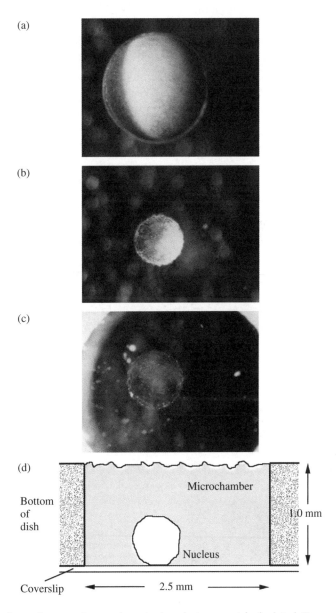

(a)

(b)

(c)

(d)

Bottom of dish

Microchamber

1.0 mm

Nucleus

Coverslip ◄——— 2.5 mm ———►

Figure 6.5 Specimen for assaying nucleocytoplasmic transport in isolated *Xenopus* oocyte nuclei. The nucleus of a stage VI *Xenopus* oocyte (a) was isolated manually (b) and, after manual purification with a fine glass needle, deposited in a microchamber (c). The microchamber (d) has a volume of 5 μl. (From ref. 4)

ovary, kept in amphibian Ringer's solution, and transfer it into a small glass dish containing mock 3. Using a stereomicroscope at ~16× total magnification and fine forceps open the oocyte and set free the nucleus (Figure 6.5(b)). Transfer the nucleus into another small glass dish containing fresh mock 3. Further purify the nucleus from adhering yolk particles by repeatedly touching the nucleus with a microcapillary. Transfer the nucleus into the microchamber (Figure 6.5(c)), strictly avoiding the nucleus touching the air–water interface.

3. Mount the microchamber on the stage of a confocal laser scanning microscope. Visualize the nucleus in through-light at low magnification (10× or 16× objective). Focus onto the largest perimeter of the nucleus. Adjust laser power, ③ wavelength, multiplier voltages, pinhole, etc. so that both the transport substrate (NLS protein) and the control substrate (Texas-red labelled 70 kDa dextran) will be optimally imaged.

4. Use a GELoader Tip® to inject ~10 µl of the transport solution into the microchamber. Start timer and scanner

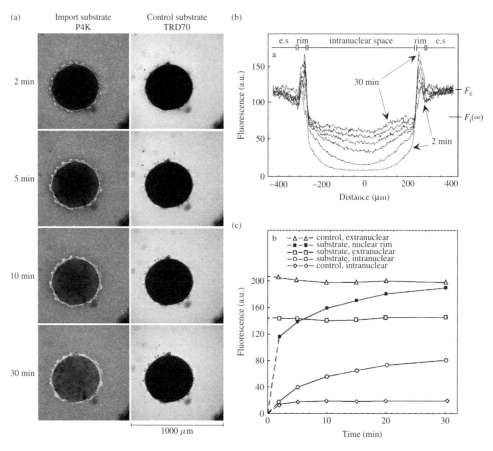

Figure 6.6 Example of a transport measurement. (a) A *Xenopus* oocyte nucleus was incubated with a solution containing a NLS-protein (P4K) and a control substrate (TRD70). Scans were taken at indicated times after transport solution addition. (b) Profiles through the nucleus showing the time development of the transport substrate fluorescence. (c) Time development of mean intra- and extranuclear fluorescence of transport and control substrate. (From ref. 4)

$(t = 0)$. Continue with scanning at intervals properly resolving transport kinetics (Figure 6.6(a)).

5. Evaluate the image series (Figure 6.6(b)), using an image processing program such as Image J (public domain Java version by W. Rasband, http://rbs.info.nih.gov/ij/) to derive F_{si}, F_{se}, F_{ci}, F_{ce}, the mean corrected intra- and extranuclear fluorescence intensities of transport or control substrate, as well as the nuclear radius R (Figure 6.6(c)). The number $N_i(t)$ of imported substrate molecules can then be calculated according to:

$$N_i(t) = (F_i/F_e)\, C_e\, V_N\, L$$

where C_e is the substrate concentration in the transport solution, V_N is the nuclear volume $(= 4/3\pi\, R^3)$ and L is Avogadro's number. In addition, intensity profiles (Figure 6.6(b)) may be obtained.

Notes

This procedure will take approximately 1 h.

① Mock 3 is an intracellular mock medium adjusted, however, to 3 μM free Ca^{++}. Prepare a fresh batch every day.

② BSA serves to balance intranuclear osmotic pressure. Dialyse stock BSA solution (100 g/l) against mock 3.

③ Set laser power as low as compatible with a sufficient image quality to avoid photobleaching.

References

1. Adam, S. A., Stern-Marr, R. and Gerace, L. (1990) Nuclear protein import in permeabilized mammalian cells requires soluble cytoplasmic factors. *J. Cell Biol.*, **111**, 807–816.
2. Forbes, D. J. Kirschner, M. W. and Newport, J. W. (1983) Spontaneous formation of nucleus-like structures around bacteriophage DNA microinjected into *Xenopus* eggs. *Cell*, **34**, 13–23.
3. Lohka, M. J and Masui, Y. (1983) Formation *in vitro* of sperm pronuclei and mitotic chromosomes induced by amphibian ooplasmic components. *Science*, **220**, 719–721.
4. Radtke,T., Schmalz,D., Coutavas,E., Soliman, T. M. and Peters,R. (2001) Kinetics of protein import into isolated *Xenopus* oocyte nuclei. *Proc. Natl. Acad. Sci. USA*, **98**, 2407–2412.
5. Keminer, O., Siebrasse, J. P. Zerf, K. and Peters. R. (1999) Optical recording of signal-mediated protein transport through single nuclear pore complexes. *Proc. Natl. Acad. Sci. USA*, **96**, 11 842–11 847.

Transport measurements in microarrays of nuclear envelope patches by optical single transporter recording

Reiner Peters

Introduction

A most powerful approach to the experimental analysis of membrane transport processes is single transporter recording. However, previously this was possible only in the case of ion channels using the electrophysiological patch clamp method [1]. In optical single transporter recording (OSTR) [2, 3] membranes are attached to microarrays of cylindrical test compartments and transport across membrane patches that may contain single transporter or transporter populations is recorded by confocal microscopy. OSTR features a very high sensitivity, single-transporter resolution, multiplexing of transport substrates and parallel data acquisition (for review, see ref. 4). Here, a protocol for the measurement of nuclear export [5, 6] is given. For a complementary method, employing isolated *Xenopus* oocyte nuclei, see *Protocol 6.10*.

Reagents and materials

Tissue culture dishes, e.g. Falcon no. 353004

3.5 mm diameter drill

Polycarbonate track etched (PCTE) membrane filters, type Cyclopore Transparent® (Whatman, Maidstone, Kent, UK), available in pore sizes of 0.1 to 8.0 μm

Eastman Instant Adhesive no. 910

Amphibian Ringer's solution: 88 mM NaCl, 1 mM KCl, 0.8 mM $MgSO_4$, 1.4 mM $CaCl_2$, 5 mM Hepes (pH 7.4)

Mock 3 intracellular medium: ① 90 mM KCl, 10 mM NaCl, 2 mM $MgCl_2$, 0.1 mM $CaCl_2$, 1.0 mM N-(2-hydroxyethyl) ethylene-diaminetriacetic acid (HEDTA), 10 mM Hepes (pH 7.3)

Transport solution: mock 3 containing 1 μM of a fluorescent protein containing a nuclear export signal (NES) such as GG-NES [5], 1 μM of the nuclear export receptor CRM1, 1 μM Ran-GTP, 10 g/l BSA ②

Piece of *Xenopus laevis* ovary

Equipment

Forceps, e.g. Dumont no. 5

GELoader Tips® (i.e. very fine pipette tips made by Eppendorf, Hamburg, Germany)

Stereomicroscope

Confocal laser scanning fluorescence microscope

Image processing program (e.g. Image J)

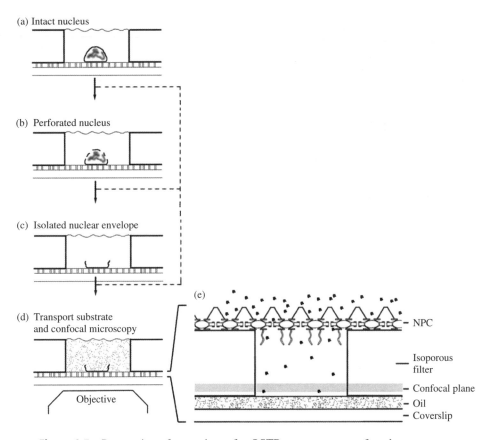

(a) Intact nucleus

(b) Perforated nucleus

(c) Isolated nuclear envelope

(d) Transport substrate and confocal microscopy

Objective

(e)

NPC

Isoporous filter

Confocal plane

Oil

Coverslip

Figure 6.7 Preparation of a specimen for OSTR measurements of nuclear transport

Procedure

1. Prepare an OSTR chamber by drilling a hole of 3.5 mm diameter into the bottom of a tissue culture dish. Apply a piece of optically clear double-sticky tape to a clean coverslip (15 mm × 15 mm). Cut a PCTE membrane filter into small pieces (5 mm × 5 mm) and place a filter piece, shiny side up, on the tape. Attach the cover slip-filter assembly to the culture dish such that the filter forms the bottom of the hole. Fill the microchamber with 12 µl of mock 3. If the pores of the filter do not fill spontaneously with mock 3, swim the chamber on an ultrasonic bath and apply a short pulse (some seconds) of ultrasound. Put a moistened piece of laboratory tissue into the chamber. Close the chamber lid.

2. Manually isolate and purify a nucleus from a *Xenopus* oocyte, as described in *Protocol 6.10*, step 2. Transfer the purified nucleus into the OSTR chamber. Attach the nucleus firmly to the PCTE filter by pressing the nucleus against the filter using a small glass rod (Figure 6.7(a)). Use a very fine steel needle or a glass microcapillary to open the nucleus (Figure 6.7(b)). Remove the nuclear contents and purify the nuclear side of the nuclear envelope by washing it three times with 15 µl of mock 3 (Figure 6.7(c)).

3. Use a GELoader Tip® to inject ∼ 15 µl of the transport solution into the microchamber (Figure 6.7(d)). The transport solution, containing 10 g/l BSA, will sink to the bottom of the

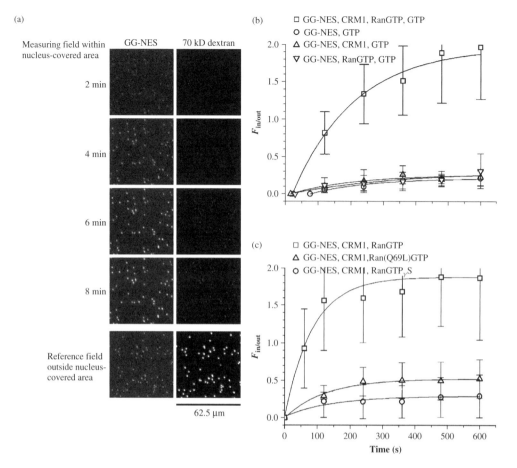

Figure 6.8 Example of an OSTR measurement of signal-dependent nuclear protein export. (a) Raw data. A transport solution containing an export complex (i.e. a 1 : 1 : 1 complex of GG-NES, CRM1 and RanGTP) and a control substrate (Texas-red labelled 70 kDa dextran) was added to the OSTR chamber at $t = 0$ min and confocal scans of the oil–filter interface of a nucleus-covered area acquired at the indicated times. After transport had reached a plateau a nucleus-free area was imaged for reference. (b) Data showing that only the complete export complex was transported. (c) Data showing that hydrolysis of Ran-bound GTP was required for export. (From ref. 5)

chamber replacing the pure mock buffer. Remove surplus buffer. Start the timer ($t = 0$).

4. Mount the microchamber on the stage of a confocal laser scanning microscope. Visualize the nucleus in through-light at low magnification (10× objective). Place the nucleus into the centre of the field of vision. Switch to a 40-fold, oil immersion objective without moving the OSTR chamber. Focus on the interface between oil and filter. Start

3. scanner, adjust microscope parameters, and continue with scanning at intervals properly resolving transport kinetics (Figure 6.8(a)). After recording transport kinetics in the nucleus-covered area, shift the nucleus out of the field of view and obtain scans of a free area of the filter for reference (Figure 6.8(a), bottom).

5. Evaluate the image series, using an image processing program such as Image J (public domain Java version by

W. Rasband, http://rbs.info.nih.gov/ij/). We have written a plug-in to that program which automatically finds the filter pores, measures their fluorescence intensity, subtracts the local background, and plots the time-dependent background-corrected fluorescence for each filter pore individually (examples are given in Figure 6.8(b)).

Notes

This procedure will take approximately 30 min.

① Mock 3 is an intracellular mock medium adjusted, however, to 3 μM free Ca^{++}. Prepare a fresh batch every day.

② Dialyse BSA stock solution (100 g/l) against mock 3.

References

1. Neher, E. and Sakmann, B. (1976). Single-channel currents recorded from membrane of denervated frog fibres. *Nature*, **260**, 779–802.
2. Tschödrich-Rotter, M. and Peters, R. (1998) An optical method for recording the activity of single transporters in membrane patches. *J. Microsc.*, **192**, 114–125.
3. Peters, R., Sauer, H., Tschopp, J. and Fritzsch, G. (1990) Transients of perforin pore formation observed by fluorescence microscopic single channel recording. *EMBO J.*, **9**, 2447–2451.
4. Peters, R. (2003) Optical single transporter recording: transport kinetics in microarrays of membrane patches. *Ann. Rev. Biophys. Biomol. Struct.*, **32**, 47–67.
5. Siebrasse J. P., Coutavas, E. and Peters, R. (2002) Reconstitution of nuclear protein export in isolated nuclear envelopes. *J. Cell Biol.*, **158**, 849–854.
6. Siebrasse J. P. and Peters, R. (2002) Rapid translocation of NTF2 through the nuclear pore of isolated nuclei and nuclear envelopes. *EMBO Rep.*, **3**, 887–892.

Cell permeabilization with Streptolysin O

Ivan Walev

Introduction

Delivery of macromolecules into the cytosol, with retention of cell viability, is increasingly being employed by cell biologists. Most of the existing methods for introducing macromolecules into the cell require considerable methodological expertise [1]. Pore-forming bacterial toxins have found widespread application as tools to study protein trafficking in eukaryotic cells because, under appropriate experimental conditions, the plasma membrane can be selectively permeabilized while the membranes of internal compartments remain intact [2, 3]. With these approaches, bacterial toxin attack is generally lethal, so the possibilities of conducting cell biological studies after permeabilization have been restricted. Recently a simple method for reversible cell permeabilization using the bacterial toxin Streptolysin O (SLO) has been described [4]. SLO is the prototype of a toxin family that have been termed oxygen-labile or sulfydryl-activatable toxin (also termed cholesterol-dependent toxins). This is because the toxin spontaneously loses activity in the presence of atmospheric oxygen, and regains activity upon reduction (e.g. with DTT). The reactivity of the single cysteine residue in the SLO-molecule is responsible for this property. However, the single cysteine residue in SLO can be replaced by alanine, without loss of activity [5]. This mutagenized SLO can be isolated from *E. coli* and is no longer prone to oxygen-dependent inactivation [6]. SLO generates very large transmembrane pores, readily allowing delivery of molecules with mass up to 150 kDa into cytosol. Reversible permeabilization of adherent and non-adherent cells requires use of low SLO concentrations. Resealed cells remain viable for days and retain their capacity to endocytose and to proliferate [4].

Reagents

Streptolysin-O (SLO): available from several biochemical suppliers ①

Hepes buffered saline solution (HBSS): 30 mM Hepes, 150 mM NaCl (pH 7.2)

Procedure

1. Suspend the cells in HBSS without Ca^{2+} and Mg^{2+}, for 5–10 min. ②

2. *Permeabilization step*: discard the supernatants and add SLO dissolved in the same medium for 10 min at 37 °C. ③ The protein or agent to be delivered to the cytosol is included at the permeabilization step.

3. *Resealing step*: add surplus ice-cold medium (without washing) containing Ca^{2+} and Mg^{2+} (final Ca^{2+} concentration 1–2 mM) and incubate the cells for at least 60 min at 37 °C. ④

Notes

① SLO preparations may contain contaminating proteases or DNases. Such contaminants may create artefacts. Overall, it is therefore worthwhile to check whether a given SLO preparation is contaminant-free. Our procedure produces contaminant-free recombinant SLO (Email to: <hgmeyer@mail.uni-mainz.de>). Because this SLO mutant does not contain a cysteine residue, activation by reduction is not necessary.

② Cells: adherent and non-adherent cells can be used. Mouse cells are more resistant to SLO. Some adherent cells do not tolerate the 5–10 min without Ca^{2+} and detach from plastic. Permeabilization can be measured light microscopically by Trypan Blue or propidium iodide staining.

③ In all cases, selection of an appropriate toxin concentration is pivotal to success. Thus, the required SLO concentration will vary depending on cell target and density, and must be determined immediately before the delivery experiment by titration. The goal is to identify the toxin concentration that causes permeabilization of 60–80% of the total cell population under the same conditions.

④ HBSS: some deviation of pH may help. Addition of serum to the medium is good, but not absolutely necessary.

References

1. Lauer, J. L. and Fields, G. B. (1997) *Methods Enzymol.*, **289**, 564–571.
2. Bhakdi, S., Weller, U., Walev, I., Martin, E., Jonas, D. and Palmer, M. (1993) *Med. Microbiol. Immunol.*, **182**, 167–175.
3. Ahnert-Hilger, G., Bader, M. F., Bhakdi, S. and Gratzl, M. (1988) *J. Neurochem.*, **52**, 1751–1758.
4. Walev, I., Bhakdi, S. C., Hofmann, F., Djonder, N., Valeva, A., Aktories, K. and Bhakdi, S. (2001) *Proc. Natl. Acad. Sci. USA*, **98**, 3185–3190.
5. Pinkney, M., Beachey, E. and Kehoe, M. (1989) *Infect. Immun.*, **57**, 2553–2558.
6. Weller, U., Müller, L., Messner, M., Palmer, M., Valeva, A., Tranum-Jensen, J., Agrawal, P., Biermann, C., Döbereiner, A., Kehoe, M. A. and Bhakdi, S. (1996) *Eur. J. Biochem.*, **236**, 34–39.

Nanocapsules: a new vehicle for intracellular delivery of drugs

Anton I. P. M. de Kroon, Rutger W. H. M. Staffhorst, Ben de Kruijff and **Koert N. J. Burger**

Introduction

Liposomes, aqueous compartments surrounded by lipid bilayers, have been widely used as transport vectors to deliver (impermeant) substances into the cytoplasm of cells. Examples include the delivery of chemotherapeutic agents, DNA for transfection, and fluorescent probe molecules (for reviews see refs 1 and 2). The molecular mechanism of the liposome–cell interaction leading to uptake of the liposome contents depends on the lipid composition of the liposome membrane and the cell type involved. Liposome–plasma membrane fusion has been reported in a number of cases (e.g. [3]). Endocytosis is considered the major route of entry for liposomes into cells (reviewed in ref. 4).

The efficiency of encapsulation of the compound of interest in the liposomes is an important determinant of the efficiency of uptake by the cell. The structural properties of liposomes allow for highly efficient encapsulation of hydrophilic compounds and lipophilic compounds in the aqueous interior and in the lipid bilayer of the liposome, respectively [1]. Compounds that do not meet either of these criteria were so far not amenable to efficient encapsulation in a lipid bilayer coat.

One such compound is the poorly water-soluble anti-cancer drug cisplatin, *cis*-diamminedichloroplatinum (II), which is commonly used in the treatment of a variety of solid tumours, including genito-urinary, head and neck, and lung tumours [5]. Encapsulation of this drug into liposomes has many advantages including the reduction of premature inactivation of this highly reactive molecule upon entry in the blood and the reduction of deleterious side-effects such as nephro-, oto- and neurotoxicity [6].

The liposomal formulations of cisplatin developed so far (see e.g. ref 7) suffer from a limited bioavailability of the drug in the tumour [8]. A key factor is likely the low water solubility (7 mM at 37 °C) and low lipophilicity of cisplatin, leading to liposomal formulations with low drug-to-lipid molar ratios (of the order of 0.02). Serendipitously, an alternative method was recently discovered, enabling the encapsulation of cisplatin in a lipid formulation with superior efficiency [9]. Our method takes advantage of the limited solubility of the drug in water, and produces cisplatin nanocapsules, nanoprecipitates of cisplatin surrounded by a single lipid bilayer, which exhibit an unprecedented drug-to-lipid ratio and an unprecedented *in vitro* cytotoxicity. The cisplatin nanocapsules may turn out to be the paradigm for encapsulating compounds with limited solubility in water and with low lipophilicity in a lipid bilayer coat.

Technical principle

Repeated freezing and thawing of a concentrated aqueous solution of cisplatin in the presence of anionic phospholipids result in the formation of cisplatin nanocapsules. The preparation of cisplatin nanocapsules requires the presence of negatively charged phospholipids and of positively charged aqua-species of cisplatin, pointing to an essential role for electrostatic interactions in the mechanism of formation. A solution of cisplatin in water, in the absence of added chloride, contains a mixture of the neutral dichloride- and dihydroxo-species of cisplatin with low solubility in water, and positively charged aqua-species of cisplatin with a much higher solubility [5]. In the model proposed for the mechanism of nanocapsule formation [9], cisplatin is concentrated in the residual fluid during freezing, forming small aggregates when the solubility limit of the dichloro-species is exceeded. As freezing proceeds, the nanoprecipitates of the dichloro-species of cisplatin become covered by the positively charged aqua-species, which have a higher solubility limit. The negatively charged membranes interact with the positively charged cisplatin aggregates and then reorganize to wrap the aggregates in a phospholipid bilayer coat. The resulting nanocapsules do not redissolve upon thawing.

Procedure

1. Cisplatin is dissolved in MilliQ water to a concentration of 5 mM, which is facilitated by incubating at 55 °C for 30 min. The solution is incubated overnight in the dark at 37 °C to ensure full equilibration.

2. Stock solutions of 1,2-dioleoyl-*sn*-glycero-3-phosphocholine (DOPC) and 1, 2-dioleoyl-*sn*-glycero-3-phosphoserine (DOPS) are prepared in chloroform (concentration ∼5 mM). The precise concentrations are determined by phosphate analysis [10].

3. Aliquots corresponding to 0.6 micromole of each phospholipid are mixed, the solvent is removed by rotatory evaporation, and the lipid film is further dried under vacuum overnight.

4. The dry lipid film is hydrated by adding 1.2 ml of the 5 mM cisplatin solution in water and incubating for 15 min at 37 °C.

5. After brief homogenization on a vortex mixer, the dispersion is transferred to a glass tube and subjected to 10 freeze–thaw cycles using ethanol/dry-ice (−70 °C) and a water bath (37 °C).

6. The resulting colloidal solution is transferred to microfuge tubes and centrifuged for 4 min at 470g (2100 rpm in an Eppendorf centrifuge) to collect the nanocapsules.

7. After removal of the supernatant, the fluffy white layer on top of the yellow pellet, corresponding to large liposomes, is removed by a micropipette. The yellow pellet containing the cisplatin nanocapsules, is resuspended in 1 ml water and centrifuged as above in order to wash away non-encapsulated cisplatin.

8. Upon resuspending the final pellet in 0.5 ml water, the nanocapsules are stored at 4 °C until use.

Alternatively, the nanocapsules are separated from contaminating liposomes by density gradient centrifugation. Briefly, the dispersion obtained after the freeze–thaw cycles is loaded on top of a step gradient consisting of 1 ml of each 1.8, 0.6 and 0.2 M sucrose in 10 mM Pipes-NaOH, 1 mM EGTA, pH 7.4. After centrifugation

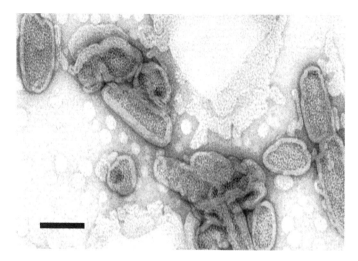

Figure 6.9 Electron micrograph of cisplatin nanocapsules visualised by negative staining. A dilute suspension of nanocapsules was transferred to a carbon-formvar coated grid, and stained with 4% (w/v) uranyl acetate for 45s. Scale bar, 100 nm

at 4 °C for 30 min at 400 000g, the pellet fraction corresponding to the nanocapsules is collected and washed as above.

Instead of DOPS, other anionic phospholipids like dioleoyl-phoshatidylglycerol (DOPG) and dioleoyl-phosphatidic acid (DOPA) can be used to prepare cisplatin nanocapsules. DOPC can be replaced by dioleoyl-phosphatidylethanolamine (DOPE) or sphingomyelin. The method is sensitive to high chloride concentrations and alkaline pH, because these conditions prevent the formation of the positively charged aqua-species. Instead of hydrating a lipid film with 5 mM cisplatin, it is also possible to add the cisplatin solution to preformed DOPC/DOPS liposomes and then start the freeze–thaw cycles. The cisplatin nanocapsules can be stored after lyophilization, and retain cisplatin upon rehydration.

Characterization

1. Encapsulation efficiency

The phospholipid content of the nanocapsules is determined by phosphate analysis [10]. The cisplatin content of the

nanocapsules is assessed by flameless atomic absorption spectrometry (NFAAS) using K_2PtCl_2 as a standard [11]. Analysis of the cisplatin nanocapsules prepared according to the above protocol, typically yields a Pt/phosphate molar ratio of 11 ± 1. Based on the size of the nanocapsules (see below), this number is estimated to correspond to an internal cisplatin concentration exceeding 0.5 M, which is far beyond the solubility limit of cisplatin and consistent with the quasi-crystalline structure of the encapsulated cisplatin [9]. The method allows encapsulation of cisplatin with an efficiency of approximately 30%.

2. Shape and size

Analysis by negative stain electron microscopy reveals bean-shaped particles consisting of an electron-dense core surrounded by a bright layer (excluding stain), corresponding to the bilayer coat (Figure 6.9). The nanocapsules have a heterogeneous size distribution, with 75% of the population having a length between 50 and 250 nm and a width of around 50 nm. Size analysis by dynamic light scattering yields consistent results. Nanocapsules of smaller size and with a narrower size distribution have been obtained by high-pressure extrusion through polycarbonate filters with a 200 nm pore size [9].

3. Delivery of contents

The cytotoxicity of the cisplatin nanocapsules towards the human ovarian carcinoma IGROV-1 cell line has been compared to that of free cisplatin. The IC50 value (the drug concentration at which cell growth is inhibited by 50%) of cisplatin administered as nanocapsules is two orders of magnitude smaller than that of the free drug [9]. The higher cytotoxicity is explained by the reduced inactivation of the drug, due to the lipid coat sequestering it from reaction with substrates in the extracellular environment. Upon binding to the cell surface or endocytic uptake of the nanocapsules, the coat is destabilized, and after membrane passage, cisplatin can exert its cytotoxic effect.

Comments

Cisplatin nanocapsules represent a new lipid formulation of cisplatin, distinct from conventional liposomal formulations in that the drug is present as a nanoprecipitate surrounded by a bilayer. This results in a drug-to-lipid ratio that exceeds that of liposomal formulations by two to three orders of magnitude and probably accounts for the typical bean-like shape of the nanocapsules. The high encapsulation efficiency of cisplatin in nanocapsules is expected to increase the bioavailability of the drug and thus improve the therapeutic index as compared to liposomal formulations of cisplatin. Data obtained so far indicate that the cisplatin nanocapsules can gain access to the cell interior via endocytosis [9]. The method for preparing cisplatin nanocapsules may be applicable to a variety of other compounds with limited solubility in water and low lipophilicity that are not efficiently encapsulated in conventional liposomes. Alternatively, compounds of interest may be co-encapsulated together with cisplatin.

Like the surface of conventional liposomes, the membrane surface of nanocapsules can be engineered to include poly (ethyleneglycol)-conjugated lipids (reviewed in ref. 2). This results in more stable nanocapsules without affecting the cytotoxicity [12]. It is expected that the technologies developed for liposomes, including the attachment of ligands or antibodies for purposes of targeting, can also be applied to nanocapsules.

Acknowledgement

Financial support by the Dutch Cancer Society (project UU2001-2493) is gratefully acknowledged.

References

1. Drummond, D. C., Meyer, O., Hong, K., Kirpotin, D. B. and Papahadjopoulos, D. (1999) *Pharmacol. Rev.* **51**, 691–743.

2. Lian, T. and Ho, R. J. Y. (2001) *J. Pharm. Sci.*, **90**, 667–680.

3. Garrett, F. E., Goel, S., Yasul, J. and Koch, R. A. (1999) *Biochim. Biophys. Acta*, **1417**, 77–88.

4. Düzgünes, N. and Nir, S. (1999) *Adv. Drug Deliv. Rev.*, **40**, 3–18.

5. Lippert, B. (1999) *Cisplatin: Chemistry and Biochemistry of a Leading Anticancer Drug.* Wiley-VCH, New York.

6. Hacker, M. P. (1991) In: *The Toxicity of Anticancer Drugs* (G. Powis and M. P. Hacker, eds), p. 82. McGraw-Hill, New York.

7. Newman, M. S., Colbern, G. T., Working, P. K., Engbers, C. and Amantea, M. A. (1999) *Cancer Chemother. Pharmacol.*, **43**, 1–7.

8. Meerum Terwogt, J. M., Groenewegen, G., Pluim, D., Maliepaard, M., Tibben, M. M., Huisman, A., ten Bokkel Huinink, W. W., Schot, M., Welbank, H., Voest, E. E., Beijnen, J. H. and Schellens, J. H. M. (2002)

Cancer Chemother. Pharmacol., **49**, 201–210.

9. Burger, K. N. J., Staffhorst, R. W. H. M., de Vijlder, H. C., Velinova, M. J., Bomans, P. H., Frederik, P. M. and de Kruijff, B. (2002) *Nat. Med.*, **8**, 81–84.

10. Rouser, G., Fleischer, S. and Yamamoto, A. (1970) *Lipids*, **5**, 494–496.

11. Burger, K. N. J., Staffhorst, R. W. H. M. and de Kruijff, B. (1999) *Biochim. Biophys. Acta*, **1419**, 43–54.

12. Velinova, M. J., Staffhorst, R. W. H. M., Mulder, W. J. M., Dries, A. S., Jansen, B. A. J., de Kruijff, B. and de Kroon, A. I. P. M. (2004) *Biochim. Biophys. Acta*, **1663**, 135–142.

Note added in proof: The molecular architecture of the cisplatin nanocapsules was recently solved (Chupin, V., de Kroon, A. I. P. M., and de Kruijff, B. (2004) *J. Am. Chem. Soc.*, **126**, 13816–13821.)

A rapid screen for determination of the protective role of antioxidant proteins in yeast

Luis Eduardo Soares Netto

Introduction

Yeast has been successfully employed as a model to study several biological processes. Indeed, Leland H. Hartwell and Paul M. Nurse received the Nobel Prize in 2001 for the discovery of genes in yeast involved in cell cycle regulation, which has had a great impact on the understanding of processes such as cancer. Some of the advantages of using yeast as a model system are: (1) this micro-organism grows very rapidly and media used to cultivate it are very simple and inexpensive; (2) it is very amenable to genetic manipulations such as gene disruption; (3) its biology is very well studied and (4) its genome has been completely sequenced and is publicly available.

There has been considerable interest in the study of antioxidant enzymes in yeast. Yeast as well as bacteria, mammals and plants possess several enzymatic systems capable of decomposing peroxides. *Saccharomyces cerevisiae* possesses five peroxiredoxins (a type of thiol-dependent peroxidase, many of them thioredoxin-dependent peroxidases); two catalases; three glutathione-dependent peroxidases and one cytochrome *c* peroxidase. Although a partial redundancy of their roles is observed, they definitively have distinct roles. For example: cytosolic thioredoxin peroxidase I (cTPxI) is important for protection of cells with dysfunctional mitochondria, whereas mitochondrial thioredoxin peroxidase I is important for respiratory-competent cells [1–3]. More recently, results have been obtained indicating that cytosolic thioredoxin peroxidase II is specifically important for cell protection against organic peroxides independently of the functional state of mitochondria [4]. One approach that has been very helpful for us is a viability test based on serial dilutions as described below. Using this approach, we were able to test mutant strains under several different conditions and found one in particular, where a gene deletion rendered cells very sensitive to oxidative stress. Several concentrations of oxidants can be tested as well as the effect of various carbon sources. Changing the carbohydrate present in the media is an interesting manipulation because yeast physiology and biochemistry are very dependent on the carbon source. For example, yeast grown under glucose get most of its ATP from glycolysis, has few mitochondria and this organelle contains few cristae. In contrast, when yeast is grown in glycerol or ethanol it only produces ATP from oxidative phosphorylation and possesses many active mitochondria [5].

It is important to emphasize that this approach can also be employed for the study of other protective roles, for example against heat stress, osmotic stress and alkylation damage. Proteins from other organisms can also be studied. In this case, it would be necessary to verify if a heterologous gene can complement the sensitivity observed for a given mutant strain.

After testing several conditions by this protocol and finding one where the mutant strain is specifically sensitive, a more accurate and quantitative assay (colony counting) can be performed. A disadvantage of colony counting assay is that it is very time consuming. Examples of this strategy are described in refs [2] and [4].

Reagents

Preparation of medium and plates

A. Synthetic medium

Per litre, add:

1.7 g Yeast nitrogen base without amino acids or ammonium sulfate (YNB-AA/AS)

5 g $(NH_4)_2SO_4$

20 g of glucose or other carbon source (2%)

This medium can support the growth of yeast with no nutritional requirements. Generally, however, yeast strains present nutritional requirements. Therefore, nutrients should be added as described in Table 6.1. Synthetic media should be autoclaved before use. Threonine and aspartic acid work better when added after autoclaving, therefore stock solutions of these amino acids should be prepared and sterilized by filtration. These amino acids can then be added to a synthetic medium in laminar flow.

B. Preparation of plates

Solid synthetic medium is prepared as described above with addition of agar

Table 6.1 Nutrients to be added with yeast strains with deficiencies in metabolic processes[a]

Nutrient	Final concentration (ug/ml)	Amount of nutrient added (mg)[b]
Adenine (hemisulfate salt)	40	40
L-arginine (HCl)	20	240
L-aspartic acid	100	100
L-glutamic acid (monosodium salt)	100	100
L-histidine	20	20
L-leucine	60	60
L-lysine (mono-HCl)	30	30
L-methionine	20	20
L-phenylalanine	50	50
L-serine	375	375
L-threonine	200	200
L-tryptophan	40	40
L-tyrosine	30	30
L-valine	150	150
Uracil	20	20

[a] These data were obtained from Ausubel *et al.* [6].
[b] The amount of nutrient was added per litre of synthetic media.

2% (20 g of agar per litre of medium). The medium is autoclaved and then left for 45–60 min at room temperature until cooled to about 50 °C, which can be felt if one can hold the flask comfortably. At this point, the medium should be poured into plates. Drugs and other compounds to be tested should also be added when the medium is at 50 °C, to avoid undesirable thermal decomposition. More details about preparation of media can be obtained from Ausubel *et al.* [6].

Procedure

Determination of hydroperoxide tolerance

1. Grow yeast cells overnight in complete synthetic media with various carbon

sources; in most cases, 2% glucose or with 2% glycerol.

2. The next day, dilute inoculate to 3×10^6 cells per ml (which corresponds to approximately 0.1 $OD_{600\ nm}$) and then cultivate until the cell density reaches approximately 3×10^7 cells per ml (which corresponds to approximately 1.0 $OD_{600\ nm}$). This takes about 10 h, which ensures that cells are well adapted to the carbon source present in the medium.

3. Dilute cell suspension to $OD_{600\ nm} = 0.2$ and make four further 1/5 dilutions of the cell suspensions.

4. Place a 10 µl droplet of each dilution into complete synthetic medium + agar with 2% glucose or the other carbon sources as described in Figure 6.10. A

control plate should contain media components alone, whereas the test plate contains peroxide (or other compound to be investigated).

5. Incubate plates for 30 or 48 h, depending on the results observed. In some cases, a differential effect can only be found after only one of these time periods.

Comments

1. In the control plate, there is no addition besides the components of the media. It is important to be very careful in the determination of $OD_{600\ nm}$ before the dilutions in order to guarantee that mutant and wild-type strains will grow equally.

2. Cells are grown in synthetic medium to avoid the reaction of peroxides with

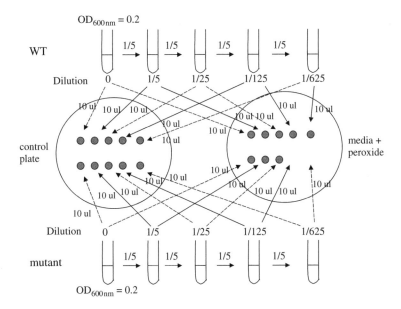

Figure 6.10 Scheme viability of yeast cells in plates after serial dilutions. Cells were submitted to four 1/5 dilutions (200 ul to 1 ml) in synthetic media; 10 µl of undiluted and 1/5, 1/25, 1/125 and 1/625 diluted suspensions were added to control plates as well as to the plates with the compounds to be tested. In control plates, the two strains grew equally which showed that their cell densities were the same in the undiluted samples. In the experimental plate, the mutant strain did not grow in the 1/125 and in the 1/625 dilutions, indicating that this gene deletion rendered cells sensitive to peroxide treatment

extracellular components present in a rich medium such as glutathione [7]. It is important to use synthetic rather than rich media such as YPD (yeast peptone dextrose) because glutathione and other substances are not very well controlled and their concentration can vary from different lots of media, which may compromise the reproducibility of the assay.

3. Peroxides, mitochondrial inhibitors or other stressful compounds can be added to the plates under several different conditions, in order to find one where deletion of a gene severely retards growth. Plates can be incubated for different intervals, which can also interfere with the results. We usually incubate plates for 30 or 48 h. It may be interesting to vary the carbon source because it provokes dramatic changes in yeast biochemistry and physiology as mentioned above.

4. It is important not to use small volumes to make the dilutions. We generally diluted 200 μl of cell suspension into 1 ml of synthetic media. In this way, errors due to manipulation of pipettes are minimized.

References

1. Kowaltowski, A. J., Vercesi, A. E., Rhee, S. G. and Netto, L. E. S. (2000) Catalases and thioredoxin peroxidase protect *Saccharomyces cerevisiae* against Ca^{2+}-induced mitochondrial membrane permeabilization and cell death. *FEBS Lett.*, **473**, 177–182.

2. Demasi, A. P. D., Pereira, G. A. G. and Netto, L. E. S. (2001). Cytosolic thioredoxin peroxidase I is essential for the antioxidant defense of yeast with dysfunctional mitochondria. *FEBS Lett.*, **509**, 430–434.

3. Monteiro, G., Pereira, G. A. G. and Netto, L. E. S. (2002) Regulation of mitochondrial thioredoxin peroxidase I expression by two different pathways: one dependent on CAMP and the other on heme. *Free Radical Biol. Med.*, **32**, 278–288.

4. Munhoz, D. C. and Netto, L. E. S. (2004) "Cytosolic thioredoxin peroxidase I and II are important defenses of yeast against organic hydroperoxide insult, *J. Biol. Chem.*, **279**, 35219–35227.

5. Pon, L. and Schatz, G. (1991) *Biogenesis of Yeast Mitochondria in the Molecular and Cellular Biology of the Yeast Saccharomyces.* Vol 1: *Genome Dynamics, Protein Synthesis and Energetics.* Cold Spring Harbor Laboratory Press, New York.

6. Ausubel, F. M., Brent, R., Kingstone, R. E., Moore, D. D., Seidman, J. A., Smith, J. A. and Struhl, K. (1994) *Saccharomyces cerevisiae.* In: *Current Protocols in Molecular Biology,* John Wiley & Sons, Inc., New York.

7. Maris, A. F., Assumpção, A. L., Bonatto, D., Brendel, M. and Henriques J. A. (2001) Diauxic shift-induced stress resistance against hydroperoxides in *Saccharomyces cerevisiae* is not an adaptive stress response and does not depend on functional mitochondria. *Curr. Genet.*, **39**, 137–149.

In vitro assessment of neuronal apoptosis

Eric Bertrand

Introduction

Apoptosis is a specific form of cell death which is triggered and executed by the cell's own machinery; as such it is considered as a kind of 'suicide'. Elimination of cells through apoptosis contributes to many important physiological processes such as development, inflammation and adult tissue homeostasis [1, 2]. Also, excessive apoptosis has been associated with a number of conditions, like neurodegenerative diseases, ischaemia or Aids. At first, the definition of apoptosis was essentially based on morphological features: cell shrinkage, condensation of cytoplasm and nucleus, clumping of chromatin, blebbing of the membrane and ultimately fragmentation in apoptotic bodies. Afterwards, it was observed that the alterations of chromatin were also accompanied by a cleavage of nuclear DNA into nucleosomal fragments; however, this type of DNA degradation is not entirely specific of apoptosis. At the biochemical level, most of the characteristic changes observed during apoptosis are mediated by the caspases, a family of cysteine proteases. The caspases cleave a number of substrates within the cell including structural proteins and enzymes, thereby modifying their function. Caspase activation is currently considered as the most reliable marker of apoptosis.

The protocols included therein should prove useful to the reader who wants to investigate whether a particular treatment induces apoptosis on neural cell cultures. Some of the issues that should be addressed in the course of such a study are: is the treatment toxic, does it induce apoptosis, which caspases are activated? Toxicity can be assessed and quantified with the TUNEL assay, while reversal of the toxic effect by a caspase inhibitor such as ZVAD indicates the involvement of apoptosis. ZVAD is a broad-range caspase inhibitor, some caspase inhibitors are more selective than ZVAD, but they are usually not entirely specific. For instance, DEVD preferentially inhibits caspase-3 but is also active against caspase 8, 7 and 10 [3]. Therefore, even if DEVD blocks the apoptosis induced by the treatment, a direct demonstration of caspase-3 activation using immunostaining is required. Finally, a very specific inhibition of a given caspase can be achieved with antisense oligonucleotides coupled to the cell permeable carrier penetratin.

Reagents

PBS: GIBCO/Invitrogen, cat. no. 14200-059

zDEVD-fmk: Tebu, cat. no. P414

zVAD-fmk: France Biochem, cat. no. 727610

Vectastain ABC-AP kit: Vector Laboratories, cat. no. AK-5200

NBT/BCIP: Pierce, cat. no. 34042

In situ Cell Detection Kit, AP: Roche Diagnostics Cat N° 1 684 809

Anti p20 antibody: Pharmingen, cat. no. 557035

J peptide: Syntem,

Penetratin peptide and oligonucleotides: Quantum, cat. no. 151300

Procedures

Immunocytochemical methods to study apoptosis at the single cell level

The protocols within this section are intended for cells that have been cultured in ELISA wells; with some adjustments on solution volumes they can be adapted to glass coverslips and other plate formats. The cells were fixed with 4% paraformaldehyde at room temperature for 30 min or overnight at 4 °C.

TUNEL staining

TUNEL stands for TdT mediated dUTP nick end labelling; this reaction labels DNA strand breaks by incorporating modified nucleotides at the 3'-OH termini. During apoptosis, nuclear DNA is degraded to nucleosomal fragments which are detected by the TUNEL reaction. In this protocol we used the alkaline phosphatase version of the *In Situ* Cell Death Detection Kit provided by Roche Diagnostics. In most cases we followed the instructions of the manufacturer, however we obtained a better specificity of the staining on primary cultures of cortical neurons by reducing the incubation times for the 1 + 2 reaction mixture (DNA labelling solution) and the converter-AP solution (alkaline phosphatase coupled antibody). Some adjustment of incubation times and other parameters might be necessary, depending on the cell type studied and the type of signal observed (fluorescence, alkaline phosphatase, horseradish peroxydase); the manufacturer gives a number of useful pieces of advice in this respect (*http://www.roche-applied-science.com*). This protocol has been used to detect apoptosis on rat and mouse cortical neurons and it should be applicable to other species and neuronal types.

1. Wash cells 3× with PBS and then permeabilize in a 0.1% sodium citrate/0.1% triton solution for 2 min on ice (at 4 °C).

2. Wash wells 2× with PBS.

3. Add 50 µl of TUNEL 1 + 2 reaction mixture (enzyme solution + label solution) from Roche Diagnostics kit for 30 min at 37 °C.

4. Wash wells 3× with PBS.

5. Incubate the cells for 15 min at 37 °C in 50 µl of converter-AP solution (alkaline phosphatase coupled antibody).

6. Prepare NBT/BCIP substrate solution for alkaline phosphatase:

   100 mM Tris pH 9.5

   10 mM NaCl

   5 mM MgCl2

   1 mM levamisole to inhibit endogenous phosphatases

7. Add NBT/BCIP according to the manufacturer's guidelines.

8. Wash wells 3× in PBS, add 100 µl of NBT/BCIP substrate solution and incubate for approximately 10 min at room temperature. Follow the course of the reaction and block it by adding 10 µl of 0.5 M EDTA in the wells when the staining is intense enough.

9. Wash wells 3× in deionized water and let them dry out for 1–2 h.

10. Count cells using a phase contrast microscope.

Fractin staining

Fractin is a cleavage product of actin after its processing by caspase-1 and caspase-3, we used an affinity-purified antibody provided by Dr Greg Cole to detect fractin. This antibody has been tested *in vitro* on mice and can also be used on human and mouse brain sections with a few adaptations.

1. Wash cells 3× with PBS, then block and permeabilize in PBS 10% FCS/ 0.2% Triton for 1h at room temperature.

2. Incubate overnight at 4 °C with purified primary antibody diluted 1/2000 in PBS 10% SVF/0.2% triton.

3. Wash wells 3× with PBS, incubate 2 h at room temperature with secondary biotinylated anti-rabbit antibody from Vectastain ABC-AP (alkaline phosphatase) kit at 1/200 in PBS 5% SVF/0.2% triton.

4. Wash wells 3× again with PBS, incubate 1 h at room temperature with A + B complex from ABC-AP Vectastain kit at 1/200 (A + B complex was preformed for 40 min in PBS).

5. Wash wells 3× with PBS, add 100 μl of NBT/BCIP substrate solution and incubate for approximately 10 min at room temperature. Follow the course of the reaction and block it by adding 10 μl of 0.5 M EDTA in the wells when the staining is intense enough.

6. Wash wells 3× with deionized water and let them dry out for 1–2 h.

p20 staining

The p20 antibody (Pharmingen) is targeted against the active (17–22 Kd) form of caspase-3 which is a cleavage product of a larger (37 Kd) propeptide; it can be used for Western blots with a few adaptations.

The immunocytochemistry procedure is exactly the same as for the fractin staining except that the dilution for the purified primary antibody is 1/500. So far, this antibody has been used on human and murine cells.

Inhibition of apoptosis

In our hands, the irreversible caspase inhibitors zVAD-fmk at 100 μM and zDEVD-fmk at 200 μM were effective at inhibiting apoptosis occurring through caspase-3 activation. We also used antisense oligonucleotides coupled to cell-permeant peptides as described by Troy *et al.* [4] at a concentration of 200 nM.

Oligonucleotide sequences directly coupled to the cell-permeant vector penetratin are commercially available from Quantum Biotechnologies. The penetratin peptide and its properties have been described in detail by Derossi *et al.* [5] and Dupont *et al.* [6]. Its amino acid sequence is: RKQIKIWFQNRRMKWKK.

The oligonucleotides used were designed by Troy *et al.* [4] and their sequences are:

Sequence for antisense caspase-1 CCTC-AGGACCTTGTCGGCCAT

Sequence for antisense caspase-3 GTTG-TTGTCCATGGTCACTTT

These sequences correspond to rat caspases; if you investigate another species you should check these sequences and modify the oligonucleotides accordingly. The caspase-3 antisense oligonucleotide has proven effective on murine cells [7].

Note that the peptide/oligonucleotide conjugate should be kept away from reducing agents as the two parts are linked by a disulfide bond. The coupling takes place

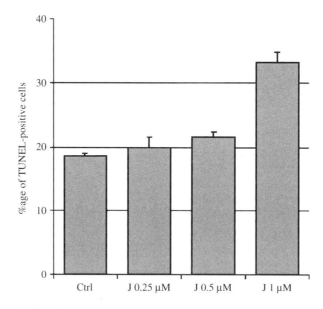

Figure 6.11 Toxicity of peptide J on primary cultures of cortical neurons (error bars represent standard deviation)

between a thiol function at the 5′ end of the nucleotide and a modified cysteine at the *N*-terminal part of the penetratin peptide. More details on the chemistry involved in the coupling can be found in Troy *et al*. [8] and Dupont *et al*. [6].

Examples

The cell-permeable peptide J has been used to induce apoptosis; it contains a cytoplasmic juxtamembrane sequence of the amyloid precursor protein (APP) linked to a penetratin vector. Our intent was to simulate an excessive generation of amyloid precursor protein (APP) juxtamembrane fragments by internalizing J in cortical neurons (more detail in Bertrand *et al*. [7]).

Peptide J RKQIKIWFQNRRMKW KKKYTSIHHG

(APP juxtamembrane sequence is underlined)

The following experiments were performed on E15 rat cortical neurons which were cultured in ELISA wells precoated with polyornithine, as in ref. 7. The treatments were applied 2 h after plating; whenever necessary the cells were incubated with ZVAD 1 h before peptide J addition.

We first wanted to determine whether this peptide was toxic at concentrations in the micromolar range. We compared the number of TUNEL-positive cells of four conditions after 24 h of culture: neurons treated with 0.25 □M, 0.5 □M, 1 □M of peptide J and untreated controls (Figure 6.11). For each condition, about 300 cells were counted on the diameter of six ELISA wells (for a total of 2000). There is no significant difference between control, J 0.25 □M and J 0.5 □M conditions (Student, $p > 0.05$), whereas the percentage of TUNEL positive cells is higher for J 1 □M than for the control (Student, $p < 0.001$). We conclude that the peptide J is toxic at 1 □M but not at the lower concentrations.

In the next experiment we tested whether the toxicity of the peptide J can be blocked by the caspase inhibitor ZVAD

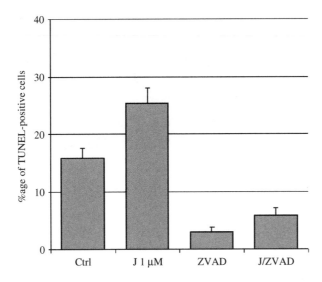

Figure 6.12 The caspase inhibitor ZVAD reduces the toxic effect of peptide J (error bars represent standard deviation)

(Figure 6.12). We studied rat E15 cortical neurons and observed the effect of ZVAD on J 1 □M treated and untreated cells, these cells were fixed after 18 h of treatment. For each condition, 300 cells were counted on the diameter of one ELISA well, and the experiment was repeated three times with similar results. The toxicity of peptide J is very effectively reduced by ZVAD which also reduces TUNEL-staining in the control ($p < 0.001$ in both cases, Khi2 test). This means that the toxic effect of peptide J is mediated by the caspases and that the TUNEL-staining in our system is caused by apoptosis.

Data on the activation of caspase-3 by APP juxtamembrane peptides can be found in Bertrand *et al.* [7], along with some *in vivo* results.

References

1. Jacobson, M. D., Weil, M. and Raff, M. C. (1997) Programmed cell death in animal development. *Cell*, **88**, 347–354.
2. Nagata, S. (1997) Apoptosis by death factor. *Cell*, **88**, 355–365.
3. Garcia-Calvo, M., Peterson, E. P., Leiting, B., Ruel, R., Nicholson, D. W. and Thornberry, N. A. (1998) Inhibition of human caspases by peptide-based and macromolecular inhibitors. *J. Biol. Chem.*, **273**, 32 608–32 613.
4. Troy, C. M., Rabacchi, S. A., Friedman, W. J., Frappier, T. F., Brown, K. and Shelanski, M. L. (2000) Caspase-2 mediates neuronal cell death induced by beta-amyloid. *J. Neurosci.*, **20**, 1386–1392.
5. Derossi, D., Chassaing, G. and Prochiantz, A. (1998) Trojan peptides: the penetratin system for intracellular delivery. *Trends Cell Biol.*, **8**, 84–87.
6. Dupont, E., Joliot, A. and Prochiantz, A. (2002) Penetratins. In: *CRC Handbook on Cell Penetrating Peptides* Ü. Langel, ed.), CRC Press.
7. Bertrand, E., Brouillet, E., Caille, I., Bouillot, C., Cole, G. M., Prochiantz, A. and Allinquant, B. (2001) A short cytoplasmic domain of the amyloid precursor protein induces apoptosis *in vitro* and *in vivo*. *Mol. Cell. Neurosci.*, **18**, 503–511.
8. Troy, C. M., Derossi, D., Prochiantz, A., Greene, L. A. and Shelanski, M. L. (1996) Downregulation of Cu/Zn superoxide dismutase leads to cell death via the nitric oxide–peroxynitrite pathway. *J. Neurosci.*, **16**, 253–261.

The mitochondrial permeability transition [1–5]

Judie B. Alimonti and **Arnold H. Greenberg**†

Introduction

Mitochondrial permeability transition (PT) occurs when the PT pore opens allowing solutes and water to enter the mitochondrial matrix, often accompanied with the loss of the mitochondrial membrane potential ($\Delta\Psi$m). Direct measurement of the opening of the PT pore is best ascertained using calcein AM. The non-fluorescent calcein AM is membrane permeable, and upon entering the cell is cleaved by esterases to become fluorescent and membrane impermeable. $CoCl_2$ is added to eliminate calcein fluorescence in the cytoplasm so that only the mitochondria will fluoresce. Calcein exits the mitochondria only upon opening of the PT pore, therefore it is an indicator of PT. Since a closed PT pore is required to maintain $\Delta\Psi$m, the loss of $\Delta\Psi$m can also be used to determine PT. However, remember that loss of $\Delta\Psi$m can occur independently of PT, therefore you must verify that it is due to PT by either adding a PT inhibitor or calcein AM. The utilization of fluorescent cationic dyes that accumulate around a membrane with a transmembrane potential allow the measurement of $\Delta\Psi$m. The stronger the membrane potential the more dye accumulates at a membrane.

The mitochondrial permeability transition: PT and $\Delta\Psi$m loss determined in cells or isolated mitochondria with confocal laser imaging [2, 4, 5]

Equipment

For cells and isolated mitochondria

Chambered glass coverslips (e.g. Nunc 12-565-103N)

Inverted confocal microscope

Pipette and tips

Polystyrene ice container and ice

Tabletop centrifuge and rotors

Reagents

For cells and isolated mitochondria

Calcein AM (1 mM stock in DMSO)

Tetramethylrhodamine (TMRM: 10 mM stock in DMSO)

Carbonylcyanide m-chlorophenylhydrazone (CCCP: 100 mM stock in DMSO)

Cyclosporin A (10 mM stock in ethanol)

For cells only

Tissue culture medium

Cell buffer: Hanks balanced salt solutions (HBSS), 10 mM Hepes pH 7.2

For isolated mitochondria only

Buffer H: 300 mM sucrose, 5 mM TES, and 200 μM EGTA, pH to 7.3 with NaOH. (TES = N-tris[hydroxymethyl]-methyl-2-aminoethanesulfonic acid).

Freshly isolated mitochondria and S100 (*see Protocols 4.7–4.10*)

Assay buffer: 220 mM sucrose, 68 mM mannitol, 10 mM Hepes-KOH, 10 mM KCl, 5 mM KH_2PO_4, 2 mM $MgCl_2$, 0.5 mM EGTA, 5 mM succinate, 2 μM rotenone, pH 7.2

Procedure

Cells

1. Seed adherent cells onto a chambered glass coverslip. ① Incubate 1–2 days in a humidified 37 °C/ 5% CO_2 incubator until the cells are in log phase.

2. Pretreat the cells with any PT inhibitors (i.e. 10 μM cyclosporin A or 50 μM Bongkrekic acid) as necessary before incubating the cells with the PT inducers for the required times. ②

3. Gently wash the cells twice and resuspend in cell buffer.

4. Load the cells with the fluorescent marker. Add 1 μM calcein AM and 1–5 mM $CoCl_2$. ③ Incubate for 10–15 min at room temperature. Wash four times then resuspend in cell buffer.

5. Imaging is performed on an inverted confocal microscope using a band by-pass filter of 488 nm for calcein, and

Nomarski optics for transmitted light images of the same cells. Cells containing mitochondria with a closed PT pore will fluoresce whereas the cells that have undergone PT have less or no fluorescence. The level of fluorescence per cell can then be quantitated using imaging software, along with the number of cells with high, medium or low levels of fluorescence. ④ If you are also using 100 nm TMRM to measure $\Delta\Psi$m in the same cells then you will have to switch the optics back and forth from 488 nm for calcein, to a band bypass filter of 568 nm for TMRM.

Isolated mitochondria

1. Add 20 μg/ml freshly isolated mitochondria in H buffer to a chambered coverslip. ⑤ Centrifuge at 1475g, 5 min, 4 °C. Wash off the excess layers of mitochondria. ⑥

2. Resuspend in assay buffer containing 100 nM TMRM, 8 μM calcein AM. Incubate 10–15 min at room temperature. Wash four times and resuspend in assay buffer with TMRM, along with the S100 if required.

3. Pretreat with any PT inhibitors (i.e. 10 μM cyclosporin A, 30 min or 50 μM Bongkrekic acid) as necessary before incubating the mitochondria with the PT inducers for the required times.

4. Imaging is performed on an inverted confocal microscope using a band bypass filter of 488 nm for calcein, and 568 nm for TMRM. Take an image before adding the PT inducer, then at various time points to follow PT and loss of $\Delta\Psi$m. ⑦

Notes

① For non-adherent cells you can cytospin them onto a glass coverslip, dry off the surrounding area and apply a non-toxic, non-reactive grease around the cells to form a well. Alternatively, you can use a polylysine-coated chambered coverslip to help the cells adhere.

② If your PT inducer acts rapidly then you can stain your cells with the fluorophores first and take a baseline picture on the confocal microscope before adding your PT inducer and following the loss of fluorescence.

③ CoCl$_2$ is toxic to the cells over extended times, therefore must be washed out. You may have to adjust the CoCl$_2$ concentration for each cell line to determine the minimum concentration of CoCl$_2$ required to quench the cytoplasmic calcein yet remain non-toxic.

④ If you induce PT before staining then you must standardize the staining and confocal settings for all of the samples. Remember to include an untreated sample for your negative control, and CCCP 50–100 μM treatment for your positive control. The PT inhibitor cyclosporin A (10–20 μM) can be added to block the PT inducers to verify your test substance works on the PT pore.

⑤ Alternatively, add 20 μg/ml freshly isolated mitochondria in H buffer to a six-well plate containing a 22 × 30 mm coverslip. Centrifuge at 1475g, 5 min, 4 °C. The coverslip can either be added to an enclosed heated stage that passes buffer over it, or dry off the surrounding area and apply a non-toxic, non-reactive grease around the mitochondria to form a well.

⑥ Mitochondria are sticky but adhere loosely to the coverslip, so you will have to gently wash off the excess multiple layers of mitochondria for easier viewing on the confocal microscope.

⑦ The confocal laser can induce PT in isolated mitochondria, therefore limit the exposure to the laser, and include an untreated mitochondria sample for the same time period as the experiment to determine the stability of the mitochondria under experimental conditions. CCCP 50–100 μM treatment is your positive control.

The mitochondrial permeability transition: measuring PT and $\Delta\Psi m$ loss in isolated mitochondria with Rh123 in a fluorometer [1, 2]

Reagents

Freshly isolated mitochondria and S100 (see *Protocols 4.7–4.10*)

Carbonylcyanide *m*-chlorophenylhydra-zone (CCCP: 100 mM stock in DMSO)

Rhodamine 123 (Rh123) (16 mM stock in DMF)

Cyclosporin A (10 mM stock in ethanol)

Assay buffer: 220 mM sucrose, 68 mM mannitol, 10 mM Hepes-KOH, 10 mM KCl, 5 mM KH_2PO_4, 2 mM $MgCl_2$, 0.5 mM EGTA, 5 mM succinate, 2 μM rotenone, pH 7.2

Equipment

Fluorometer

96-well non-fluorescent black plates

Pipettes and tips

5% CO_2/37 °C humidified incubator

Procedure

1. Load freshly isolated mitochondria in assay buffer in triplicate onto a 96-well plate with a final mitochondrial concentration between 0.75 and 1 mg/ml and a minimum volume of 0.1 ml/well. Also add the S100 if required.

2. The positive control for $\Delta\Psi m$ loss is CCCP (50–100 μM final). Pretreatment with PT inhibitors like cyclosporin A (10–20 μM, 30 min) should be used to verify that the $\Delta\Psi m$ loss is due to PT. Add the PT inducers for the required time at 37 °C.

3. Add the fluorescent dye Rh123 (5 μM). ① The fluorometer requires an excitation/emission filter of 485/535 nm. Take readings during the linear range (5–20 min) of the accumulation of Rh123. ② ③

Notes

① As Rh123 accumulates around respiring mitochondria the fluorescence decreases since Rh123 fluorescence is quenched at high concentrations, so the higher the $\Delta\Psi m$ the lower the fluorescence. This is unique to Rh123.

② The results can be expressed as the percentage of maximum $\Delta\Psi m$ loss, using the CCCP sample as the maximum loss. The mitochondria will deteriorate with time and the background increases with longer incubations.

③ Alternatively, you can incubate the mitochondria with Rh123 until the quenching is complete. You then add your PT inducer and watch for the increase in fluorescence.

The mitochondrial permeability transition: measuring PT and $\Delta\Psi$m loss in cells and isolated mitochondria on the FACS [1, 4, 5]

Equipment

FACS and FACS tubes

Pipettes and tips

Polystyrene ice container and ice

5% CO_2/37 °C humidified incubator

Tabletop centrifuge

Reagents

Freshly isolated mitochondria and S100 (see *Protocols 4.7–4.10*)

Rh123 (16 mM stock in DMF)

JC-1 (10 mg/ml stock in DMSO)

$DiOC_6$ (3) (100 μM stock in DMSO)

For cells: Hanks balanced salt solution (HBSS), 10 mM Hepes pH 7.2

For mitochondria: Assay buffer: 220 mM sucrose, 68 mM mannitol, 10 mM Hepes-KOH, 10 mM KCl, 5 mM KH_2PO_4, 2 mM $MgCl_2$, 0.5 mM EGTA, 5 mM succinate, 2 μM rotenone, pH 7.2

Procedure

Cells

1. Treat the cells with the PT inhibitors and inducers for the required times. ① Wash the cells twice with HBSS,

10 mM Hepes pH 7.2. Resuspend at 1×10^6 cells in $\frac{1}{2}$ ml HBSS, 10 mM Hepes pH 7.2.

2. Label the cells with the Rh123 (2 μM) for 10 min at room temperature. ② Read on a FACS in FL1.

Mitochondria

1. Pre-incubate 50 μg/ml of freshly isolated mitochondria in $\frac{1}{2}$ ml assay buffer with PT inhibitors (10–20 μM cyclosporin A) for 30 min before adding the PT inducers for the required time. ①

2. Label the mitochondria with the Rh123 (5 μM) for 10 min at room temperature. ② Read on a FACS in FL1.

Notes

① The positive control for $\Delta\Psi$m loss is CCCP (50–100 μM final). There will be a shift in fluorescence with a change in $\Delta\Psi$m.

② Other dyes that can be utilized include $DiOC_6$ (3) (25 nM : mitochondria or 40 nM : cells), or JC-1 (5 μg/ml : mitochondria or 1 μM : cells). JC-1 fluoresces green when it is a monomer but red when it aggregates about a mitochondria with high $\Delta\Psi$m.

Measuring cytochrome *c* release in isolated mitochondria by Western blot analysis [2, 4]

Equipment

SDS-PAGE apparatus

96-well U-bottom tissue culture plate

Pipette and tips

5% CO_2/37 °C humidified incubator

Reagents

Freshly isolated mitochondria and S100 (see *Protocols 4.7–4.10*)

Assay buffer: 220 mM sucrose, 68 mM mannitol, 10 mM Hepes-KOH, 10 mM KCl, 5 mM KH_2PO_4, 2 mM $MgCl_2$, 0.5 mM EGTA, 5 mM succinate, 2 μM rotenone, pH 7.2

Cyclosporin A (10 mM stock in ethanol)

PT inducers

Cytochrome *c* antibody (Pharmingen)

Procedure

1. Add 1 mg/ml freshly isolated mitochondria in assay buffer to a 96-well U-bottom plate with or without S100, in a final volume of 60 μl.

2. Pretreat with any PT inhibitors (i.e. 10 μM cyclosporin A, 30 min) as necessary before incubating the mitochondria with the PT inducers at 37 °C for the desired time period, usually within 0–4 h.

3. Centrifuge the plate at 760*g*, 5 min. Add 30 μl of supernatant to 7.5 μl of 5× SDS loading buffer. Boil for 5 min then add 10 μl to a standard 15% SDS-PAGE gel for Western blotting. Use antibodies recognizing cytochrome *c* to verify cytochrome *c* release into the supernatant and mitochondrial dysfunction. ①

Note

① Cytochrome *c* release is a measure of mitochondrial dysfunction but not necessarily PT. Pre-incubating with PT pore inhibitors can confirm that the release of cytochrome *c* may be due to PT.

PROTOCOL 6.20

Protein import into isolated mitochondria [3, 4]

The influence of anti-apoptotic proteins such as Bcl-2 on the inhibition of PT can be examined with mitochondria over-expressing Bcl-2. Bcl-2^{high} mitochondria can be obtained by isolating them from cells over-expressing Bcl-2, or by loading the isolated mitochondria with Bcl-2 using a protein import protocol.

Reagents

Freshly isolated mitochondria and S100 (see *Protocols 4.7–4.10*)

Buffer 1: 250 mM sucrose, 10 mM Hepes-KOH, 2 mM K$_2$HPO$_4$, 5 mM Na succinate, 1 mM ATP, 0.08 mM ADP, 1 mM DTT, pH 7.5

Buffer 2: 20 mM Hepes-KOH, 10 mM KCl, 2.5 mM Mg Cl$_2$, 1 mM EDTA, 1 mM EGTA, 1 mM DTT, pH 7.5

Buffer 3: 250 mM sucrose, 10 mM Hepes-KOH, 1 mM DTT, pH 7.5

Assay buffer: 220 mM sucrose, 68 mM mannitol, 10 mM Hepes-KOH, 10 mM KCl, 5 mM KH$_2$PO$_4$, 2 mM MgCl$_2$, 0.5 mM EGTA, 5 mM succinate, 2 μM rotenone, pH 7.2

Bcl-2 protein

Equipment

37 °C water bath

Eppendorfs

Pipettes and tips

Microcentrifuge

Procedure

1. 50 μl 1 mg/ml freshly isolated mitochondria in buffer 1 is added to 50 μl buffer 2 containing 0.125–0.5 μg/ml Bcl-2 protein (higher concentrations may be toxic). Incubate 30 min at 37 °C.

2. Layer 100 μl onto a 500 μl cushion of buffer 3 and centrifuge at 14 500g, 5 min, 4 °C.

3. Resuspend pellet in 10 μl of assay buffer. The mitochondria can then be used in the confocal, fluorometer, cell-free (Western), or FACS protocols.

References

1. Susin, S. A., Larochette, N., Gueskens, M. and Kroemer, G. (2000) Quantitation of the mitochondrial transmembrane potential in cells and isolated mitochondria. *Meth. Enzymol.*, **322**, 205–208.
2. Alimonti, J. B., Shi, L., Baijal, P. K. and Greenberg, A. H. (2001) Granzyme B induces BID-mediated cytochrome *c* release and mitochondrial permeability transition. *J. Biol. Chem.*, **276**, 6974–6982.
3. Goping, I. S., Gross, A., Lavoie, J. N., Nguyen, M., Jemmerson, R., Roth, K., Korsmeyer, S. J. and Shore, G. C. (1998) Regulated targeting of BAX to mitochondria. *J. Cell Biol.*, **143**, 207–215.

4. Vande Velde, C., Cizeau, J., Dubik, D., Alimonti, J., Brown, T., Israels, S., Hakem, R.
and Greenberg, A. H. (2000) BNIP3 and genetic control of necrosis-like cell death through
the mitochondrial permeability transition pore.
Molec. Cell. Biol., **20**, 5454–5468.

5. Castedo, M., Ferri, K., Roumier, T., Meti-
vier, D., Zamzami, N. and Kroemer, G. (2002)
Quantitation of mitochondrial alterations associated with apoptosis. *J. Immunol. Meth.*, **265**,
39–47.

Formation of ternary SNARE complexes *in vitro*

Jinnan Xiao, Anuradha Pradhan and **Yuechueng Liu**

Introduction

SNARE (soluble NSF attachment protein receptor) proteins have been shown to play essential roles in vesicular transport [1]. In nervous systems, v (vesicle)-SNARE and t (target)-SNAREs interact with each other to form a tight ternary complex. The assembly of such a ternary complex allows the opposing membranes to be pulled into close proximity, permitting synaptic vesicle fusion to proceed and release neurotransmitters [2]. The neuronal SNARE core complex includes v-SNARE synaptobrevin (also known as VAMP), t-SNAREs syntaxin and SNAP-25. While both synaptobrevin and syntaxin are membrane proteins with a single C-terminal transmembrane domain, SNAP-25 attaches to membranes via post-translational palmitoylation of its cysteine residues.

Many studies involving vesicular trafficking require reconstitution of the SNARE core complex *in vitro*. For instance, GST-fusion protein pull-down assay is often used to investigate proteins implicated in regulation of membrane fusion via their interactions with the SNAREs. This short protocol will focus on the isolation of recombinant SNAREs and the reconstitution of the SNARE core complex *in vitro* (see Figure 6.13).

Equipment

Tabletop microcentrifuge and 1.5 ml microcentrifuge tubes

Figure 6.13 An example of reconstituted SNARE ternary complexes analysed by SDS-PAGE and stained with Coomassie Blue. The apparent molecular weights for the SNAREs are: 60 kDa (EST-SyntaxinΔTM), 32 kDa (6 × His-SNAP-25), and 20 kDa (Synaptobrevin). Prestained molecular weight marker is shown on the left. The molecular weights for the markers are (in kDa): 106, 78, 50, 32, 26, 20

37 °C Incubator, with shaker

Sonicator (e.g. Virtis Virsonic 50)

Rotating shaker (e.g. Labquake Shaker)

Reagents

LB bacterial culture media

IPTG (e.g. Sigma, GIBCO)

Glutathione-agarose (e.g. Pierce Chemicals)

Phosphate-buffered saline (PBS)

Thrombin cleavage and capture kit (e.g. Novagen)

6 × His-tagged SNAP-25, in a bacterial expression plasmid vector (e.g. Invitrogen's pTrcHis, Qiagen's pQE)

Syntaxin with deleted transmembrane domain (syntaxinΔTM) in pGEX-2T

Synaptobrevin in pGEX-2T

Procedure

1. Prepare 6 × His-SNAP-25, GST-syntaxinΔTM and GST-synaptobrevin. The procedure for GST fusion protein purification has been described in detail by Smith and Johnson [3]. For affinity purification of 6 × His-tagged proteins using Ni-NTA agarose, follow the manufacturers' instructions (e.g. Qiagen's The QIAexpressionist). Dialyse the purified His-SNAP-25 against PBS overnight at 4 °C with at least three buffer changes. The purified proteins should be kept at 4 °C for short-term (<5 days) storage. For long-term storage, the proteins can be kept in −20 °C freezer (avoid repeated freezing/thawing). The GST-syntaxinΔTM agarose is kept at 4 °C and it is stable for at least 2 weeks.

2. Add 5 units of biotinated thrombin to ~1 ml slurry of GST-synaptobrevin in PBS containing 0.1% Triton X-100. Mix the sample well by gently reversing the microcentrifuge tube several times and put the tube onto a rotating shaker. Incubate overnight at 4 °C. ①

3. Centrifuge for 2 min in a microcentrifuge at maximum speed at 4 °C. Collect the supernatant containing synaptobrevin. Wash the agarose beads once with PBS by centrifugation at 4 °C. Use ~0.4–0.5 ml PBS which is about the same volume as the agarose beads.

Combine the supernatants. However, if one wants a more concentrated synaptobrevin preparation, it will be better to keep them separated, since the first supernatant usually has more synaptobrevin.

4. Add at least 30 µl streptoavidin-agarose to the supernatant and incubate for 20–30 min at room temperature. Again, use a rotating shaker for continuous mixing of the sample.

5. Centrifuge the sample for 5 min at room temperature and at maximum speed. Collect the supernatant and discard the pellet. Dialyse the sample against PBS overnight at 4 °C with three buffer changes.

6. Aliquot 100 µl of His-SNAP-25 (0.2 mg/ml), 100 µl of synaptobrevin (0.15 mg/ml) and 100 µl agarose slurry of GST-syntaxinΔTM (~0.5 mg/ml) and mix in a 1.5 ml microcentrifuge tube by gently reversing the tube several times. ②

7. Place the mixture on a rotator and incubate overnight at 4 °C. ③

8. Centrifuge the sample in a microcentrifuge at maximum speed for 2 min at 4 °C. Discard the supernatant (or keep it for further experiments). Wash the agarose resin with 1 ml PBS three times at 4 °C by centrifugation.

9. The agarose beads containing ternary complexes can be boiled in SDS sample buffer and directly analysed by SDS-PAGE, or immunoblot analysis using specific antibodies against the SNAREs. Alternatively, the complexes can be eluted with 5 mM reduced glutathione for further experiments.

Notes

This procedure will take approximately 48 h (not including the preparation of

His-SNAP-25, GST-syntaxinΔTM, and GST-synaptobrevin).

① Triton X-100 is helpful in releasing the synaptobrevin from the GST agarose.

② The molar ratio for the ternary complex *in vivo* is 1 : 1 : 1.

③ Incubation time can be shorter, e.g. 1 h at room temperature. However, overnight incubation at 4 °C seems to produce more complexes.

References

1. Jahn R. and Südhof T. C. (1999) *Ann. Rev. Biochem.*, **66**, 863–911.
2. Brunger A. T. (2001) *Curr. Opin. Struct. Biol.*, **11**, 163–173.
3. Smith, D. B. and Johnson, K. (1988) *Gene*, **67**, 31–40.

In vitro reconstitution of liver endoplasmic reticulum

Jacques Paiement and Robin Young

Introduction

In vitro reconstitution is among the most important and widely used strategies in studying membrane traffic [1]. We describe here an *in vitro* assay which allows study of the reconstitution of rat liver endoplasmic reticulum. Membrane derivatives of liver endoplasmic reticulum (ER) are incubated under specific conditions leading to membrane fusion and the formation of large ER membrane complexes. Following incubation the reconstituted ER membranes are studied by electron microscopy. Depending on the type of ER derivative (i.e. rough or smooth?) used in the reconstitution assay the assembly products will exhibit different structural properties and can be distinguished based on morphological criteria. Examples are described below.

Reagents

ATP (100 mM) to be stored in aliquots in powder form in a dessicator at $-80\,°C$ (High grade. Sigma Chemicals cat. no. A-7699)

Complete™ Protease Inhibitor Cocktail (1 tablet/ml = 50 × concentrated) stored at $-20\,°C$ (Roche Applied Science cat. no. 1697498)

Creatine Phosphokinase (1 U/μl) to be stored in aliquots in powder form in a dessicator at $4\,°C$ (Roche Applied Science cat. no. 127566)

DTT (250 mM) to be stored in aliquots in powder form in a dessicator at $4\,°C$ (Sigma Chemicals cat. no. D-0632)

GTP (25 mM) to be stored in aliquots in powder form in a dessicator at $-80\,°C$ (High grade. Sigma Chemicals cat. no. G-8877)

$MgCl_2$ (125 mM), filtered on Whatman no. 42 ashless filter to remove any particulate matter in solution

Phosphocreatine (50 mM) to be stored in aliquots in powder form in a dessicator at $4\,°C$ (Roche Applied Science cat. no. 621714)

Sucrose-Imidazole (0.25 mM sucrose, 3 mM imidazole, pH 7.4), filtered on Whatman no. 42 ashless filter to remove any particulate matter in solution

tER derived membranes (see 'Fractionation of rough and smooth microsomes from rat liver homogenates', *Protocol 6.22* for preparation)

Tris-HCl buffer (200 mM, pH 7.4), filtered on Whatman no. 42 ashless filter to remove any particulate matter in solution

Equipment

Polypropylene conical microcentrifuge tubes (1 ml)

Incubation bath, with shaker (37 °C)

Procedure ①

1. Prepare test samples and all stock solutions on ice (ensure all solutions are at least 4 °C prior to preparation of medium).

2. Prepare stock solutions of DTT, creatine phosphokinase and phosphocreatine. Remove protease inhibitor stock from −20 °C and put on ice to thaw.

3. The following reagents are added to each tube to produce a final incubation mixture of 250 μl:

   • 125 μl Tris buffer to produce final concentration of 100 mM,

   • 10 μl MgCl$_2$ to produce final concentration of 5 mM,

   • 1.8 μl creatine phosphokinase to produce final concentration of 7.3 U/ml,

   • 10 μl phosphocreatine to produce final concentration of 2 mM,

   • 10 μl DTT to produce final concentration of 0.1 mM,

   • 5 μl protease inhibitor cocktail to produce final concentration of 1×.

4. Take membranes out of −80 °C and place on ice.

5. Add sucrose-imidazole (amount added is calculated to bring final incubation volume to 250 μl, after addition of all reagents).

6. Take GTP and ATP out of −80 °C and keep on ice.

7. Prepare stock ATP and add 5 μl to each tube to produce a concentration of 2 mM.

8. Prepare stock GTP and add 10 μl to each tube to produce a concentration of 1 mM.

9. Verify membranes are thawed and add 150 μg of membrane protein to each tube.

10. Incubate at 37 °C for 90 min with constant oscillation, with additional manual agitation every half-hour.

11. At end of 90 min add 5 and 10 μl of ATP and GTP respectively. ②

12. Continue incubation for an additional 90 min at 37 °C with constant oscillation, with additional manual agitation of individual tubes every half-hour.

Comments

The incubation medium described here contains factors that will stimulate fusion of membranes of classical rough microsomes [2, 3], fusion of membranes of low-density rough microsomes [4 7], fusion of membranes of smooth microsomes [4–7] and fusion of membranes of the nuclear envelope [8]. The reconstituted elements formed as a consequence of membrane fusion have been characterized by a variety of biochemical, histochemical and immunocytochemical procedures [2–8]. The factors involved in the transformation of various ER membrane subdomains are summarized in a review [9].

Figure 6.14 shows *in vitro* reconstituted transitional endoplasmic reticulum (tER). Derivatives of liver tER were incubated using the specific incubation conditions described above. Following incubation the membrane sample was fixed in suspension and processed as described previously [7]. The two domains of the tER are recognizable based on morphological criteria. The

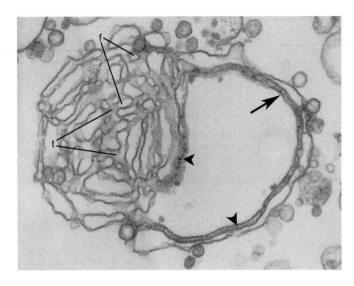

Figure 6.14 Micrograph of reconstituted hepatocyte ER showing a membrane complex consisting of branching and anastomosing smooth tubules in continuity with peripheral rough ER cisternae. The arrow points to closely apposed membranes of two rough ER cisternae. One cisterna is dilated and contains the arrow and the arrowheads, the other cisterna is flattened and tightly bound at the periphery of the dilated cisterna. Ribosomes are mainly observed between the two cisternae. Arrowheads point to ribosomes. f, fenestrations; t, smooth tubes.

rER domain consists of parallel cisternae limited by ribosome-studded membranes and the sER domain consists of a network of interconnecting tubules limited by smooth membranes devoid of attached ribosomes.

Notes

This procedure will take approximately 4 h.

① Similar cell-free incubation conditions for reconstitution of tER have previously been summarized and include variations in concentration of certain reagents and other parameters of incubation [4–6].

② This will compensate for potential nucleotide hydrolysis during incubation by membrane nucleotidases.

References

1. Mellman, I. and Warren, G. (2000) *Cell*, **100**, 99–112.

2. Paiement, J. and Bergeron, J. J. M. (1983) *J. Cell Biol.*, **96**, 1791–1796.

3. Paiement, J., Beaufay, H. and Godelaine, D. (1980) *J. Cell Biol.*, **86**, 29–37.

4. Roy, L., Bergeron, J. J. M., Lavoie, C., Hendriks, R., Gushue, J., Fazel, A., Pelletier, A., Morré, D. J., Subramaniam, V. N., Hong, W. and Paiement, J. (2000) *Mol. Biol. Cell*, **11**, 2529–2542.

5. Lavoie, C. Chevet, E., Roy, L., Tonks, N. K., Fazel, A., Posner, B. I., Paiement, J. and Bergeron, J. J. M. (2000) *Proc. Nat. Acad. Sci.*, **25**, 13 637–13 642.

6. Lavoie, C., Paiement, J., Dominguez, M., Roy, L., Dahan, S., Gushue, J. N. and Bergeron, J. J. M. (1999) *J. Cell Biol.*, **146**, 285–299.

7. Lavoie, C., Lanoix, J., Kan, F. W. K. and Paiement, J. (1996) *J. Cell Sci.*, **109**, 1415–1425.

8. Paiement, J. (1984) *Exp. Cell Res.*, **151**, 354–366.

9. Paiement, J. and Bergeron, J. (2001) *Biochem. Cell Biol.*, **79**, 587–592.

Asymmetric incorporation of glycolipids into membranes and detection of lipid flip-flop movement

Félix M. Goñi*, Ana-Victoria Villar, F.-Xabier Contreras and Alicia Alonso

Introduction

Assemblies of glycosphingolipids and cholesterol are believed to form microdomains ('rafts') with specific functions in membrane traffic and signal transduction. Flask-shaped 60 nm invaginations of the cell plasma membrane, termed caveolae, may be the sites at which these domains cluster and self-stabilize [1, 2]. Rafts and caveolae recruit specific membrane proteins which are implicated in cell signalling [3, 4].

These domains have a particular asymmetric disposition as glycosphingolipids and glycosylphosphatidylinositol (GPI)-linked proteins locate preferentially in the outer leaflet [1, 2, 5, 6]. While artificial lipidic vesicles (liposomes) have been extremely useful in the understanding of multiple aspects of membrane structure and function, no simple technology was available for the reliable preparation of liposomes with asymmetrically distributed lipids. In view of the structural and functional importance of rafts and caveolae, we have recently described the preparation of liposomes with GPI and/or glycosphingolipids located preferentially in the outer monolayer [7].

Cell membrane asymmetry, however, is not a static phenomenon. Instead, it arises from a series of concerted transmembrane movements, leading to a time-invariant distribution of bilayer components. Lipid asymmetry in particular is known to be altered in physiological or pathological events such as recognition by phagocytes, blood coagulation, or apoptosis [8]. The collapse of lipid asymmetry is often known as lipid 'scrambling'.

Sphingomyelinases cleave the sphingophospholipid sphingomyelin yielding phosphorylcholine and ceramide [9, 10]. In turn, ceramide may alter a number of membrane properties: it increases lipid order, gives rise to ceramide-rich separate domains, and destabilizes the lamellar structures, inducing membrane permeabilization and membrane fusion [4, 11–14].

In this context, we have developed an assay to test transmembrane ('flip-flop') lipid movement induced by sphingomyelinase via ceramide formation [15]. Both the preparation of asymmetric vesicles and the flip-flop assay are described below.

A. Asymmetric incorporation of glycolipids into liposomal membranes

Materials

GM3 monosialoganglioside (e.g. Matreya Inc. or other supplier) ①

Neuraminidase (EC 3.2.1.18) from *Clostridium perfringens* (Sigma)

FITC-dextran (fluorescein isothiocyanate-dextran, average molecular weight 4400 Da) (Sigma)

Egg lipids (sphingomyelin, phosphatidylcholine, phosphatidylethanolamine) are purchased from Lipid Products, UK

Cholesterol, 8-aminonaphthalene-1, 3, 6-trisulfonic acid (ANTS) (Molecular Probes Inc.)

p-Xylenebis (pyridinium bromide) (DPX) (Molecular Probes Inc.)

Liposomes (large unilamellar vesicles obtained by extrusion through polycarbonate filters 100–400 nm pore diameter [16]) ②

Procedure

Asymmetric incorporation of glycolipids

1. Incorporate GPI or GM3 to liposomes prepared by extrusion, by drying the glycolipid from organic solvent and resuspend in a volume of methanol equivalent to 5% of the vesicle suspension volume. ③

2. Add vesicles to the methanolic glycolipid solution with vortex mixing.

3. Incubate the vesicles for 15 min [7]. ④

Representative results

The mild character of the bilayer perturbation caused by glycolipids during their incorporation under our conditions is demonstrated by an experiment in which fluorescent molecules (FITC-dextran, molecular weight 4400 Da) are entrapped in the vesicles. These molecules are self-quenching and their fluorescence increases upon dilution [17]. As shown in Figure 6.15, neither methanol nor GPI in methanolic solution is able to release FITC-dextran molecules from the liposomes. As expected, addition of 0.025% (w/v) Triton X-100 releases the whole vesicular contents. Similar experiments with GM3 instead of GPI also fail to show any vesicle leakage secondary to ganglioside insertion (not shown). When ANTS-DPX, the molecular weight of which is one order of magnitude below that of FITC-dextran 4400, is entrapped in the vesicles, incorporation of GPI is accompanied by partial release of the dyes, while GM3 is completely inactive in this respect [7].

The asymmetric insertion of GM3 into the outer monolayer of liposomes is shown in the following experiment. Large unilamellar vesicles of either phosphatidylethanolamine : phosphatidylcholine : cholesterol (2 : 1 : 1) or phosphatidylethanolamine : phosphatidylcholine (2 : 1) are added to GM3 ganglioside in methanolic solution (5 mole % of total lipid). After vortexing and incubation at room temperature, the vesicles (0.3 mM) are treated with neuraminidase (0.16 U/ml), an enzyme known to degrade the glycosidic part of gangliosides, releasing sialic acid. In this preparation, lipids other than GM3 are not hydrolysed by neuraminidase. Furthermore, neuraminidase cannot cross the membrane and reach the inner liposomal compartment. Neuraminidase is assayed in 10 mM Hepes, pH 5.6, 39 °C, with continuous stirring. Total lipid concentration is 0.3 mM. Enzyme concentration is 0.16 U/ml. Aliquots are removed from the reaction mixture at regular intervals and extracted with chloroform/methanol/hydrochloric acid (200/100/1, by volume). Sialic acid is assayed in the aqueous phase with the resorcinol method [18]. The enzyme activity is shown as a function of time in Figure 6.16. After 60 min, when c.75% of the ganglioside appears to have been cleaved,

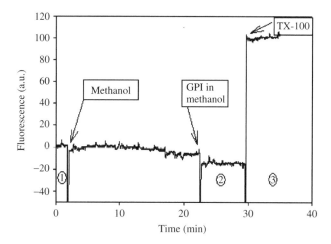

Figure 6.15 Asymmetric incorporation of GPI into the outer monolayer of liposomes, and its effect on the permeability barrier properties of the vesicles. Lipid composition was phosphatidylethanolamine : phosphatidylcholine : cholesterol 2 : 1 : 1 mole ratio. Liposomes were large unilamellar vesicles containing entrapped FITC-dextran (average molecular weight 4400 Da). The various reagents were added as indicated by the arrows. The encircled figures at the bottom correspond to the times at which average diameters were measured by quasi-elastic light scattering. Average particle diameters were: 1, 131.9 ± 1.1 nm; 2, 129.7 ± 2.3 nm; 3, complete heterogeneity (polydispersity 1.0). Taken from ref. 7, with permission

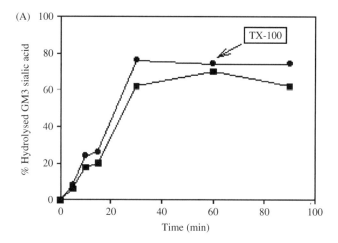

Figure 6.16 Neuraminidase cleavage of GM3 that had been asymmetrically inserted into the outer monolayer of liposomes. Lipid composition as in Figure 6.15. Total lipid concentration was 0.3 mM. GM3 concentration was 0.015 mM. Enzyme concentration was 0.16 U/ml. Experiment started in the absence of detergent, Triton X-100 (0.025% w/v) added at the time indicated by the arrow. Taken from ref. 7, with permission

Triton X-100 (0.025% w/v final concentration) is added in order to terminate the membrane barrier effect (Triton X-100 under these conditions permeabilizes the liposomal membranes to large molecules, see Figure 6.15). No further hydrolysis occurs during the following 30 min. This is interpreted as an indication that most, if not all, of the GM3 molecules were originally located in the outer monolayer and were accessible to neuraminidase. Independent experiments had shown that the lack of further hydrolysis after membrane permeabilization by Triton X-100 was not due to a detergent-induced enzyme inactivation.

B. Detection of lipid transmembrane (flip-flop) movement

Materials

Sphingomyelinase (E.C. 3.1.4.12) from *Bacillus cereus* (Sigma) ⑤

Other materials, liposome preparation and asymmetric incorporation of lipids are performed as described in section A, above. When required, neuraminidase (0.16 U/ml) is added to the hydration buffer. In this case, non-entrapped neuraminidase is removed by gel filtration through Sephadex G-75

Anti-neuraminidase IgG ⑥

Procedure

Ganglioside flip-flop translocation in lipid bilayers

1. Treat LUV composed of SM : PE : Ch (2 : 1 : 1 mole ratio) with GM3 ganglioside in methanol to obtain vesicles containing GM3 (~10 mole % of total lipid) located exclusively on the outer leaflet, as described above. ⑦

2. Add sphingomyelinase to the suspension of these vesicles to induce SM hydrolysis, that reaches equilibrium after *c*.20 min, when *c*.40% of SM has been hydrolysed (Figure 6.17(a)).

3. Addition of Triton X-100 after 60 min causes membrane disruption, but SM hydrolysis does not go beyond 50%

30 min after detergent addition, i.e. 90 min after sphingomyelinase addition. ⑧

4. Remove aliquots of the vesicle suspension at fixed times after sphingomyelinase addition, and analyse for the GM3 product of neuraminidase activity, sialic acid. GM3 is hydrolysed almost in parallel with SM, except that no saturation was observed (Figure 6.17(b)). ⑨

Comment

It is necessary to rule out the possibility of neuraminidase coming out from the vesicles as a result of sphingomyelinase activity. Ceramides increase membrane permeability, and efflux of molecules up to 40 kDa has been observed in sphingomyelin-treated vesicles [13]. Neuraminidase has a molecular mass of 70 kDa. In order to clarify this aspect, a control experiment can be performed in which any extravesicular neuraminidase would be neutralized by a specific antibody. With this aim, a polyclonal antineuraminidase antibody was raised in rabbits, and it was checked that the purified antineuraminidase IgG at 50 μg/ml completely abolished neuraminidase activity at 0.16 U/ml. The same concentration of IgG had no effect on sphingomyelinase activity. When the experiment described in Figure 6.17 is repeated with 50 μg/ml antineuraminidase antibody in the medium, the results are exactly the same (data not shown). It can be concluded that GM3 hydrolysis is catalysed by neuraminidase inside the vesicles, thus sphingomyelinase activity causes GM3 to flip to the inside lipid monolayer.

Notes

① Glycosylphosphatidylinositol (GPI) is purified from rat livers as described by Varela-Nieto *et al.* [19].

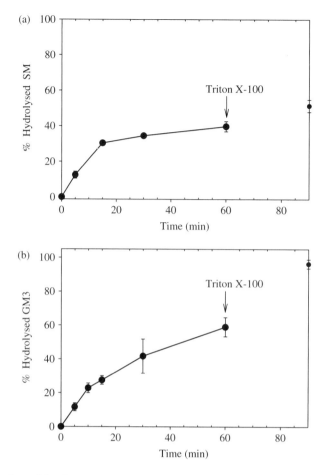

Figure 6.17   Flip-flop of GM3 ganglioside induced by sphingomyelinase activity, in large unilamellar vesicles. (a) Time-course of sphingomyelin hydrolysis by external sphingomyelinase. (b) GM3 ganglioside hydrolysis by entrapped neuraminidase. Average values ± SEM. Vesicle composition was sphingomyelin : phosphatidylethanolamine : cholesterol (2 : 1 : 1, mole ratio) + 10 mole % GM3 on the outer leaflet only [15]

② Several lipid compositions based on sphingomyelin, phosphatidylcholine and phosphatidylethanolamine with or without cholesterol have been used with similar results from the point of view of glycolipid incorporation. When required, liposomes are prepared containing entrapped fluorescent dyes, such as ANTS-DPX [20], or FITC-dextrans [17]. In these cases the fluorescent probes are included in the lipid hydration buffer during liposome preparation, and the non-entrapped probes are removed from the vesicles in a Sephadex G-75 separation column.

③ Glycolipids have been used at molar fractions of up to 0.1 of total lipid.

④ In our experience this incubation is allowed to proceed at room temperature because the lipids used in liposome preparation are of natural origin, and their mixtures are expected to be in the fluid state at room temperatures. Other lipid compositions may require incubation at higher temperatures.

⑤ Sphingomyelinase is used in the presence of 2 mM *o*-phenantroline in order to inhibit traces of contaminant phospholipase C activity. Previous studies had shown that *o*-phenantroline does not affect sphingomyelinase activity.

⑥ Raise antibody in rabbits and purify using a Hi-trap protein G affinity column. The crude antibody (1 ml) is passed through the column and the latter is washed three times with 20 mM phosphate, pH 7 buffer. Antineuraminidase IgG is eluted from the affinity column with a 0.1 mM glycine-HCl, pH 2.7 buffer.

⑦ The vesicles contain neuraminidase and, when kept at 4 °C, the vesicles are stable for at least 12 h. No significant amount of GM3 is hydrolysed in this period of time.

⑧ Previous experiments had shown that this Triton X-100 concentration did not inhibit sphingomyelinase or neuraminidase.

⑨ After addition of Triton X-100, virtually all GM3 is cleaved by neuraminidase. The data suggest that, as a consequence of sphingomyelinase activity, GM3 is flipping to the inner leaflet, thus becoming accessible to neuraminidase.

References

1. Simons, K. and Ehehalt, R. (2002) *J. Clin. Invest.*, **110**, 597–603.
2. Razani, B., Woodman, S. E. and Lisanti, M. P. (2002) *Pharmacol. Rev.*, **54**, 431–467.
3. Kenworthy, A. (2002) *Trends Biochem. Sci.*, **27**, 435–437.
4. Kolesnick, R. N., Goñi, F. M. and Alonso, A. (2000) *J. Cell. Physiol.*, **184**, 285–300.
5. Simons, K. and Ikonen, E. (1997) *Nature*, **387**, 569–572.
6. Hooper, N. M. (1998) *Curr. Biol.*, **8**, R114–R116.
7. Villar, A. V., Alonso, A., Pañeda, C., Varela-Nieto, I., Brodbeck, U. and Goñi, F. M. (1999) *FEBS Lett.*, **457**, 71–74.
8. Bevers, E. M., Comfurius, P., Dekkers, D. W. C. and Zwaal, R. F. A. (1999) *Biochim. Biophys. Acta*, **1439**, 317–330.
9. Barenholz, Y., Roitman, A. and Gatt, S. (1966) *J. Biol. Chem.*, **241**, 3731–3737.
10. Goñi, F. M. and Alonso, A. (2002) *FEBS Lett.*, **531**, 38–46.
11. Basañez, G. and Edwards, K. (1997) *Biophys. J.*, **72**, 2630–2637.
12. Goñi, F. M. and Alonso, A. (2000) *Biosci. Rep.*, **20**, 443–463.
13. Montes, L. R., Goñi, F. M. and Alonso, A. (2002) *J. Biol. Chem.*, **277**, 11 788–11 794.
14. Ruiz-Argüello, M. B., Goñi, F. M. and Alonso, A. (1998) *J. Biol. Chem.*, **273**, 22 977–22 982.
15. Contreras, F. X., Villar, A. V., Alonso, A., Kolesnick, R. N. and Goñi, F. M. (2003) **278**, 37169–37174 .
16. Mayer, L. D., Hope, M. H. and Cullis, P. R. (1986) *Biochim. Biophys. Acta*, **858**, 161–168.
17. Ostolaza, H., Bartolomé, B., Ortiz de Zárate, I., de la Cruz, F. and Goñi, F. M. (1993) *Biochim. Biophys. Acta*, **1147**, 81–88.
18. Wybenga, L. E., Epand, R. F., Nir, S., Chu, J. W. K., Sharon, F. J., Flanagan, T. D. and Epand, R. M. (1996) *Biochemistry*, **35**, 9513–9518.
19. Varela-Nieto, I., Alvarez, L. and Mato, J. M. (1993) *Handb. Endocr. Res. Tech.*, **20**, 391–405.
20. Ellens, H., Bentz, J. and Szoka, F. C. (1986) *Biochemistry*, **25**, 4141–4147.

PROTOCOL 6.24

Purification of clathrin-coated vesicles from rat brains

Brian J. Peter and Ian G. Mills

Introduction

Clathrin-coated vesicles (CCVs) are endocytic and trafficking organelles which are highly enriched in the coat protein clathrin, as well as AP complexes and a number of other proteins involved in vesicle budding and endocytosis. There are other protocols published detailing their purification [1, 2], but this protocol is quick (1 day) and useful for purifying small amounts of CCVs sufficient for Western blots, EM, or further clathrin purification.

Reagents

Frozen rat brains ①

HKM buffer (25 mM Hepes pH 7.4, 125 mM potassium acetate, 5 mM magnesium acetate, 1 mM dithiothreitol)

HKM buffer with 12.5% Ficoll and 12.5% sucrose, and HKM buffer made up with deuterium oxide (D_2O) with 8% sucrose ②

Protease inhibitor cocktail III (Calbiochem)

Equipment

Low-speed refrigerated centrifuge and rotor (e.g. Sorvall SS34 rotor or equivalent)

Potter-Elvehjem homogenizer (with both 45 and 15 ml tubes)

Preparative ultracentrifuge and rotors (e.g. Beckman type 70 Ti, type 45 Ti, SW 40 Ti + tubes, or equivalent)

Procedure

1. Crush 10 rat brains into small pieces while still frozen; this can be done in a sealed bag immediately after removing brains from freezer, or under liquid nitrogen with a mortar and pestle. ③

2. Make up to 40 ml with HKM and add Calbiochem protease cocktail set III.

3. Homogenize brains in a 50 ml potter-Elvehjem homogenizer (10–15 strokes). Avoid frothing.

4. Spin at 7000 rpm ($5800g_{max}$) for 20 min in an SS34 rotor.

5. Collect supernatant with a pipette. Be careful to minimize carryover of the loose lipid pellet.

6. Ultracentrifuge at 45 000 rpm ($208\,000g_{max}$) for 40 min in a type 70 Ti rotor.

7. Resuspend pellet in 10 ml of HKM. Homogenize pellet in a 15 ml homogenizer.

8. Add an equal volume of HKM containing Ficoll (12.5% w/v) and sucrose (12.5% w/v). Mix by inversion to ensure homogeneity.

9. Spin in a type 70 Ti rotor at 25 000 rpm ($64\,000g_{max}$) for 20 min.

10. Dilute the supernatant 1:5 in HKM.

11. Centrifuge at 35 000 rpm ($142\,000g_{max}$) for 60 min in a type 45 Ti rotor.

12. Resuspend pellet in 15 ml of HKM and homogenize.

13. Leave on ice for about 1 h. ④

14. Spin at 13 000 rpm ($17\,000g_{max}$) for 10 min in a type 70 Ti rotor to sediment insoluble material.

15. Layer supernatant over a cushion of 8% (w/v) sucrose made up with D_2O and HKM. Spin supernatant at 25 000 rpm ($80\,000g_{max}$) in a swing-out rotor (SW40) for 2 h.

16. Resuspend the pellet in HKM to yield a cloudy suspension. Fractions can be snap frozen in liquid nitrogen and stored at $-70\,°C$ or used immediately.

Notes

This procedure will take approximately 9 h. All steps should be done at $4\,°C$.

① Brains can be purchased from Harlan Seralab Loughborough, UK and should be stored at $-80\,°C$. (Fresh brains can also be used.)

② All buffers should be ice-cold during the procedure.

③ Alternatively, mince the brains with scissors after adding the cold buffer.

④ For clathrin purification, steps 13 and 14 can be omitted, as they decrease final protein yield. Pellets should be resuspended in 800 mM Tris-HCl pH 8.0 and incubated in rotating tubes overnight at $4\,°C$. Pellet the vesicles at 50 000 rpm in a Beckman TLA100 rotor (20 min spin). The supernatant is highly enriched in clathrin. Clathrin can be purified away from AP complexes and other proteins by gel filtration or ion exchange chromatography.

References

1. Pearse, B. M. (1983) Isolation of coated vesicles. *Methods Enzymol.*, **98**, 320–326.
2. Keen, J. H., Willingham, M. C. and Pastan, I. H. (1979) Clathrin-coated vesicles: isolation, dissociation and factor-dependent reassociation of clathrin baskets. *Cell*, **16**, 303–312.

Reconstitution of endocytic intermediates on a lipid monolayer

Brian J. Peter and **Matthew K. Higgins**

Introduction

Lipid monolayers have been used for many years as templates for the formation of two-dimensional crystals of soluble proteins (reviewed in ref. 1) and, more recently, membrane proteins [2]. The principle of the assay is that phospholipids, when placed onto an aqueous droplet, adopt a conformation in which the hydrophobic tails point towards the air while the hydrophilic head groups contact the solution. Proteins of interest interact with the head groups and are concentrated in a two-dimensional array. A hydrophobic electron microscope grid interacts with the lipid tails, allowing the monolayer to be removed from the droplet and studied in the electron microscope. The lipid monolayer composition can be tailored to simulate the inner leaflet of the plasma membrane, and thus can be used in conjunction with purified proteins to reconstitute early stages of endocytosis. While the fluidity and flexibility of a lipid monolayer are not the same as those of a lipid bilayer with aqueous solution on either side, this technique can nonetheless be useful for studying the properties of different proteins and their interactions with clathrin. The formation of clathrin coats can be observed using negative stain electron microscopy, while the use of platinum shadowing can reveal the degree of invagination of these coats.

For examples, see refs 3 and 4. A gallery of images obtained with this technique can be view on the web (http://www2.mrc-lmb.cam.ac.uk/groups/hmm/epsin/EM/).

Reagents

Chloroform

Cholesterol (Avanti), dissolved to 10 mg/ml in Chloroform

Clathrin (purified)   HKM buffer (25 mM Hepes pH 7.4, 125 mM potassium acetate, 5 mM magnesium acetate, 1 mM dithiothreitol) ①

Methanol

Phosphatidylinositol and Phosphatidylinositol-4,5-bisphosphate (Avanti polar lipids), dissolved to 1 mg/ml in 3 : 1 chloroform: methanol. ②

Phosphatidylserine, phosphatidylcholine and phosphatidylethanolamine (Sigma) dissolved in chloroform to 10 mg/ml.

Purified AP180, epsin or other clathrin- and lipid-binding protein.

Uranyl acetate, 2% w/v (Biorad), with 0.0025% w/v polyacrylic acid (Sigma) in water ③

Equipment

Carbon and collodion-coated gold electron microscopy grids (e.g. G204G from Agar Scientific Ltd, coated first with collodion or formvar, and then with a thin layer of evaporated carbon)

Forceps for handling EM grids – self-locking spring forceps are especially useful

Humid chamber, or covered container with a wet paper towel inside

Parafilm

Teflon block with 60 μl wells allowing for side injection (see Figure 6.18)

Transmission electron microscope

Vacuum evaporator, 0.2 mm diameter piece of platinum wire (TAAB Laboratories), 1 mm thick tungsten wire (also TAAB) (necessary for platinum shadowing)

Whatman filter paper or similar, for blotting EM grids

Procedure

1. Make up a lipid mixture containing 10% cholesterol, 40% PE, 40% PC and 10% PtdIns(4,5)P$_2$ to a final concentration of 0.1 mg/ml in a 19:1 mixture of chloroform : methanol (methanol is necessary to maintain PtdIns(4,5)P$_2$ solubility). This mixture should be made on the day the monolayer is made. If stored, it should be stored under argon at $-80\,^{\circ}$C in a glass vial with a glass or Teflon lid, for not longer than 3 days.

2. Arrange Teflon block in humid chamber, and fill wells of Teflon block with HKM buffer. Fill the wells with 40–60 μl of buffer, such that the total volume in the well will be 60 μl after injection of protein samples. ④

3. Carefully pipette (or inject with Hamilton syringe) 1 μl of lipid mixture on to the buffer in the well. As a negative control, inject pure chloroform without any lipid (this will test for lipid dependence of any structures

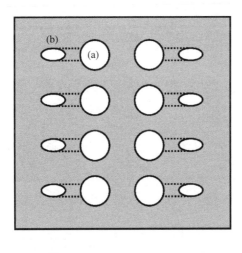
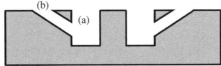

Figure 6.18   Top (above) and cross-section (below) views of the Teflon block used for monolayer formation. The block contains eight wells for processing samples in parallel. Main buffer well (a) should be 4 mm in diameter × 5 mm deep (or deeper) for a 60 μl sample volume. The monolayer is formed on top of a buffer droplet in well (a), and proteins and buffer are injected later through the side port (b)

seen, such as whether clathrin baskets form in solution instead of clathrin coats on the monolayer surface). ⑤

4. Incubate at room temperature for 60 min. The chloroform should evaporate, leaving a monolayer of lipid on the surface of the buffer.

5. Carefully place one EM grid, carbon side down, onto the top of each buffer droplet. Grids should not glow discharged before use as a hydrophobic carbon film is required to adhere to the hydrophobic lipid tails of the monolayer.

6. Gently inject proteins into the side injection well. Final protein concentrations in the well should be 0.5–2 μM for the AP180/epsin/adaptor protein,

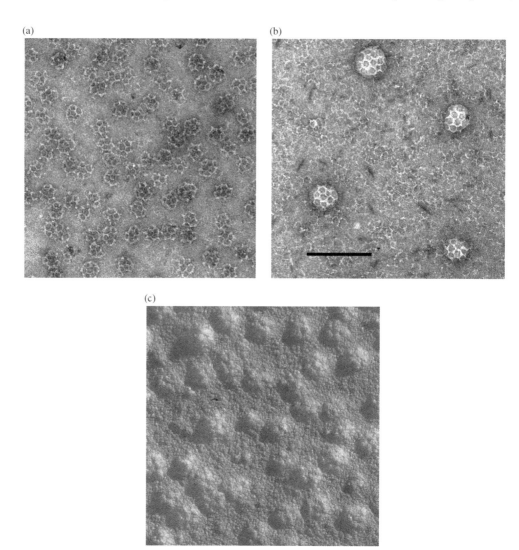

(a)

(b)

(c)

Figure 6.19 Electron micrographs of endocytic intermediates formed with the monolayer assay: (a) structures formed by incubation of AP180 and clathrin; (b) structures formed by incubation of epsin1 and clathrin; (c) structures formed by incubation of AP180, AP2 complex, and clathrin after shadowing with platinum. Scale bar in (b) = 300 nm and applies to all panels

and 30–500 nM for the clathrin. It is often useful to try several concentrations, to account for differences in protein activity.

7. Incubate 60 min at room temperature.

8. Prepare uranyl acetate stain. Lay a fresh piece of Parafilm on the bench, and place two 30 μl drops side by side on the Parafilm for each grid which will be stained. Lay out a piece of Whatman filter paper to blot buffer and stain from grids. Lay out a second piece of filter paper on which to set grids to air-dry.

9. Gently inject approximately 30 μl of buffer into the side injection port; this will raise the grid up above the surface of the Teflon block. Immediately grab the grid with forceps and lift it vertically off of the droplet.

10. Blot the grid briefly by touching it to the filter paper, then touch it to the first stain droplet and blot immediately. Touch the grid to the second stain droplet, leave for 30 s, and blot briefly. This leaves a film of stain on the surface of the grid in which the protein is embedded. If the grids will be platinum shadowed, hold the grid to the filter paper for several seconds to ensure that the entire grid surface dries. Lay the grid on another piece of filter paper to dry.

11. For negative stain EM, grids can be examined in the EM immediately (see Figure 6.19). They are also stable for several weeks, at least, at room temperature.

12. If platinum shadowing is required, set up the vacuum evaporator with a 2 cm long piece of platinum wire coiled tightly round a piece of 1 mm thick tungsten wire. Place the grids to be shadowed on a rotary platform at an angle of 10° to the line between the platform centre and the platinum coil. Create a vacuum in the evaporator. With a shield between the grids and the wire, turn on the current to the tungsten wire. When the platinum wire melts, remove the shield and the platinum will evaporate onto the grids. For rotary platinum shadowing, start rotation of the platform immediately before removing the shield. For single angle shadowing, the platform can remain stationary during the evaporation; 1–2 min of evaporation is usually sufficient, but trials may need to be done to account for differences in evaporators.

Notes

This procedure will take approximately 3 h. All reagents should be of the highest purity available, and buffers should be filtered before use.

① Clathrin should be dialysed into HKM buffer and centrifuged for 20 min at $100\,000g_{max}$ (e.g. 45 000 rpm in a Beckman TLA100 rotor) immediately prior to use to remove aggregates.

② Lipid stocks were stored at $-80\,°C$.

③ The addition of polyacrylic acid helps to prevent the stain from precipitating and forming uranyl acetate crystals.

④ It is important to ensure that the surface of the droplet is either flat or slightly concave. Overfilling the Teflon wells leads to a convex droplet surface, upon which the monolayer does not form properly. Also, filling the wells with too little volume (less than 35 μl, or depending on the geometry of the block) can lead to an uneven surface near the side injection port.

(5) Trace lipid contamination (e.g. PtdIns $(4,5)P_2$) on the Teflon block can result in misleading negative controls. The Teflon block should be rinsed with hot water, then ethanol, and finally, soaked overnight in a mixture of chloroform/methanol to remove any protein or lipid residue. Hamilton syringes are also susceptible to trace lipid contamination.

References

1. Chiu, W., Avila-Sakar, A. J. and Schmid, M. F. (1997) Electron crystallography of macro-molecular periodic arrays on phospholipid monolayers. *Adv. Biophys.*, **34**, 161–172.

2. Levy, D., Mosser, G., Lambert, O., Moeck, G. S., Bald, D. and Rigaud, J. L. (1999) Two-dimensional crystallization on lipid layer: a successful approach for membrane proteins. *J. Struct. Biol.*, **127**(1), Aug, 44–52.

3. Ford, M. G., Mills, I. G., Peter, B. J., Vallis, Y., Praefcke, G. J., Evans, P. R. and McMahon, H. T. (2002) Curvature of clathrin-coated pits driven by epsin. *Nature* **419**(6905), 361–366.

4. Ford, M. G., Pearse, B. M., Higgins, M. K., Vallis, Y., Owen, D. J., Gibson, A., Hopkins, C. R., Evans, P. R. and McMahon, H. T. (2001) Simultaneous binding of PtdIns(4,5)P_2 and clathrin by AP180 in the nucleation of clathrin lattices on membranes. *Science*, **291**(5506), 1051–1055.

PROTOCOL 6.26

Golgi membrane tubule formation

William J. Brown, K. Chambers and A. Doody

Introduction

Membrane tubules emanate from a variety of organelles including the Golgi complex, endosomes and lysosomes [1]. These tubules are generally 60–80 nm in diameter, can extend for many microns in length, and are believed to play a role in various intracellular membrane trafficking pathways. Although recent studies strongly suggest a role for a cytoplasmic Ca^{2+}-independent phospholipase A_2 enzyme in initiating the formation of Golgi and endosome tubules [2, 3], the exact molecular mechanisms of tubule formation have not been elucidated. Progress in this area has been advanced by the development of an *in vitro* reconstitution system that was used to demonstrate that tubulation of Golgi membranes was absolutely dependent on cytosolic protein [4, 5]. This assay system should allow for the further characterization and identification of cytosolic factors that are required for tubules to form. This system should also provide a means to identify molecular differences between the tubules and the main body of the organelles from which they grow (see Figure 6.20).

Reagents

Carbon- and formvar-coated electron microscope grids (300 mesh copper)

Extract of bovine brain cytosol (BBC) and isolated rat liver Golgi complexes,

prepared as described [3]. ① (See also *Protocol 4.23*)

2% w/v Phosphotungstic acid (PTA), pH 7.4 (stored at 4 °C)

Stock solutions: 10 mM Hepes, pH 7.4 (stored at −20 °C), 10 mM ATP (stored at −20 °C), 100 mM $MgCl_2$ (stored at −20 °C), 25 mM Tris containing 50 mM KCl (pH 7.4)

Equipment

Eppendorf 500 µl microfuge tubes, or similar

Self-closing, fine-tipped forceps

Transmission electron microscope

Water bath (37 °C)

Procedure

1. Thaw out all solutions and reagents and put on ice. Do not thaw out by running under hot water.

2. Pick up formvar/carbon-coated grids with the self-closing forceps. Hold the grids on the very edge. Be sure the grids look uniformly coated. Use a piece of white paper as a background.

3. Place the forceps on the white paper with coated grid side up. Use a box, e.g. the top of a pipette tip box, to cover the ends of the forceps to keep the grids from accumulating dust.

4. Place 500 µl microfuge tubes in an appropriate holder.

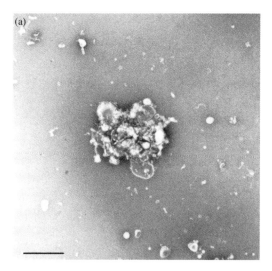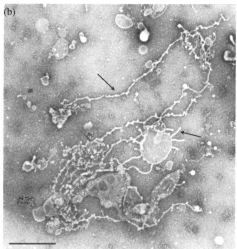

Figure 6.20   (a) shows an example of a negatively stained Golgi stack that was incubated with BSA as a negative control. These control Golgi stacks generally appear somewhat round, with a few vesicles and possibly some short tubules. (b) shows a negatively stained Golgi stack that was incubated with BBC to induce tubule formation. Under these conditions, numerous tubules, 60–80 nm in diameter and ranging up to several microns in length, can be seen (arrows). Scale bars = 1 μm

5. Prepare Mg-ATP solution in the proportions below and put on ice. You will need 1 μl/reaction.

   80 μl Tris/KCl stock buffer

   10 μl ATP

   10 μl MgCl₂

6. Prepare the tubulation buffer containing BBC. You will need 20 μl/reaction (sample). When using BBC as a positive control, its final concentration following dilution at this step should be ~3 mg/ml. For negative controls, we use bovine serum albumin (BSA) at an equivalent concentration. For solutions with a high protein concentration (e.g. BBC, which is often prepared and stored at ~20 mg/ml), use the following proportions: 85 μl 1 Tris/KCl, 1 μl Mg-ATP, 1 μl Hepes buffer, 13 μl BBC.

7. Add 20 μl of Golgi suspension to each microfuge tube.

8. Add 20 μl of tubulation buffer/BBC solution to the Golgi tubes. Add the tubulation buffer/BBC solution very carefully to the Golgi suspension in three aliquots over ~15 s. Gently mix.

9. Place tubes in 37 °C water bath and incubate for up to 30 min.

10. Take 10 μl of a sample and puddle onto the grid that is being held by a forceps. Cover with the box top. Let the suspension sit on the grid for 15 min at room temperature.

11. Carefully add 10 μl of 2% PTA to the grid three times over 5 s. Do not let the puddle spill over the edges of the grid.

12. Blot off fluid from the edge of the grid with a moist kimwipe.

13. Add 10 μl of PTA to the grid and blot off fluid as in step 12. Repeat once more.

14. Let grids dry and then store until EM observation.

Notes

Preparation of BBC (or cytosolic preparations from other tissues/cells) will take 2 days. Preparation of rat liver Golgi complexes will take 1 day. The tubulation assay will take approximately 2 h.

① Cytosolic extracts from other tissues can also be used. These reagents generally give the best results if freshly prepared, but they can be stored for several weeks at $-80\,^{\circ}$C and still give good results. However, they degrade with time, so use them as soon as possible.

References

1. Brown, W. J., Chambers, K. and Doody, A. (2003) Phospholipase A2 (PLA2) enzymes in membrane trafficking:mediators of membrane shape and function. *Traffic*, **4**, 214–221.
2. De Figueiredo, P., Drecktrah, D., Katzenellenbogen, J. A., Strang, M. and Brown, W. J. (1998) *Proc. Nat. Acad. Sci. USA*, **95**, 8642–8647.
3. De Figueiredo, P., Doody, A., Polizotto, R. S., Drecktrah, D., Wood, S., Banta, M., Strang, M. and Brown, W. J. (2001) *J. Biol. Chem.*, **276**, 47361–47370.
4. Banta, M., Polizotto, R. S., Wood, S. A., de Figueiredo, P. and Brown, W. J. (1995) *Biochemistry*, **34**, 13359–13366.
5. Polizotto, R. S., de Figueiredo, P. and Brown, W. J. (1999) *J. Cell Biochem.*, **74**, 670–683.

Tight junction assembly

C. Yan Cheng and Dolores D. Mruk

Introduction

When mammalian epithelial cells, such as Sertoli cells, keratinocytes and MDCK cells (Madin Darby canine kidney cells), are cultured *in vitro* on a suitable substratum (e.g., Matrigel for Sertoli cells) in chemically defined medium (e.g., F12/DMEM, Ham's F12 Nutrient Mixture/Dulbecco's Modified Eagle's medium for Sertoli cells) in either culture dishes or bicameral units, they assemble tight junctions (TJs) within days (for reviews, see refs 1–3). As such, this represents a useful *in vitro* system to investigate the biology and regulation of TJ dynamics (for reviews, see refs 3 and 4). There are several techniques available in the literature to monitor the assembly of epithelial cell TJs quantitatively or semi-quantitatively, which include restricted diffusion of [³H]-inulin, [¹²⁵I]-BSA or FITC-dextran across the cell epithelium and maintenance of non-equilibrium of the fluid level between the apical and basal compartments of the bicameral unit [5, 6] (see Figure 6.21). While these techniques are helpful to monitor TJ assembly, they are either tedious, require the use of radioactive isotopes, or expensive equipment (such as a cytofluorometer to quantify FITC). We present herein the use of a Millicell Electrical Resistance System (ERS) (Millipore Corp.) to quantify the transepithelial electrical resistance (TER) across the cell epithelium. This equipment is not expensive to purchase, the maintenance cost is also minimal (except for the electrodes which need periodic replacement), and yet an investigator can quantitatively assess the assembly, maintenance and/or regulation of TJs with precision and reproducibility. This technique has been tested vigorously against the other currently available methodologies [5–8], and it was shown to be the best technique to quantify TJ assembly *in vitro*.

Reagents and cell cultures

All reagents and enzymes used should be standard laboratory reagents prepared with Milli-Q quality water (such as PureLab PLUS from USFilter). Primary cultures of Sertoli cells are obtained from 20-day-old rat testes using procedures as earlier described [5, 6, 9] and cultured on Matrigel-coated (Matrigel : F12/DMEM, 1 : 7) bicameral units (Millicell-HA 0.45 μm-pore size culture plate insert composed of mixed esters of cellulose with 12 mm diameter; cat. no. PIHA 012 50, Millipore, Bedford, MA). Each treatment group should have at least triplicate culture inserts. Matrigel-coated bicameral units should be prepared 24 h prior to cell plating and stored in a 35 °C CO_2 incubator (95% air/5% CO_2, v/v) as described to allow the Matrigel to completely dry, which is essential to

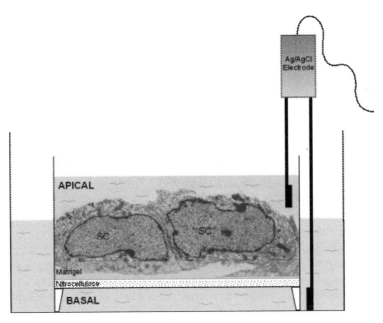

Figure 6.21 This is an electron micrograph that shows two Sertoli cells of an epithelium cultured on a Matrigel-coated bicameral unit (nitrocellulose based) on day 4. Ag/AgCl electrodes (one placed into the apical and the other into the basal compartment of the bicameral unit) are used to quantify the electrical resistance (in ohms) when an electrical current, ~20 mA, is sent across the cell epithelium. Also note the non-equilibrium of the level of F12/DMEM between the two compartments when TJs are formed between cells. SC, Sertoli cell nucleus. This figure is adapted from Lee and Cheng [10] and used with permission from the publisher.

obtain TER readings with minimal intra-experimental variations. Care should also be taken to avoid trapping air bubbles in the Matrigel [6].

Equipment

Ag/AgCl electrodes (cat. no. MERS STX 01, Millipore) – electrodes need to be replaced after ~12 months of routine use when visible loss of Ag/AgCl is detected. The electrodes are EVON chopstick electrodes. Each leaflet has an outer and inner electrode; the outer electrode is a small silver pad for passing current through the nitrocellulose membrane of the bicameral unit, whereas the inner electrode is a small Ag/AgCl voltage sensor

CO_2 humidified incubator at 35 °C

Millipore Millicell ERS system (cat. no. MERS 000 01)

Centrifuge set at room temperature

Shaking water bath with variable temperature control

Standard laboratory microscope and transmission electron microscope (TEM)

Procedure for TER measurement

1. Sertoli cells isolated from seminiferous tubules immediately after their isolation are plated immediately on Matrigel-coated bicameral units at $0.75–1 \times 10^6$ cells/cm^2 and designated time 0. Each culture insert is placed in a well of a 24-well dish with triplicate inserts for the control and each treatment group. Great care should be taken to avoid trapping air bubbles between Sertoli cells and the nitrocellulose membrane, which have a tendency to form on

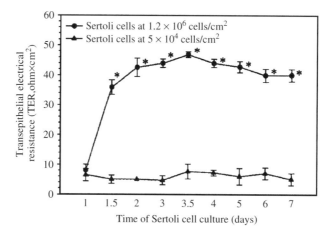

Figure 6.22 The assembly of the Sertoli cell TJ-permeability barrier *in vitro* as monitored by TER measurement across the Sertoli cell epithelium. In brief, Sertoli cells were cultured on Matrigel-coated bicameral units at either 1.2×10^6 or 5×10^4 cells/cm^2. The TJ-barrier formed between Sertoli cells when they were cultured at 1.2×10^6 cells/cm^2 as manifested by a steady rise in the resistance across the cell epithelium, yet it failed to form at the lower cell density due to the lack of cell proximity. This figure is adopted from Mruk *et al.* [6] and used with permission from the publisher. *, $p < 0.01$ when compared to its corresponding control.

the periphery of the bicameral unit. Also, it is important that cells are uniformly plated to avoid uneven clustering of cells across the epithelium. Cells should not be disturbed thereafter for at least 2–4 h. The same volume of F12/DMEM (~ 500 μl) should be used for the apical and basal compartments of the bicameral unit. The daily replacement of media should be of the same volume. By 4–5 h after plating, the first TER reading should be recorded and daily thereafter as follows:

- To test the instrument, switch the mode to RESISTANCE and turn the power ON. Press the TEST R button. With the RANGE SWITCH in the 2000 V position, the meter will read approximately 1000, whereas in the 20 k range, the meter will read 1.00. It is now ready for use.

- To test the electrodes, insert the end-plug at the end of the electrode cable into the front panel of the EVOM. Place the electrodes

into F12/DMEM. Turn the mode to VOLTS. Press the power switch ON. The reading will be ~ 1–2 mV. After 15 min, adjust this voltage to 0 mV at the adjustment labelled ZERO V with a screwdriver.

2. The range should be set to 2000 ohm. Switch the mode to RESISTANCE. Dip one electrode inside the bicameral unit on top of the cell layer (shorter end) and the other electrode (longer end) into the external solution (see Figure 6.22). The electrode should be held straight upright, as shown in Figure 6.22, and not touching the bicameral unit nor the side of the 24-well dish. However, the longer end of the electrode should rest on the bottom of the 24-well dish during measurement (see Fig. 6.21). Press the MEASURE R button for 2–3 s to send an electrical current, ~ 20 mA, across the cell epithelium. A steady reading of the resistance should result denoting the tightness of the TJ-barrier. Cells should be fed immediately thereafter

and incubated at $35\,^{\circ}C$ in a humidified incubator (95% air/5% CO_2) until the next TER reading is taken.

- Do not rinse the electrodes between recordings until readings within the triplicate set of a treatment group are completed. The electrodes can then be dipped into empty wells with F12/DMEM alone to remove treatment reagents by limiting dilutions.

- Caution must be exercised to ensure that the cell layer is not touched with the electrodes when taking measurements.

- If the depth at which the electrodes are immersed into the external and internal media varies between different bicameral units, slight differences in your measurements will result.

- Dishes containing bicameral units must adjust to room temperature for best measurements (~20 min).

- To sterilize the electrodes, soak them in 70% alcohol (v/v with Milli Q water). Electrodes must then be equilibrated in F12/DMEM for at least 2 h before use.

- The investigator must ensure that bicameral units do not tilt to the side of the 24-well dish but instead sit centred in the 24-well dish when TER measurements are taken.

- A total of four readings should be taken per bicameral unit at 12-,3-, 6- and 9-o'clock positions and averaged.

- Data will be obtained and analysed as follows:
  –Measure background (bicameral unit + media without cells)
  –Measure unknown (bicameral unit + media + cells)
  –RESISTANCE × SURFACE AREA = ohm × cm^2

3. TER readings must be taken on a daily basis. If a reagent is being tested to monitor its effects on TJ assembly, it should be removed once an effect is identified to examine if the perturbed TJ-barrier can be resealed to confirm the specificity. Figure 6.22 shows the typical result from an experiment to assess the assembly of Sertoli cell TJ-permeability barrier *in vitro*. It is noted that the TJ-barrier assembles by 3–4 days when the TER reaches a plateau level.

References

1. Gumbiner, B. and Simons, K. (1986) *J. Cell Biol.*, **102**, 457–468.
2. Grima, J., Wong, C. S. C., Zhu, L. J., Zong, S. D. and Cheng, C. Y. (1998) *J. Biol. Chem.*, **273**, 21 040–21 053.
3. Cheng, C. Y. and Mruk, D. D. (2002) *Physiol. Rev.*, **82**, 825–874.
4. Lui, W. Y., Mruk, D. D., Lee, W. M. and Cheng, C. Y. (2003) *Biol. Reprod.*, **68**, 1087–1097.
5. Grima, J., Pineau, C., Bardin, C. W. and Cheng, C. Y. (1992) *Mol. Cell. Endocrinol.*, **89**, 127–140.
6. Mruk, D. D., Siu, M. K. Y., Conway, A. M., Lee, N. P. Y., Lau, A. S. N. and Cheng, C. Y. (2003) *J. Androl.*, **24**, 510–523.
7. Lui, W. Y., Lee, W. M. and Cheng, C. Y. (2001) *Endocrinology*, **142**, 1865–1877.
8. Chung, N. P. Y. and Cheng, C. Y. (2001) *Endocrinology*, **142**, 1878–1888.
9. Cheng, C. Y., Mather, J. P., Byer, A. L. and Bardin, C. W. (1986) *Endocrinology* **118**, 480–488.
10. Lee, N. P. Y. and Cheng, C. Y. (2003) *Endocrinology*, **144**, 3114–3129.

Reconstitution of the major light-harvesting chlorophyll *a/b* complex into liposomes

Chunhong Yang, Helmut Kirchhoff, Winfried Haase, Stephanie Boggasch and **Harald Paulsen**

Introduction

The major light-harvesting chlorophyll (Chl) *a/b* protein complex of photosystem II (LHCIIb) is the most abundant membrane protein which comprises more than 50% of the total Chl in plants [1]. It is also a very important protein in the biosphere, because LHCIIb has a key function in the conversion of solar energy to biochemical energy in plants. The main function of LHCIIb is to absorb the solar energy and to transfer excited electronic energy to the reaction centre of photosystem (PS) II under moderate illumination, and to adjust energy transfer between PS II and PS I under strong illumination [2]. LHCIIb can be assembled *in vitro* from its recombinant apoprotein and purified pigments in detergent solution [3, 4].

This protocol presents methods for reconstituting either native (Procedure A) or recombinant LHCIIb (Procedure B) into artificial liposomes made of thylakoid lipids.

Reagents

Bio-Beads SM-2 Adsorbent (Bio-RAD)

Chloroform

Ficoll (Amersham Biosciences) in Rb, 10%, 20%, 30%, 40%

Lipids (synthetic or isolated from plant thylakoids): phosphatidylglycerol (PG), digalactosyl diacylglycerol (DGDG), monogalactosyl diacylglycerol (MGDG), sulfolipid (SL)

Liquid nitrogen

Reconstitution buffer for Procedure A (Rb): 10 mM Tricine pH 7.8 (KOH)

$2\times$ Reconstitution buffer for Procedure B (Rb2): 20 mM Tris HCl (pH 7.5), 20 mM NaCl, 20 mM β-Mecaptoethanol

Triton X-100

Equipment

Extruder (LiposoFast, Avestin)

Polycarbonate membranes (pore diameter: 100 nm)

Rotary evaporator

Sonicator set up with a micro-tip, and a bath sonicator

Ultracentrifuge

UV-Vis spectrophotometer

Procedure A

1. Mix 300 nmol lipids dissolved in chloroform in a glass tube with a glass stopper (47 mol% MGDG; 27 mol% DGDG; 39 mol% SL; 14 mol% PG, stoichiometries for grana thylakoids, see ref. 5)

2. Evaporate the chloroform phase carefully under an N_2 stream and then in a rotary evaporator at 40 °C for 45 min (lipids form a thin film on the glass wall).

3. For the following steps the air in the glass tube is removed by N_2. Add 1.35 ml Rb saturated with N_2.

4. Sonicate the suspension three times for 10 s in an ice bath at low sonication power (the suspension becomes opalescent).

5. Add 0.15 ml 5 mM Triton X-100 under constant slow stirring. Incubate for 10 min under slow constant stirring.

6. Add 5 µl LHCIIb preparation (about 2 mM Chl in 0.37% (w/v) Triton X-100). See Figure 6.25 for different results depending on the amount of LHCIIb added. Incubate for 20 min under slow constant stirring. Take a UV spectrum (200–350 nm) of the undiluted solution and put this aliquot back in the reconstitution batch. (LHCIIb is isolated according to ref. 6 with modifications given in ref. 7.)

7. Add 8 mg Bio-Beads and incubate under constant stirring overnight at 4 °C. Take a spectrum as in step 6.

8. Remove the solution from Bio-Beads with a pipette and put it in a fresh glass tube. Add 16 mg Bio-Beads and incubate for 1 h at room temperature (RT). Take a spectrum as in step 6.

9. Repeat step 8. Quantity of removed Triton X-100 is calculated from the difference spectra steps 7–9 minus step 6, according to the following equation:

$$[Triton\ X\text{-}100] = (OD_{275} - OD_{300})$$

$$* 1.46\ [mM]$$

OD_{275}: maximum absorption of Triton X-100, A_{275} at 1 cm pathlength; OD_{300}: non-specific absorption, A_{300} at 1 cm pathlength.

10. Freeze (in liquid nitrogen) and thaw at RT twice.

11. Put the sample on a Ficoll step gradient (10, 20, 30, 40%) and ultracentrifuge at 100 000g for 18 h.

12. Collect the green 30% Ficoll step band in an Eppendorf cup and fill up the cup with Rb. Centrifuge at about 20 000g for 10 min at 7 °C (removal of Ficoll).

13. Suspend the pellet in 1.2 ml Rb and centrifuge again. Suspend the pellet in a small volume of Rb (LHCIIb-liposomes).

14. Characterization: lipid analysis by 2D thin layer chromatography [8], absorption spectra, 77 K fluorescence spectra and electron microscopy.

Procedure B

1. Prepare 100 µg recombinant LHCIIb trimer complex according to refs 3 and 4.

2. Mix 1 mg lipids in chloroform at a stoichiometry similar to that of natural thylakoid lipids (without MGDG, PG: 21.4%; DGDG 61.9%; SL 16.7%) in a glass tube.

3. Evaporate the chloroform phase in a rotary evaporator at 40 °C so that a thin lipid film is formed on the inner glass wall.

4. Add 100 µl Rb2 and 50 µl bidest water, incubate at RT for 30 min.

5. Vortex the mixture vigorously until a homogenous suspension is formed.

6. Sonicate the lipid suspension for 30 s with 5 s interval between every 10 s (the solution becomes opalescent).

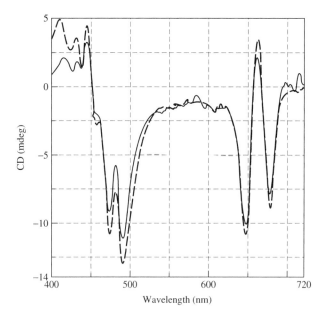

Figure 6.23 Circular dichroic absorption spectra of LHCIIb proteoliposomes and LHCIIb in 0.05% (w/v) Triton X-100 solution. (solid line: LHCIIb trimer in Triton X-100 solution; dashed line: reconstituted LHCIIb trimer in proteoliposome)

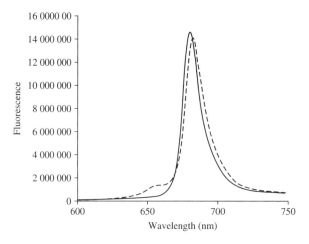

Figure 6.24 77 K Fluorescence emission spectra of LHCIIb proteoliposomes and LHCIIb complexes in 0.05% (w/v) Triton-X 100 solution upon excitation of Chl *b* at 470 nm. (solid line: LHCIIb trimer in Triton X-100 solution; dashed line: reconstituted LHCIIb trimer in proteoliposome)

7. Freeze (in liquid nitrogen) and thaw (at 25 °C) the solution twice.

8. Force the lipid solution through a polycarbonate membrane (pore diameter 100 nm) for about 20 times by using an extruder (LiposoFast).

9. Add Triton X-100 solution so that the end concentration of Triton X-100 is 1.5 mM, and add distilled water to 200 µl.

10. Add 100 µg LHCIIb trimer complex in maximum 100 µL 0.05% (w/v)

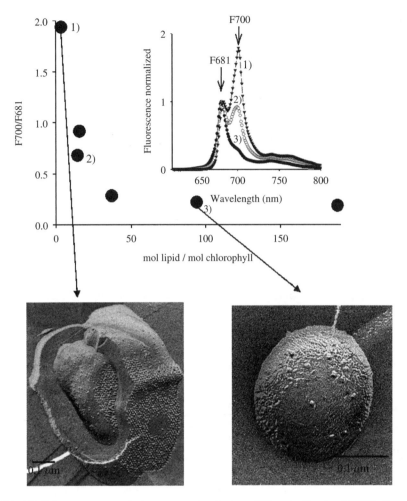

Figure 6.25 77 K fluorescence emission spectra (excitation: 480 nm) of native LHCIIb liposomes and electron microscopic pictures of freeze-fractured LHCIIb liposomes. The emission ratio F700/F681 is a measure of the LHCIIb aggregation level. If the molar lipid : Chl ratio for the reconstitution is lower than about 40, LHCIIb begins to aggregate (increase in F700/F681). The aggregated samples show a hexagonal arrangement of the proteins (see left-hand EM picture). This is clear evidence for a random protein orientation within the liposomes [3]. At high lipid : Chl ratios the protein is diluted enough so that no aggregation occurs (right-hand EM picture)

Triton X-100 to the liposome solution dropwise under sonication in a bath sonicator.

11. Vortex the mixture for 1 min, incubate at RT for 30 min.

12. Add 10 mg Bio-Beads, incubate at RT for 1 h.

13. Centrifuge shortly, collect the supernatant.

14. Add 10 mg new Bio-Beads to the supernatant and incubate at 4 °C overnight.

15. Repeat step 13 and add 20 mg new Bio-Beads to supernatant, incubate at RT for 1 h.

16. Repeat step 15.

17. Centrifuge the tube shortly and collect supernatant.

18. Analyse the proteoliposomes by spectroscopy.

Results

1. Circular dichoism (CD) in the visible range shows that the pigment organization is very similar in LHCIIb proteoliposomes as it is in detergent solubilized LHCIIb (Figure 6.23).

2. Low-temperature fluorescence spectroscopy can be used to check whether an LHCIIb preparation is intact and functional, i.e. whether Chl *b* quantitatively transfers its excitation energy to Chl *a*. Low-temperature fluorescence upon excitation of Chl *b* at 470 nm shows that most fluorescence is emitted by Chl *a* (680 nm) and only very little by Chl *b* (660 nm) (Figure 6.24), demonstrating that only very little Chl *b* has been disconnected from Chl *a*.

3. 77 K fluorescence emission spectra of native LHCIIb liposomes and electron microscopic pictures of freeze fractured LHCIIb liposomes show that LHCIIb complexes begin to aggregate when the reconstitution has been done at a low lipids : Chl ratio (see Figure 6.25).

Note

The whole procedure will take ~2 days.

References

1. Kühlbrandt, W., Wang, D. N. and Fujiyoshi, Y. (1994) *Nature*, **367**, 614–621.
2. Allen, J. F. (1994) *Physiol. Plant.*, **92**, 196–205.
3. Hobe, S., Prytulla, S., Kühlbrandt W. and Paulsen H. (1994) *EMBO J.*, **13**, 3423–3429.
4. Rogl, H., Kosemund, K., Kühlbrandt, W. and Collinson, I. (1998) *FEBS Lett.*, **432**, 21–26.
5. Kühlbrandt, W., Thaler, T. and Wehrli, E. (1983) *J. Cell Biol.*, **96**, 1414–1424.
6. Burke, J. J., Ditto, C. L. and Arntzen, C. J. (1978) *Arch. Biochem. Biophys.*, **187**, 252–263.
7. Duchene, S. and Siegenthaler P. A. (2000) *Lipids*, **35**, 739–744.
8. Christie W. W. and Dobson G. (1999) *Lipid Technology*, **11**, pp. 64–66.

Reconstitution of photosystem 2 into liposomes

Julie Benesova, Sven-T. Liffers and Matthias Rögner

A. Preparation of liposomes

Reagents

Buffer (20 mM MES, pH 6.5; 10 mM $MgCl_2$; 10 mM $CaCl_2$)

Chloroform

Phosphatidylcholine in chloroform (25 mg/ml)

Phosphatic acid in chloroform (5 mg/ml) [1]

Equipment

Cryostat

Glass container with round bottom (50 ml)

Polycarbonate filter (Ø 10 mm, pore diameter 0.2 μm)

Rotavapor-R

Ultrasonication apparatus, with sonication tip

Waterbath (7 l)

Procedure

1. Cool Rotavapor-R down to 4 °C.

2. Mix the lipid stock solutions in the glass with the round bottom.

3. Rotate for 120 rpm at 40 °C, 30 min, vacuum at −600 mbar.

4. Rotate for 120 rpm at 40 °C, 1 h, vacuum at −1200 mbar.

5. Resuspend the obtained lipid layer carefully in 3 ml ether and 1 ml buffer.

6. Use the ultrasonic tip to make the suspension homogeneous.

7. Rotate for 120 rpm at 27.4 °C, vacuum at −225 mbar until the film is completely detached from the glass wall.

8. Add 2 ml of the buffer.

9. Rotate for 120 rpm at 27.4 °C, vacuum at −1200 mbar to remove ether completely.

10. Filter the liposome solution after about 90 min through the polycarbonate filter.

B. Reconstitution of photosystem 2 [1–3]

Introduction

The importance of this photosynthetic membrane protein complex is based on its role as a light-driven water-splitting nano machine. It is the first component of the photosynthetic electron transport chain and transfers the electrons extracted from water to plastoquinone. The size of the membrane-integral part of this complex is in the case of dimers (about 500 kDa) $190 \times 100 \times 40$ Å. Beyond this membrane-embedded part, the donor side of this complex extends up to 55 Å out of the membrane into the lumen and the acceptor side up to 10 Å into the stromal region [2]. The lumen-exposed part performs the water-splitting process while the membrane-embedded part harbours the components of the electron transport chain leading to light-driven charge separation.

Reconstitution of the PS2 complex into liposomes facilitates various studies of structure/function relationships of PS2 including its assembly, parameters for its orientation in the membrane, the process of reversible mono-/dimerization and the impact of the lipid composition on its function.

Reagents

1 M Octylglucopyranosid (OGP): 146 mg/ 0.5 ml buffer (= 30%)

Buffer (20 mM MES, pH 6,5; 30 mM CaCl$_2$; 10 mM MgCl$_2$; 1 M glycin-Betain

Liposome suspension (see above)

110–190 µg Chl PS2 (depending on activity of PS2)

Equipment

Biobeads (BioRad)

Cryostat

Water cooling by self-made glass cooler

Pretreatment of biobeads

- Stir 20–30 g Biobeads in 150–200 ml methanol for 1 h at RT in a beaker.

- Remove the supernatant by Pasteur pipette (connected to aspirator), after short centrifugation.

- Add 200 ml water and stir for 1 h.

- Remove water as above.

- Add again 200 ml water and stir for 30 min.

- Keep in 30–50 ml in the fridge (stable for several weeks!).

Procedure

1. Mix 110–190 µg Chl PS2 (dependent on activity, see above) and 180 µl liposome suspension; add buffer to a final volume of 900 µl.

2. Add 1 M OGP stock solution in small parts under mild stirring until the milky suspension is clear (start with 66 µl 1 M OGP = 2% OGP final conc.; if still cloudy, add in steps of 0.5%)

3. Add buffer up to a final volume of 1 ml and stir slowly in the dark at 20 °C.

4. Add 80 mg Biobeads and incubate for 1 h; repeat this step.

5. Add 160 mg Biobeads and incubate again for 1 h with slow stirring (see above).

References

1. Rigaud, J.-L., Pitard, B. and Levly, D. (1995) Reconstitution of membrane proteins into liposomes: application to energy-transducing membrane proteins. *Biochim. Biophys. Acta*, **1231**, 223–246.
2. Zouni, A., Witt, H. T., Kern, J., Fromme, P., Krauss, N., Saenger, W. and Orth, P. (2001) Crystal structure of photosystem 2 from *Synechoccus elongatus* at 3.8 Å resolution. *Nature*, **409**, 739–743.
3. Kruip, J., Karapetyan, N. V., Terekhovas, I. V. and Rögner, M. (1999) *In vitro* oligomerization of a membrane protein complex: liposome-based reconstitution of trimeric photosystem I from isolated monomers. *J. Biol. Chem.*, **274**, 18 181–18 188.

Golgi – vimentin interaction *in vitro* and *in vivo*

Ya-sheng Gao and **Elizabeth Sztul**

Introduction

Intermediate filaments constitute one of the three major cytoskeletal systems of eukaryotic cells. Intermediate filaments, especially those in epithelial cells, play a key role in maintaining the physical integrity of cells [1]. In cells of mesenchymal origin and in cultured cells, the intermediate cytoskeleton is predominantly composed of vimentin. The dynamics of the vimentin cytoskeleton appear to be regulated by various cellular proteins, including proteins associated with the Golgi complex [2]. To study the interactions of vimentin with the Golgi complex, biochemical and light microscopy based approaches are most direct and easiest to follow. Importantly, biochemical approaches allow testing of compounds that might regulate the vimentin–Golgi interaction by removing or introducing such compounds into the assay. Light microscopy approaches allow the examination of the vimentin–Golgi interactions within the architectural context of the cell.

1. Biochemical assay for vimentin binding to purified Golgi membranes

Reagents

BSA (10 mg/ml in PBS)

ECL kit (PIERCE Chemical Co. or other vendors)

HKM buffer (25 mM Hepes, 0.115 M potassium acetate, 2.5 mM magnesium chloride, pH 7.4)

HRP conjugated secondary antibodies (Amersham Pharmacia Biotechnologies or other vendors)

Isolated stacked Golgi membranes [3] purified through additional up-flotation (protocol below)

Mouse anti-GM130 (Transduction Laboratories) and goat anti-vimentin (ICN Biomedicals) antibodies

Nycodenz solution (60% in HKM) from Life Technologies

Protein Assay Kit (from Bio-Rad Laboratories)

Purified vimentin (unpolymerized) from cytoskeleton, Inc. or produced as described in ref. 2

Sucrose solution (2 M, in HKM buffer)

Equipment

Densitometer

Eppendorf tubes

Equipment for SDS-PAGE (minigel) and immunoblotting

Nitrocellulose membrane (Micron Separations Inc. or other vendors)

Pipettes

Preparative ultracentrifuge equipped with SW55 Ti and SW41 Ti rotors (Beckman Coulter) or equivalents

Spectrophotometer

Ultra-Clear centrifugation tubes compatible with the above rotors (Beckman Coulter or equivalent)

Vortex mixer

X-ray film and X-ray developing machine

Procedure 1

Part 1: Purification of Golgi membranes

Golgi membrane fractions isolated by density-gradient centrifugation are likely to be contaminated by cytosolic proteins (such as vimentin subunits, vimentin or Golgi interacting proteins) that may interfere with the experiment. Therefore, isolated Golgi membranes need to be separated from the soluble contaminants by up-flotation in Nycodenz gradient.

1. Isolate stacked Golgi membranes from rat liver ([3]; see also *Protocol 4.26*).

2. Make 60% (w/v) Nycodenz in HKM and prechill at 4 °C.

3. Mix stacked Golgi membranes with prechilled 60% Nycodenz solution to make 40% Nycodenz final concentration (total volume should be less than 3 ml).

4. Load the Golgi mixture at the bottom of a 5 ml Ultra-Clear centrifugation tube.

5. Overlay the Golgi mixture with 1 ml of prechilled 36% and 1 ml of 32% Nycodenz-HKM.

6. Centrifuge at 135 000*g* for 7 h at 4 °C in a swinging bucket rotor.

7. Fractionate from top of the gradient by pipetting 0.5 ml samples.

8. Take a portion of each fraction, process for SDS-PAGE, and transfer the gel to nitrocellulose filter.

9. Immunoblot the nitrocellulose filter with antibodies against the Golgi marker protein GM130.

10. Pool GM130 containing fractions – this represents purified Golgi membranes.

11. Quantify protein concentration in the Golgi preparation using Protein Assay Kit with BSA as standard.

12. Use the membrane preparation freshly, or freeze in liquid nitrogen and store at −80 °C.

Part 2: Association of vimentin with purified Golgi membranes

1. Commercially obtained vimentin should be made up in 10 mM Tris-HCl (pH 7.4) and 0.1% β-mercaptoethanol. For long-term storage, freeze at −80 °C [2, 3].

2. Adjust protein concentration of the Golgi membranes to about 0.25 mg/ml with cold HKM buffer before use.

3. Adjust the concentration of vimentin to about 2 mg/ml with 10 mM Tris-HCl, pH 7.4 and 0.1% β-mercaptoethanol.

4. Add 0.05–0.075 mg Golgi and 0.01 mg vimentin to an Eppendorf tube on ice, mix briefly.

5. Let the tube sit on ice for 3 h with intermittent mixing.

Part 3: Separation of membrane-bound from non-bound vimentin

1. Add sufficient cold 2 M sucrose to the vimentin–Golgi membrane mixture to make 1.6 M sucrose and load the solution (less than 3 ml) at the bottom of a 5 ml Ultra Clear centrifugation tube.

2. Overlay with 1 ml prechilled 1.3 M and 1 ml of 0.8 M sucrose made in HKM.

3. Centrifuge at 135 000*g* for 7 h at 4 °C in a swinging bucket rotor.

4. Fractionate from top of the gradient by pipetting 0.5 ml samples.

5. Take a portion of each fraction, process for SDS-PAGE, and transfer the gel to nitrocellulose filter.

6. Immunoblot the nitrocellulose filter with antibodies against the Golgi marker protein GM130 and against vimentin (fractions that contain Golgi with associated vimentin will be in the middle of the gradient).

7. Quantify the amount of vimentin recovered with Golgi membranes by densitometry. A proportion of vimentin will fractionate with Golgi membranes (Figure 6.26).

Notes

1. This procedure can be applied to test whether a protein (or another compound) regulates the vimentin–Golgi membrane interaction [4]. The modified procedure to test whether a protein binds to Golgi membranes to promote/inhibit the vimentin–Golgi membrane interaction is listed below.

   (1) Add the protein (purified or commercially available) to isolated stacked Golgi membranes and incubate at 4 °C for 2 h.

   (2) Add cold 2 M sucrose (in HKM buffer) to the mixture to make 1.6 M sucrose and load (less than 3 ml) at the bottom of a 5 ml centrifugation tube.

   (3) Overlay the sample with 1 ml ice-cold 1.3 M and 1 ml 0.8 M sucrose in HKM.

(4) Centrifuge at 135 000*g* for 7 h at 4 °C in a swinging bucket rotor.

(5) Fractionate from top of the gradient by pipetting 0.5 ml samples.

(6) Take a portion of each fraction, process for SDS-PAGE, and transfer the gel to nitrocellulose filter.

(7) Immunoblot the nitrocellulose filter with antibodies against the Golgi marker protein GM130.

(8) Pool GM130 containing fractions – this represents purified Golgi membranes to which the tested protein has bound.

(9) Quantify protein concentration in Golgi preparation using Protein Assay Kit (Bio-Rad Laboratories) with BSA as standard.

(10) Use the membrane preparation for vimentin interaction assay as described above.

2. When incubating vimentin with the membranes, it is desirable to keep the concentration of vimentin below 0.05 μg/ml to maintain vimentin in its subunit form. Polymerized vimentin is likely to aggregate with membranes. This makes separation of membrane-bound vimentin from non-bound vimentin difficult.

3. To keep vimentin in its subunit form, vimentin stock solution should not contain more than 5 mM salt.

Procedure 2: Immunofluorescence analysis of vimentin–Golgi interaction

In cells, vimentin intermediate filaments, microtubules and actin filaments are linked together by cross-linking proteins [5]. Conventional immunofluorescence procedures cannot dissect the role of vimentin intermediate filaments in Golgi binding

(a)

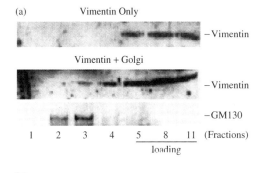

(b)

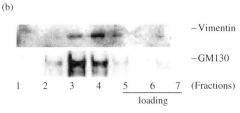

Figure 6.26 Binding of vimentin to Golgi membranes *in vitro*. (a) Recombinant vimentin alone, or recombinant vimentin incubated with purified Golgi membranes were loaded at the bottom of a sucrose density gradient and subjected to equilibrium centrifugation. Fractions were collected from the top of the gradient. An equivalent amount of each fraction was processed by SDS-PAGE, transferred to NC and immunoblotted with anti-vimentin and anti-GM130 antibodies. Vimentin remains in the load when centrifuged alone, but is recovered in a lower density fraction containing the Golgi marker GM130 when incubated with Golgi elements. (b) Fractions 3 and 4 from panel A were combined, loaded at the bottom of a sucrose density gradient and subjected to another round of equilibrium centrifugation. Fractions were collected and analysed as in (a). Vimentin remains associated with Golgi membranes. Reproduced from *The Journal of Cell Biology*, 2001, **152**(5), 877–893, by copyright permission of the Rockefeller University Press.

from the function of microtubules that interact with the Golgi at the microtubule organizing centre. To dissociate the roles of microtubules and vimentin filaments, we describe a procedure that disrupts the microtubular cytoskeleton in cells, allowing the visualization of bona fide vimentin–Golgi interactions [2].

Reagents

Antibodies against vimentin (ICN Biomedicals) and GM130 (Transduction laboratories) or antibodies against another Golgi marker protein

Blocking solution (3% normal goat serum in PBS-T) and anti-fade mounting solution (9 : 1 glycerol/PBS with 0.1% *q*-phenylenediamine)

Fluorophore conjugated secondary antibodies (Molecular Probes)

Methanol, prechilled at $-20\,^{\circ}\mathrm{C}$

PBS-T (PBS, 0.05% Tween-20)

Phosphate-buffered saline (PBS)

Nocodazole (10 mg/ml in DMSO) (Sigma-Aldrich)

Round glass coverslips, 12 mm (Fisher Scientific)

Equipment

Cell culture facility

Fluorescence imaging system

Image processing software (Photoshop or equivalent)

Procedure

Plate cells (any mammalian cell line) on glass coverslips and culture at $37\,^{\circ}\mathrm{C}$ till ∼70% confluent.

1. Make nocodazole ① at 10 μg/ml in culture medium from 10 mg/ml DMSO stock solution and replace normal medium with nocodazole medium. Incubate cells at $37\,^{\circ}\mathrm{C}$ for 5 h.

2. Rinse cells with PBS and immediately transfer coverslips into $-20\,^{\circ}\mathrm{C}$ prechilled methanol for 15 min. ②

3. Move coverslips to a humid box and wash cells with PBS-T (at room temperature) for 5 min (three times in total).

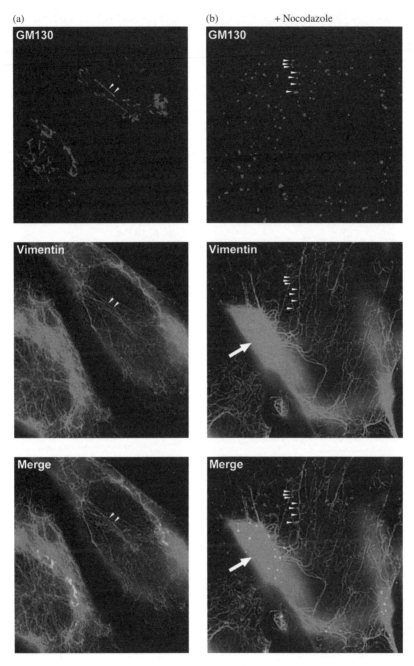

Figure 6.27 Golgi elements and vimentin filaments interact *in vivo*. MFT-6 cells were untreated (a) or treated with 10 μg/ml nocodazole for 5 h at 37 °C to disrupt microtubules (b). Cells were then labelled with polyclonal anti-GM130 and monoclonal anti-vimentin antibodies. Extended Golgi elements align with vimentin filaments in untreated cells (arrowheads in (a)). In cells with disrupted microtubules, Golgi elements associate with vimentin filaments (arrowheads in B). Arrow points to a mass of collapsed vimentin. Reproduced from *The Journal of Cell Biology*, 2001, **152**(5), 877–893, by copyright permission of the Rockefeller University Press.

4. Follow immunofluorescence procedure as described (check procedure part 1/1, *Protocol 6.26*). ③

Notes

① Nocodazole (or similar reagents) disrupt microtubules, resulting in the collapse of the majority of vimentin filaments to a peri-nuclear region. However, a fraction of peripheral vimentin filaments remains. Immunofluorescence with antibodies against Golgi proteins allows the visualization of individual Golgi fragments aligned on the vimentin filaments (Figure 6.27).

② Alternative fixatives (3% formaldehyde) and permeabilization reagents (0.07% Triton X-100 in PBS) can be used.

③ Individual filaments (or filament bundles) have weak fluorescence, and anti-vimentin antibodies must be used at high concentrations. In addition, images may need to be enhanced with software to desired brightness [6].

References

1. Fuchs, E. and Weber, K. (1994) *Ann. Rev. Biochem.*, **63**, 345–382.
2. Gao, Y.-S. and Sztul, E. (2001) *J. Cell Biol.*, **152**, 877–893.
3. Taylor, R. S., Jones, S. M., Dahl, R. H., Nordeen, M. H. and Howell, K. E. (1997) *Mol. Biol. Cell*, **8**, 1911–1931.
4. Gao, Y.-S., Vrielink, A., MacKenzie, R. and Sztul, E. (2002) *Euro. J. Cell Biol.*, **81**, 391–401.
5. Gurland, G. and Gundersen, G. G. (1995) *J. Cell Biol.*, **131**, 1275–1290.
6. Franke, W. W., Grund, C., Kuhn, C., Jackson, B. W. and Illmensee, K. (1982) *Differentiation*, **23**, 43–59.

Meinolf Thiemann and **H. Dariush Fahimi**

Introduction

The microtubular system plays a vital role for the distribution, translocation and determination of the shape of cell organelles. Peroxisomes are ubiquitous membrane-bounded organelles with essential functions in lipid metabolism and detoxification of reactive oxygen species. Their association to microtubules has been shown and the motor-dependent transport of peroxisomes along microtubules demonstrated [1–4]. The molecular mechanisms of these binding and motility events, however, are far from being understood. Hence we have developed a semi-quantitative *In vitro* peroxisome–microtubule binding assay for investigating the physicochemical characteristics of this binding process and to screen for the proteins involved [5].

The peroxisome–microtubule binding assay presented here could also be adapted to investigate the interaction of other membrane-bounded organelles with microtubules and to identify the proteins mediating it.

Reagents

MAP-free tubulin from bovine brain (Tebu, Frankfurt/Main, Germany)

Taxol (= paclitaxel) was obtained from Sigma (Taufkirchen, Germany)

(All chemicals should be of analytical grade and be prepared with high quality deionized water)

Equipment

Microcentrifuge

Shaker

Standard equipment for SDS-PAGE, Western blotting and enhanced chemiluminescence

Scanner and software for densitometric quantitation of immunoreactive protein bands

Water bath

96-Well microtiterplates (flat bottom microtiterplates with high binding capacity from Costar, Cambridge, MA, USA, proved to be well suited for the attachment of microtubules)

Procedure

The protocol of the microtubule–peroxisome binding assay is summarized in Figure 6.28.

1. Polymerize microtubules by incubating glycerol stabilized MAP-free tubulin in P/G/T-buffer (35 mM **P**ipes, 5 mM $MgSO_4$, 1 mM EGTA, 0.5 mM EDTA, 1 mM DTT, 1 mM **G**TP, 20 μM **t**axol, pH 7.4) at 37 °C for 15 min.

2. Perform the binding assay (Figure 6.28) at a constant temperature of 25 °C in 96-well microtiterplates.

3. Coat wells with microtubules by incubating 50 μl of the microtubule

**Microtubule–peroxisome
binding assay**

**Binding of peroxisomes to microtubules
immobilized to a microtiterplate**

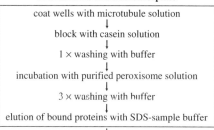

**Detection and quantitation
of peroxisomes bound to microtubules**

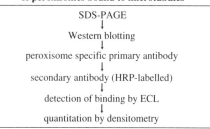

Figure 6.28 Schematic summary of the protocol of the microtubule–peroxisome binding assay which is carried out in microtiter plates. The binding of peroxisomes to microtubules is quantitated by the detection of a peroxisomal marker protein in rat liver (e.g. urate oxidase) by Western blotting

solution per well (final concentration 0.25 mg protein/ml) for 45 min. For controls take 50 μl P/G/T-buffer per well.

4. Block non-specific binding to the microtiterplate surface by 45 min incubation with 0.3 ml/well casein (2.5 mg protein/ml in P/G/T-buffer).

5. Wash once with 0.3 ml/well P/G/T-buffer.

6. Isolate purified peroxisomes by differential cell fractionation of rat liver homogenate according to a protocol described previously [5,6]. Add 40 μl of the peroxisomal preparation ① – adjusted to 0.1 mg protein/ml with homogenization buffer (250 mM sucrose, 5 mM MOPS, 0.1% ethanol,

1 mM EDTA, 0.2 mM PMSF, 1 mM 6-aminocaproic acid, 5 mM benzamidine, 10 μg/ml leupeptin, 0.2 mM DTT, pH 7.4) – to each well and incubate for 30 min.

7. Wash three times with 50 μl/well P/G/T-buffer.

8. Elute bound proteins and organelles in 20 μl SDS-sample buffer by pipetting five times up and down.

9. Perform SDS-PAGE with the samples by using 10% polyacrylamide gels.

10. Transfer proteins from the gel electrophoretically to a PVDF-membrane.

11. Perform immunocomplexing by using an antibody to a specific peroxisomal protein, e.g. urate oxidase. ②

12. Detect immunocomplexes with a horseradish peroxidase-labelled secondary antibody, employing enhanced chemiluminescence.

13. Quantitate the intensity of immunoreactive bands by densitometry.

Examples of data

In vitro binding of peroxisomes to microtubules can be demonstrated by negative staining electron microscopy [7] (Figure 6.29). Purified rat liver peroxisomes specifically associate to taxol-stabilized microtubules reconstituted from bovine brain tubulin. Binding was observed at the entire length of microtubules and only very few unbound peroxisomes were seen. Although this morphological assay employs different experimental conditions than the biochemical binding assay described here, both assay formats point out the specificity of the *In vitro* binding of peroxisomes to microtubules. The biochemical assay was saturable for both, the concentrations of peroxisomes and

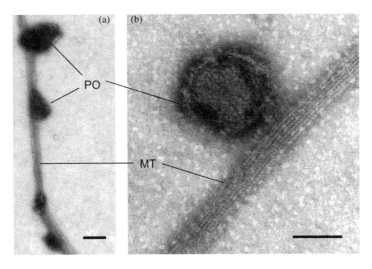

Figure 6.29 *In vitro* binding of peroxisomes to microtubules as demonstrated by negative staining electron microscopy. Peroxisomes (0.1 mg/ml) were incubated with microtubules (0.25 mg/ml) for 1 h in Eppendorf tubes. After fixation of the peroxisome–microtubule complexes with 0.25% glutaraldehyde for 5 min, a drop of this mixture was placed on a formvar-coated 200-mesh grid for 2 min. Finally, a washing step was followed by the staining of the preparation with 2% phosphotungstic acid for 3 min. Negative staining electron microscopy (EM) reveals the specificity of the association of highly purified rat liver peroxisomes (PO) to taxol-stabilized microtubules from bovine brain (MT). Bars: 0.25 μm (a), 0.05 μm (b)

microtubules (Figure 6.30). The optimal assay conditions comprised a tubulin concentration of 0.25 mg/ml and a peroxisome concentration of 0.1 mg/ml. As an example, an immunoblot incubated with an antibody to urate oxidase and serving for the quantification of peroxisomes to microtubules is shown (Figure 6.30(c)). The binding assay showed excellent reproducibility and high specificity and provided a powerful tool for the evaluation of binding conditions and the various factors involved.

Employing this assay, the binding of peroxisomes to microtubules was characterized and a CLIP-like protein was shown to be involved in the binding process [5].

Notes

This procedure will take approximately 24 h. Store tubulin and peroxisomes at −80 °C and avoid repeated thawing and freezing.

① It is essential that peroxisomes are not aggregated. To ensure the removal of peroxisomal aggregates, briefly centrifuge the peroxisomal preparation (10 s, 2000*g*) prior to use in the assay.

② An antibody to urate oxidase [8] was shown to be an excellent marker for the binding of peroxisomes to microtubules in this assay. Urate oxidase is exclusively localized to the crystalline cores of rat liver peroxisomes, which are not released from the organelle by washing procedures. Furthermore, isolated urate oxidase cores do not bind to microtubules.

References

1. Schrader, M., Burkhardt, J. K., Baumgart, E., Lüers, G., Spring, H., Völkl, A. and Fahimi, H. D. (1996) *Eur. J. Cell Biol.*, **69**, 24–35.
2. Rapp, S., Saffrich, R., Anton, M., Jäckle, W., Ansorge, W., Gorgas, K. and Just, W. (1996) *J. Cell Sci.*, **109**, 837–849.

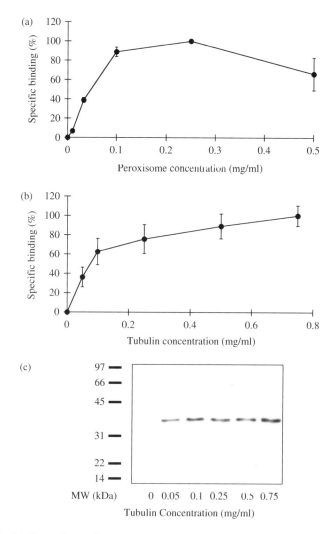

Figure 6.30 The binding of peroxisomes to microtubules depends on the concentrations of peroxisomes and tubulin. The *In vitro* binding assay was performed and binding was analysed as described. To optimize the protocol of the binding assay, the concentrations of peroxisomes (a) and tubulin (b,c) were varied. (c) As an example, the immunoblot, generated according to the standard protocol of the binding assay and incubated with an antibody to urate oxidase, is shown. Data are presented as mean ± SE from six experiments. Reproduced by permission of the *European Journal of Biochemistry* [5]

3. Wiemer, E. A. C., Wenzel, T., Deerinck, T. J., Ellisman, M. H. and Subramani, S. (1997) *J. Cell Biol.*, **136**, 71–80.

4. Schrader, M., King, S. J., Stroh, T. A. and Schroer, T. A. (2000) *J. Cell Sci.*, **113**, 3663–3671.

5. Thiemann, M., Schrader, M., Völkl, A., Baumgart, E. and Fahimi, H. D. (2000) *Eur. J. Biochem.*, **267**, 6264–6275.

6. Völkl, A., Baumgart, E. and Fahimi, H. D. (1996) In: *Subcellular Fractionation – a Practical Approach* (J. Graham and D. Rickwood, eds), pp. 143–167. Oxford University Press, Oxford, UK.

7. Schrader, M., Thiemann, M. and Fahimi, H. D. (2003) *Micr. Res. Tech.*, **11**, 171–178.

8. Völkl, A., Baumgart, E. and Fahimi, H. D. (1988) *J. Histochem. Cytochem.*, **36**, 329–336.

Detection of cytomatrix proteins by immunogold embedment-free electron microscopy

Robert Gniadecki and **Barbara Gajkowska**

Introduction

The simple view of the cell as the membrane sack containing organelles floating in the amorphous cytoplasm has over the last 30 years been replaced by a concept of a complicated three-dimensional cellular structure comprising spatially separated yet functionally interconnected compartments. The basic scaffold of most cells is afforded by the cytoskeleton (comprising actin microfilaments, intermediate filaments and the microtubules) that not only determines rigidity and motility of the cells but also provides support for the organelles.

However, our understanding of cellular structure is still incomplete. No imaging technique is alone able to visualize the intrinsic structure of the cell. Immunofluorescence microscopy with labelled antibodies against various cytoskeletal components turned out to be a very powerful technique that accounts for much of the progress that has been made in the understanding of the three-dimensional cell structure. However, the major drawbacks are low resolution and the fact that only known proteins, against which the labelled probes are available, may be investigated.

Electron microscopy seems to be ideally suited for studying cell ultrastructure. However, this technique has three major limitations:

1. The optical properties of the resins used for embedding are similar to those of proteins, hence most proteinaceous structures remain unresolved and the cytoplasm in electron microscopic images seems to be quite homogeneous.

2. In classic electron microscopy only some filaments are visualized, and only those that happen to lie at the surface of the section.

3. Aldehyde fixation, which is a necessary step during the preparation of the material, cross-links proteins and lead to the emergence of artificial structures.

Some of these limitations may be overcome by the use of the embedment-free electron microscopy (EFEM). It this technique the material is temporarily embedded in the mounting medium for cutting, but the embedding material is removed before observation. Although fixation is still necessary, no resin is present at the moment of observation in the microscope. The group of S. Penman should be credited with the development and popularization of EFEM [1, 2]. The main use of EFEM has been the study of the nuclear matrix. Penman *et al.* successfully visualized the non-chromatin nuclear matrix as a complicated three-dimensional network comprised of 10 nm 'core' filaments decorated with globular ribonucleoprotein structures and dense bodies [3]. The outline of the EFEM protocol is as follows:

1. The cells are extracted with non-ionic detergent that dissolves the membranes and allows the soluble proteins to diffuse away.

2. DNA is removed by digestion with nuclease.

3. The remaining structure (cellular scaffold or cytomatrix) is temporarily embedded in diethylene glycol distearate (DGD) and sectioned.

4. Before viewing in the electron microscope the DGD is removed so that intact scaffolds are observed.

A recent EFEM protocol may be found in Nickerson *et al.* [4]. In our laboratory we have successfully adopted this protocol for the study of diverse cell types such as normal or transformed keratinocytes and cancer cell lines (C6, U-373, MG [glioma lines], COLO 205 [colorectal adenocarcinoma], PA-1 [ovarian cancer cells]) [5–8].

The material processed for EFEM may also be used for immunostaining with gold-labelled antibodies. This provides an elegant combination of the clarity and resolution of EFEM with the power of immunocytochemistry. It is important to underline that only proteins associated with the cytomatrix will be identified. Soluble proteins will be lost during the preparation of the scaffolds and remain undetected. We have successfully used single-, double- and triple immunostaining in combination with EFEM for the study of the cellular distribution of apoptosis-related proteins bax, bid, vdac-1 and caspase-8 [5, 8].

Reagents

Chemicals

Aminoethylbenzenesulfonyl fluoride (AEBSF, Pefablok SC, Roche Applied Science)

Ammonium sulfate (Sigma)

Bovine serum albumin (Sigma)

Diethylene glycol distearate (Polysciences Inc., Warrington, PA)

DNase I, RNase free (Roche Applied Science)

Donkey serum (Sigma)

EGTA (Sigma)

Ethanol (100%)

Glutaraldehyde, EM grade, 50% stock solution (Sigma)

HMDS (hexamethyldisilazane) (Sigma)

Hydrochloric acid (HCl, 1 M)

Isobutanol (Sigma)

Magnesium chloride (Sigma)

Osmium tetroxide (Sigma)

Paraformaldehyde (Merck, Hohenbrunn, Germany)

PBS (phosphate-buffered saline, without calcium and magnesium, pH 7.2) (Sigma)

PIPES (Sigma)

Poly-L-lysine (Sigma)

Propylene oxide (Sigma)

Sodium cacodylate trihydrate, ultra pure (Sigma)

Sodium chloride (Sigma)

Sodium hydroxide (NaOH), 1 M solution

Sucrose (Sigma)

Triton X-100 (Sigma)

Tween 20 (Sigma)

Vanadyl riboside complex (VRC) (Fluka)

Water, double distilled

Consumables

Carbon-coated copper or nickel grids (can be coated with poly-L-lysine for increased adhesiveness)

Embedding capsules BEEM (LKB, USA)

Equipment

Electron microscope (e.g. JEOL 1200 EX at 80 kV)

Ultramicrotome (e.g. LKB Ultracut from Reichert, Germany) equipped with diamond knives (glass knives are reported to work for fine EFEM, but we have no experience of our own) carbon evaporator

Laboratory tabletop centrifuge with cooling

pH meter

Forceps

Pasteur and automatic pipettes

Scalpel or razor blades

1. Cell extraction

Reagents

Extraction buffer 1: 10 mM PIPES, pH6.8, 300 mM sucrose, 100 mM NaCl, 3 mM MgCl$_2$, 1 mM EGTA, 0.5% Triton X-100, 20 mM vanadyl riboside complex (VRC), 1 mM AEBSF

Remarks: PIPES is dissolved first and pH adjusted with NaOH. Then sucrose, NaCl, MgCl$_2$ and EGTA are dissolved sequentially at room temperature (RT) under constant magnetic stirring. At this stage the buffer can be sterile-filtrated and stored for a couple of weeks in a refrigerator or frozen at $-20\,^\circ$C for up to 3 months. Other components (Triton X-100, VRC and AEBSF) are added before use. Triton X-100 is added v/v with a 1 ml syringe under constant, slow stirring.

Extraction buffer 2: 10 mM PIPES, pH6.8, 250 mM ammonium sulfate, 300 mM sucrose, 3 mM MgCl$_2$, 1 mM EGTA, 0.5% Triton X-100, 20 mM VRC, 1 mM AEBSF

Remarks: Prepared as extraction buffer 1 above; in this buffer NaCl is replaced by ammonium sulfate.

DNA digestion buffer: 10 mM PIPES, pH6.8, 300 mM sucrose, 50 mM NaCl, 3 mM MgCl$_2$, 1 mM EGTA, 0.5% Triton X-100, 20 mM vanadyl riboside complex, 1 mM AEBSF, 200 units/ml RNase free DNase I.

Remarks: Prepared as extraction buffer 1. DNase I is added before use.

High salt buffer: 10 mM PIPES, pH 6.8, 300 mM sucrose, 2 M NaCl, 3 mM MgCl$_2$, 1 mM EGTA, 20 mM VRC, 1 mM AEBSF

Remarks: PIPES, sucrose, MgCl$_2$ and EGTA are added and dissolved in this sequence. pH is adjusted with NaOH. Then add NaCl to the buffer and dissolve at room temperature. VRC and AEBSF are added before use.

Procedure 1

Adherent cells may be extracted in monolayers attached to thick glass coverslips or grown in chamber slides with a glass bottom. Alternatively, the cells may be gently trypsinized or scrapped by the rubber policeman. Suspended cells are washed by centrifugation at 600 g, 5 min at 4$\,^\circ$C between steps. Coverslips with attached cells are moved between different solutions. We got best results with scrapped or trypsinized cells in suspensions. Since some loss of material is inevitable due to multiple centrifugations, start with a large amount of cells ($>$5 million). Most steps should be performed at 4$\,^\circ$C. Precool all the buffers. After the last extraction step we are left with cellular scaffolds ready for temporary embedding and sectioning.

1. Wash the cells with PBS, twice at $4\,^{\circ}C$.

2. Extract in 2 ml extraction buffer 1 at $4\,^{\circ}C$ for 5 min with occasional agitation. At this step the membranes are opened by Triton X-100 and soluble proteins diffuse from the cell to the outside.

3. Extract in 2 ml extraction buffer 2 at $4\,^{\circ}C$ for 5 min with occasional agitation. Ammonium sulphate in this buffer will extract some of the histones (mainly histone 1) and many cytoplasmic proteins [4].

4. Incubate in the DNA digestion buffer for 1 h at $32\,^{\circ}C$. Stir gently once every 7–10 min. At this step DNA is digested and removed from the nuclei. According to Nickerson *et al.* [4] the completeness of DNA removal should be monitored in pilot experiments by e.g. assessment of $[^{3}H]$thymidine release. This is probably necessary in the experiments where ultimate control is needed during the preservation of the structure of nuclear matrix. We found that satisfactory results are obtained for a wide range of different cells with standard conditions.

5. Extract in the high salt buffer at $4\,^{\circ}C$ for 5 min. High salt concentration removes the histones and most of the high mobility group proteins from the nucleus and stabilizes the cellular scaffolding.

2. Preparation for electron microscopy

Reagents

Buffered glutaraldehyde: 2.5% glutaraldehyde in extraction buffer 1
Remarks: Dissolve glutaraldehyde from the stock solution in extraction buffer 1, shortly before use. Keep in refrigerator at $4\,^{\circ}C$. Fixative older than 6 h should be discarded.

Sodium cacodylate buffer: 0.1 M sodium cacodylate, pH 7.2
Remarks: Prepare stock 0.2 M solution of sodium cacodylate in distilled water and adjust pH with 0.2 M HCl. The solution is stable for several months when stored at $4\,^{\circ}C$. Before use dilute 1 : 1 (vol/vol) with distilled water.

Buffered osmium tetroxide: 1% osmium tetroxide in cacodylate buffer
Remarks: Mix osmium tetroxide stock solution with 0.2 M sodium cacodylate stock solution and dilute with distilled water to bring the concentration of sodium cacodylate to 0.1 M. The solution is stable for several months at $4\,^{\circ}C$.

Procedure 2

1. Fixation. We fix routinely in 2.5% buffered glutaraldehyde at $4\,^{\circ}C$ for 40 min. Remember that cell pellet should ideally be approximately 1 mm^3 for further embedding.

2. Wash the pellet or cells on the coverslip with sodium cacodylate buffer for 5 min, $4\,^{\circ}C$.

3. Post-fix in buffered osmium tetroxide for 30 min at $4\,^{\circ}C$.

4. Wash in cacodylate buffer for 30 min.

5. Dehydrate the material by transfer through the ethanol series: 50, 70, 80, 90, 96% two changes for 5 min each. End with three changes of dehydrated 100% ethanol for 5 min each. At this step the material may be left overnight at $4\,^{\circ}C$. After transfer through 80% ethanol the pellets are hard enough to be transferred to BEEM capsules.

6. Transfer the cells to a 1 : 1 (v/v) mixture of ethanol : isobutanol for 5 min at room temperature, and then to pure isobutanol, twice for 5 min each time. Place the samples in the oven at $60\,^{\circ}C$ and allow the solvents

to evaporate (15 min). The aim is to remove the rest of ethanol, which is not miscible with DGD.

7. Prewarm a mixture of 1 : 1 (v/v) DGD : isobutanol in the oven at 60 °C and pour over the cells. Leave the samples in the oven until butanol evaporates (approximately 30 min) and infiltrate the cells with molten DGD (60 °C) twice, 60 min each.

8. Place the samples at room temperature to solidify. For cells on coverslips, peel the coverslip from DGD; the cells will remain embedded in DGD as a monolayer. For samples in the capsules, cut away the capsule and trim the block to a surface of 1–2 mm.

9. Cut with glass or diamond knives at angle of 10° on the microtome. Thickness between 0.1 and 1 μm can be obtained, but we had best results with 0.3 μm sections. The interference colour of the sections is the same as for Epon sections.

10. Collect the sections on the carbon-coated copper grinds that have additionally been coated with poly-L-lysine.

11. Remove DGD by immersing the grids for 1 h in toluene at room temperature.

12. Transfer the grids to a 1 : 1 (v/v) solution of ethanol : toluene for 5 min, and then to three changes of 100% ethanol (each 10 min). At this step the sections may be processed for immunogold staining, as described in Procedure 3 below. If immunocytochemistry is not required, immerse the grids for 5 min in a 1 : 1 (v/v) mixture ethanol : HMDS followed by three changes of HMDS, 10 min each. Air-dry.

13. The sections may optionally be carbon-coated to increase the robustness of the samples.

3. Staining with antibodies

Reagents

Buffered paraformaldehyde/glutaraldehyde: 3.7% paraformaldehyde and 0.1% glutaraldehyde in extraction buffer 1

Comment: Prepared freshly, stored at 4 °C for up to 3 h.

PTA buffer (PBS-Tween-albumin): 0.05% (v/v) Tween 20, 0.1% (w/v) bovine serum albumin in PBS

Comments: Place a vessel with PBS on a magnetic stirrer and dissolve the required amount of Tween 20. Switch the stirrer off, sprinkle albumin pulver on the fluid surface, switch the stirrer on and stir slowly until albumin is dissolved (about 1 h). Sterile filtrate. PTA buffer can be stored at 4 °C for a week or frozen for several months.

Glutaraldehyde cacodylate buffer: 2.5% glutaraldehyde dissolved in 0.1 M sodium cacodylate buffer, pH 7.2

Buffered osmium tetroxide: prepared as for section 2, above

Antibodies

Primary mouse, rabbit and goat antibodies have been used. An antibody that worked for the post-embedding immunogold electron microscopy is also likely to work for EFEM. We used the following secondary, gold-labelled antibodies: donkey anti-rabbit (4 nm gold), donkey anti-goat (12 nm), donkey anti-mouse (20 and 30 nm) (Jackson ImmunoResearch, West Grove, PA). We dilute antibodies in PTA buffer or in PBS. PTA buffer gives lower background staining.

Procedure 3

We have developed the following post-embedding protocol, where the sections are stained after DGD embedding. An alternative pre-embedding protocol can be found in ref. 4. Post-embedding staining gives results with a very low background, but some epitopes may be masked during prior embedding steps.

1. Cell extraction: performed as above in Procedure 1 (steps 1–5).

2. Fixation. Pure glutaraldehyde is a heavily cross-linking agent and may mask some epitopes. If problems are encountered we had good results after fixing with buffered paraformaldehyde/glutaraldehyde for 3 h at 4 °C. This solution should be prepared freshly from the EM-grade reagents. Osmium post-fixation (steps 7–9, Procedure 2) is omitted.

3. Embedding in DGD, sectioning, removal of DGD: performed as above in section 2, steps 10–17.

4. Rehydrate the material by immersing in a series of alcohols: 100, 96, 90, 80, 70, 50, 30%, 2 × 5 min at room temperature (RT).

5. Rinse the grids with the material by floating them on PBS for 10 min followed by PTA buffer 10 min, RT. All incubation/washing steps are performed by placing the grids on a large drop of fluid placed on Parafilm or similar hydrophobic membrane, specimen side down.

6. Block with 5% normal donkey serum in PTA buffer for 10 min, RT.

7. Incubate with the first antibody. Depending on the type of antibody, incubation times from 2 h to overnight are appropriate. For short times incubation at 37 °C rather than at 4 °C gives better results. All antibody dilutions are made in PTA buffer.

8. Rinse in PTA buffer three times, 5 min each, RT.

9. If double or triple staining is performed incubate with additional primary antibodies.

10. Rinse in PTA buffer three times, 5 min each at RT, followed by additional blocking with 5% normal donkey serum in PTA buffer for 10 min.

11. Incubate with gold-labelled secondary donkey antibodies diluted in PTA. Usually 1–2 h at 37 °C would suffice.

12. Rinse in PBS, 3 × 10 min, RT.

13. Rinse in sodium cacodylate buffer, three times, 5 min, RT.

14. Fixed in glutaraldehyde cacodylate buffer for 1 h at 37 °C.

15. Post-fix in buffered osmium tetroxide for 30 min followed by a rinse for 30 min in sodium cacodylate buffer at RT.

16. Dehydrate in ethanol 50–100% as described above (Procedure 2, step 5).

17. Transfer to the mixture 1:1 (v/v) 100% ethanol : HMDS for 10 min followed by three changes of HMDS, 10 min each. Air-dry.

18. The grids containing the dried samples can be lightly carbon coated in a Kinlay carbon evaporator and stored in a dessicator.

Results

Images of cell scaffolds after cell extraction and EFEM (Procedures 1 and 2) should reveal a continuous meshwork of filaments in the cytoplasmic and nuclear areas (Figure 6.31). The thickness and arrangement of filaments depend on cell type and state of differentiation. For example in C6 glioma the cell line cytomatrix comprises 10–15 nm branched fibrils, 5 nm microfilaments and short

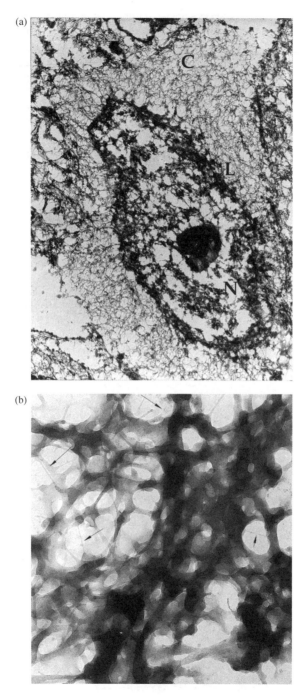

Figure 6.31 EFEM of extracted human keratinocyte showing cytomatrix and nuclear matrix. (a) Low magnifications with a meshwork of filaments in the cytoplasmic field (C), nuclear lamina (L) and the coarse nuclear matrix (N). (b) High magnification showing filaments of different diameter in the cytomatrix. Arrows: ultrathin 1 nm filaments

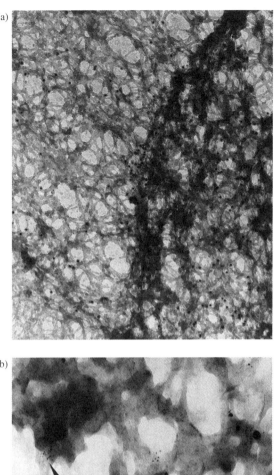

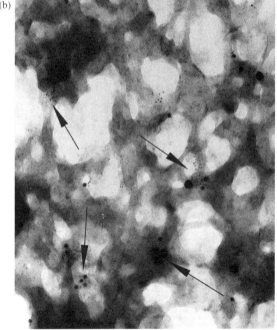

Figure 6.32 Post-embedding immunogold cytochemistry on extracted COLO 20 cells. (a) Anti-Bax staining during apoptosis showing concentration of Bax at the nuclear lamin and in the nucleus. (b) Triple immunogold labelling (anti-Bax: 4 nm gold particles, anti-Bid: 12 nm gold, anti-VDAC-1: 20 nm gold) of the nuclear matrix of COLO 25 cell during apoptosis. Note clusters of gold particles of different sizes (arrows)

(40–200 nm) very thin (1–3 nm) fibrils [6]. A similar structure of the cytomatrix was seen in proliferating human keratinocytes, but during cell differentiation the 5 and 1 nm filaments disappeared and were replaced by a tight meshwork of 15 nm fibrils [7]. The nuclear matrix consists of a meshwork of 10–30 nm filaments attached to a compact nuclear lamina. Ultrathin 1 nm filaments are also present in the nuclear matrix. The globular portion of the nuclear matrix corresponds to ribonucleoprotein.

Examples of post-embedding immunogold labelling of extracted matrices are shown in Figure 6.32. Details of antibody concentration and incubation tomes may be found in our recent publications [5, 8]. Labelling with up to three antibodies is feasible.

References

1. Capco, D. G., Krochmalnic, G. and Penman, S. (1984) A new method of preparing embedment-free sections for transmission electron microscopy: applications to the cytoskeletal framework and other three-dimensional networks. *J. Cell Biol.*, **98**, 1878–1885.
2. Penman, S. (1995) Rethinking cell structure. *Proc. Nat. Acad. Sci. USA*, **92**, 5251–5257.
3. Fey, E. G., Krochmalnic, G. and Penman S. (1986) The nonchromatin substructures of the nucleus: the ribonucleoprotein (RNP)-containing and RNP-depleted matrices analyzed by sequential fractionation and resinless section electron microscopy. *J. Cell Biol.*, **102**, 1654–1665.
4. Nickerson, J. A., Krochmalnic, G. and Penman, S. (1998) Isolation and visualization of the nuclear matrix, the nonchromatin structure of the nucleus. In: *Cell Biology: A Laboratory Handbook* (J. E. Celis, ed.), 2nd edn, vol. 2, pp. 194–192. Academic Press.
5. Gajkowska, B., Motyl, T., Olszewska-Bądarczuk, H., Gniadecki, R. and Koronkiewicz, M. (2000a) Structural association of Bax with nuclear matrix and cytomatrix revealed by embedment-free immunogold electron microscopy. *Cell Biol. Int.*, **24**, 649–656.
6. Gajkowska, B., Cholewinski, M. and Gniadecki, R. (2000b) Structure of cytomatrix and nuclear matrix revealed by embedment-free electron microscopy. *Acta Neurobiol. Exp.*, **60**, 147–158.
7. Gniadecki, R., Olszewska, H. and Gajkowska, B. (2001) Changes in the ultrastructure of cytoskeleton and nuclear matrix during HaCaT keratinocyte differentiation. *Exp. Dermatol.*, **10**, 71–79.
8. Gajkowska, B. and Wojewódzka, U. (2002) A novel immunoelectron embedment free electron microscopy reveals association of apoptosis-regulating proteins with subcellular structures. *Histochem. J.*, **34**, 441–446.

Tubulin assembly induced by taxol and other microtubule assembly promoters

Susan L. Bane

Introduction

Substances that affect microtubule assembly and dynamics have important therapeutic applications, particularly in cancer chemotherapy. Taxol was the first example of an effector molecule of microtubule dynamics that promotes rather than inhibits *in vitro* microtubule polymerization. The success of Taxol as an anticancer agent has sparked a tremendous interest in the discovery of other molecules that promote microtubule assembly [1].

Microtubule assembly promoters (MT-APs) generally are screened by assessing the ability of the molecule to promote microtubule assembly *in vitro*. Activity is most commonly assessed as an I_{50} value, which is the concentration of MTAP required to promote tubulin to assemble to 50% of its maximum. This is a useful and easily obtainable value; however, some care should be taken in its interpretation. ①

Three procedures follow:

Procedure 1. *Tubulin assembly induced by Taxol*. This is the basic procedure for measuring tubulin polymerization by assembly-promoting molecules such as Taxol. It can also be used to prepare Taxol-microtubules for other studies.

Procedure 2. *Potency measurements of MTAPs using absorption spectroscopy*. The basic procedure for Taxol assembly is applied to potency measurements of new substances. Absorption spectroscopy is convenient when a small number of molecules are to be screened.

Procedure 3. *Potency measurements of MTAPs using fluorescence spectroscopy*. Procedures for measuring potency are performed in a 96-well plate [2]. This procedure is well suited to screening a larger number of molecules.

Reagents

Purified tubulin, free of microtubule-associated proteins ② ③

PMEG buffer: 0.1 M PIPES, 1 mM $MgSO_4$, 2 mM EGTA, 0.1 mM GTP, pH 6.9. The buffer must be kept cold ($4\,^\circ C$) to prevent GTP hydrolysis. GTP-containing buffers are discarded after 2 weeks ④

Taxol, which is available from several commercial sources

Sephadex G-50 in PMEG buffer

Dimethylsulfoxide (DMSO), reagent grade or better

MTAPs to be tested (for Procedures 2 and 3)

$4', 6$-diamidino-2-phenylindole (DAPI) (for Procedure 3)

Equipment

Gel filtration column and fraction collector

Procedures 1 and 2: Absorption spectro-photometer with thermostatted cell holder (Multicell transporter desirable)

Quartz cells for spectrophotometer. A 0.4×1.0 cm cell requires about 0.5 ml solution for each polymerization in a typical instrument

Procedure 3: Plate reader capable of monitoring emission at 450 nm (excitation at 360 nm).

96-well plates

Procedures

Procedure 1 Taxol-induced tubulin assembly

1. Prepare a stock solution of Taxol of approximately 1 mM in DMSO.

2. Equilibrate tubulin in cold PMEG buffer using Sephadex G-50. ⑤

3. Set up absorption spectrophotometer to measure absorption at 350 nm for 15 min at a cuvette temperature of 37 °C.

4. Add buffer followed by tubulin to the cuvette to yield a tubulin concentration of 10 μM in a total volume of 0.5 ml. Equilibrate solution to 37 °C and begin measuring absorption at 350 nm.

5. Add an aliquot of the Taxol solution to yield a final Taxol concentration of 10 μM (and a DMSO concentration of ≤3% vol/vol ⑥), mix rapidly and gently, taking care not to introduce bubbles into the cuvette.

6. Collect data until a plateau is reached.

Procedure 2 Potency of MTAPs using an absorption spectrometer

1. Prepare 1 mM stock solution of Taxol in DMSO and stock solution(s) of the MTAP(s). ⑦

2. Prepare tubulin and absorption spectrometer as above. The rest of this protocol assumes that the absorption instrument is capable of measuring six cuvettes simultaneously.

3. Add PMEG buffer followed by tubulin to absorption cell to yield a tubulin concentration of 10 μM in a total volume of 0.5 ml. Equilibrate the solutions to 37 °C and begin measuring absorption at 350 nm.

4. To the first cuvette add an aliquot of the Taxol solution to yield a final Taxol concentration of 10 μM. To the other five cuvettes, add aliquots of the MTAP to yield the chosen final concentration of MTAP. ⑥ Mix each cuvette rapidly and gently, taking care not to introduce bubbles into the solution.

5. Continue measuring the absorption of the solution with time until a steady state is reached in the slowest sample.

6. Calculate the relative extent of assembly by subtracting the absorption value before addition of Taxol or MTAP from the absorption value of the plateau for each sample.

7. Determine the fraction assembled in each sample by dividing the change in absorption for the Taxol sample into the change of absorption of the other samples.

8. Plot the fraction assembled versus MTAP concentration and determine the I_{50}.

Procedure 3 Potency of MTAPs using a multi-well plate and a fluorescence plate reader

1. Prepare 1 mM stock solution of Taxol in DMSO and stock solution(s) of the MTAP(s) in DMSO. ⑦

2. Prepare a 1 mM stock solution of DAPI in water. ⑧

3. Equilibrate tubulin in cold PMEG buffer using Sephadex G-50. (5)

4. Prepare the tubulin-DAPI solution for pipetting into the plate. The final concentration of DAPI and tubulin in each well will be 10 and 5 μM, respectively. (9) Mix DAPI and buffer first, then add tubulin and mix gently. Add 194 μl of this solution to each well.

5. Add the MTAP in DMSO (6 μl) to a produce a final volume of 200 μl. Prepare duplicates of each MTAP concentration. Quickly mix the solutions in the wells using a multi-channel pipettor, taking care to avoid introduction of bubbles into the wells.

6. Also prepare duplicates of two types of controls: a 'no assembly' control, which contains 5 μM tubulin, 10 μM DAPI, 6 μl of DMSO but no MTAP, and a 'complete assembly' control, which contains 5 μM tubulin, 10 μM DAPI and 10 μM Taxol.

7. Initiate the plate reader (excitation = 360 nm, emission = 450 nm (10)) and collect data until a plateau is reached in the wells. (11)

8. Subtract the average fluorescence value for the 'no assembly' wells from the average of the two fluorescence values of each plateau.

9. Divide the average fluorescence value for the 'complete assembly' control into the average fluorescence value for each of the MTAP concentrations to determine the fraction assembled.

10. Plot fraction assembled versus concentration of MTAP and determine the I_{50} value.

Examples of data

Figure 6.33 shows an example of **Procedure 2** in which tubulin assembly is induced by increasing concentrations of baccatin III, an MTAP. The arrow indicates the point at which the baseline

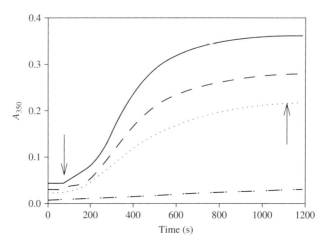

Figure 6.33 Tubulin assembly monitored on an absorption spectrometer according to **Procedure 2**. Assembly was initiated after a baseline was collected for about 75 s (arrow). The absorption value for the baseline is subtracted from the absorption value of the plateau (arrow) to evaluate the extent of assembly. In this illustration, assembly of 10 μM tubulin was induced by baccatin III at concentrations of 6 μM (solid curve), 3 μM (dashed curve) and 1.5 μM (dotted curve). A control containing no baccatin III is shown in the dot–dash curve. The extent of assembly of the Taxol control is not shown in this illustration

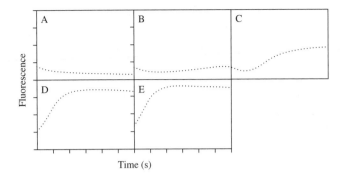

Figure 6.34 Tubulin assembly in a 96-well plate monitored by fluorescence according to **Procedure 3**. Fluorescence was collected over a period of 100 min at a temperature of 32 °C. The *x* and *y* scales of each panel are equal. Panel A is the 'no assembly' control, containing 5 μM tubulin without MTAP. The concentration of MTAP in panels B–E was 0.5, 2.0, 4.0 and 5.0 μM, respectively. The 'complete assembly' control and duplicates of each well were also run (data not shown)

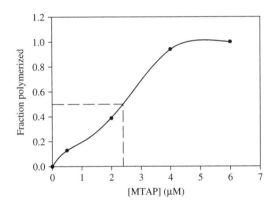

Figure 6.35 The I_{50} for the MTAP tested in Figure 6.36 is determined graphically to be 2.3 μM

was taken, and the second arrow indicates where the absorption reading for the plateau was taken.

Figure 6.34 shows an example of **Procedure 3** in which tubulin assembly is induced by increasing concentrations of a Taxol analogue.

Figure 6.35 shows a plot of the fraction of tubulin assembled as a function of MTAP concentration and the I_{50} value.

Notes

① Assembly promotion is a function of both the affinity of the MTAP for the receptor site (K_a) and the critical concentration of tubulin in the presence of saturating concentrations of the MTAP (C_{crit}). A structural alteration of Taxol can affect its K_a, C_{crit} or both parameters. For more detailed discussions, see refs 3 and 4.

② If only a few measurements are anticipated, the protein can be purchased from Cytoskeleton, Inc. If many measurements are to be done, it is more cost effective to purify the protein from mammalian brain, described in ref. 5.

③ These assays can also be performed using microtubule protein, which is tubulin plus microtubule-associated proteins (MAPs). MAPs lower the critical concentration of tubulin. Therefore, if microtubule protein is used, a lower protein concentration may be necessary to prevent polymerization of the 'no assembly' controls.

④ It is convenient to prepare stock solutions of the individual components: 0.4 M PIPES, pH 6.9, prepared from the free acid and 5 M NaOH; 100 mM EGTA, pH 6.9, prepared from the free acid and 1 M NaOH; 50 mM MgSO₄, and 50 mM GTP. PIPES stock is stored at 4 °C, GTP stock is stored at −20 °C or below, MgSO₄ and EGTA stocks are stored at 25 °C.

⑤ If a low-speed centrifuge is available, tubulin can be quickly equilibrated into PMEG buffer by the method of Penefsky [6], in which 1 ml syringes are used instead of a single column.

⑥ It is known that DMSO induces tubulin to assemble at concentrations ≥8% vol/vol [7]. We have found that DMSO at a concentration of ≤3% does not affect tubulin polymerization under these assay conditions.

⑦ We find it convenient to make serial dilutions of the stock solutions so that an equal volume of DMSO is added to each well.

⑧ This solution may be frozen in small aliquots for storage at −20 °C. The DAPI solutions are not subjected to a second freeze–thaw cycle.

⑨ We use a lower tubulin concentration in the 96-well plates than in the absorption spectrometer to decrease the mass of tubulin consumed in each experiment.

⑩ Appropriate cutoff filters can be used if the plate reader lacks monochrometers.

⑪ A plate reader equipped with a temperature-controlling unit is ideal. In the absence of such an accessory it is important to maintain a constant temperature in the 96-well plate. Lower temperatures will decrease the rates of the assembly reactions.

References

1. Altman, K.-H. (2001) *Curr. Opin. Cell Biol.*, **5**, 424–431.
2. Barron, D. M., Chatterjee, S. K., Ravindra, R., Roof, R., Baloglu, E., Kingston, D. G. I. and Bane, S. (2003) *Anal. Biochem.*, **315**, 49–56.
3. Chatterjee, S. K., Barron, D. M., Vos, S. and Bane, S. (2001) *Biochemistry*, **40**, 6964–6970.
4. Andreu, J. M. and Barasoain, I. (2001) *Biochemistry*, **40**, 11 975–11 984.
5. Williams, R. C. Jr and Lee, J. C. (1982) *Methods Enzymol.*, **85**, 376–385.
6. Penefsky, H. S. (1979) *Methods Enzymol.*, **56**, 527–530.
7. Robinson, J. and Engelborghs, Y. (1982) *J. Biol. Chem.*, **257**, 5367–5371.

Vimentin production, purification, assembly and study by EPR

John F. Hess, John C. Voss and **Paul G. FitzGerald**

Introduction

The hallmark of all intermediate filament (IF) proteins is the central rod domain, a region long predicted to form an alpha helical coiled coil. Although full-length IF proteins have not been crystallized, the alpha helical coiled nature of subdomains within the central rod has been confirmed by crystallization of vimentin fragments [1, 2] and EPR spectroscopy [3]. We use the following methods to produce, purify, spin label and then assemble filaments for EPR studies of vimentin structure and assembly. The process of spin labelling a protein requires that a single cysteine be substituted at the position where the spin label is required. Therefore, we have removed the single endogenous cysteine present in human vimentin and substituted cysteines where required. All of our constructs contain a serine at position 328, the original location of the endogenous cysteine. Thus, vimentin cys[342] contains a cysteine at position 342 and a serine at position 328.

Procedures

A. Production of vimentin in E. coli

Expression of vimentin in *E. coli* is performed using a pET7 expression plasmid generously provided by Roy Quinlan (Durham University, UK). This construct is introduced into any of several *E. coli* BL21 (DE3) strains for IPTG induction of vimentin expression followed by isolation and purification of IBs (inclusion bodies).

Reagents and equipment

For bacterial growth:

$37\,^\circ$C Shaking incubator for liquid culture

$37\,^\circ$C Incubator for bacterial plates

Bacterial culture media including antibiotics

IPTG (isopropyl-B-D-galactopyranoside)

For preparation of bacterial lysates and gel electrophoresis:

Microcentrifuge

SDS gels and apparatus for electrophoresis

Protein standards

Centrifuge and rotors

Procedure

General plasmid transformation and bacterial growth protocols can be found in references such as *Molecular Cloning* [4]. Protocols related to bacterial induction using IPTG can be obtained from companies such as Novagen, Invitrogen, Stratagene and others.

1. Transform competent *E. coli* BL21 (DE3) (Novagen, Invitrogen or other

suppliers), with plasmid DNA and plate on LB plates containing 100 μg of ampicillin (amp)/ml.

2. Pick three or four colonies and grow in 2 ml of LB-amp until slightly cloudy (OD 600 ~0.3–0.4, usually ~4 h). Then 1 ml of each culture is removed and transferred to a fresh tube and 1 μl of 1 M IPTG is added. The original tube, with 1 ml remaining, is retained for use as an uninduced control. The IPTG induced culture is returned to a shaking incubator for 4 h. Following induction, 500 μl of both induced and uninduced bacteria are harvested by low-speed centrifugation (5000 rpm for 2 min) and the supernatants discarded. Then 100 μl of TE (10 mM Tris pH 8, 1 mM EDTA) is added and the bacteria resuspended by vigorous vortexing. Then 35 μl of 4× SDS-PAGE gel loading dye is added and the solution gently mixed; 15 μl of this bacterial lysate is sufficient to run on a typical SDS-PAGE minigel. If the solution is too viscous to pipette, a 1 ml syringe fitted with a 25-gauge needle can be used to shear the bacterial genomic DNA. Coomassie blue staining and destaining of the gel will reveal which colony (or colonies) provided the best expression. From the original uninduced culture, the best expressing colony should be streaked out onto a fresh LB-amp plate.

3. Day 2: select a single isolated colony and grow in 11 ml of LB-amp. This culture is grown for 6–8 h, until very cloudy, and two glycerol stock tubes are prepared by mixing 500 μl aliquots of bacteria with an equal volume of sterile 50% glycerol and stored at −80 °C. The remainder of the culture (~10 ml) is stored at 4 °C overnight.

4. Day 3: use the entire 10 ml culture from the refrigerator to inoculate 500 ml of

LB-amp supplemented with 10 ml of 20% glucose. When the culture reaches an A_{600} of 0.8, IPTG is added to 0.5 mM. Induction is allowed to proceed for 6 h, and the bacteria harvested by centrifugation in a swinging bucket rotor at 4000 rpm for 15 min. The use of a swinging bucket rotor produces a thin pellet of bacteria on the bottom of the 250 ml bottle, instead of a thick pellet in the corner of a fixed angle rotor. This pellet geometry aids resuspension of the bacteria; complete and thorough resuspension is essential for a clean inclusion body preparation. Following centrifugation, the bacterial pellet is frozen overnight.

B. Isolation and purification of inclusion bodies (IBs) [3, 5]

Reagents and equipment

Centrifuge and rotors (for harvesting bacterial cultures and washing of IBs, e.g. Sorvall RC5C, HS4 rotor, SS34 rotor)
Lysozyme
DNase I
RNase A

Procedure

1. Thaw bacterial pellets in centrifuge bottles by the addition of a GET buffer (50 mM glucose, 25 mM Tris pH 8, 10 mM EDTA) containing 10 mg/ml egg white lysozyme (Sigma L-6876), 10 ml per centrifuge bottle (20 ml per 500 ml culture). The bottle is vigorously vortexed to resuspend the pellet and create a homogeneous suspension. The cell suspension is transferred to a disposable 50 ml screw-cap tube using a Pasteur pipette. Any remaining clumps are broken up by rapid pipetting.

2. Incubate the bacterial suspension in a 37 °C water bath for ~15–20 min.

During this time, the suspension should change from a homogeneous brown/grey/beige colour into a more clumped suspension. This indicates completion of the lysozyme digestion and beginning of bacterial lysis.

3. Produce complete lysis by the addition of an equal volume of 20 mM Tris pH 7.5, 0.2 M NaCl, 1 mM EDTA 1% deoxycholic acid, 1%NP-40. The bacterial solution and the lysis buffer should be gently rocked back and forth several times over several minutes to mix the solutions; the solution should rapidly turn viscous (more of a gelatinous blob than a suspension). This indicates lysis and release of bacterial genomic DNA.

4. Add magnesium chloride (1 M stock), DNase I (Sigma D-5025, stock solution 10 mg/ml) and RNase A (Sigma R4875), 10 mg/ml stock solution) to final concentrations of 10 mM ($MgCl_2$) and 10 µg/ml (RNase and DNase). The bacterial lysate is again mixed by rocking back and forth, then incubated at 37 °C. Over the course of 15–30 min, digestion of the bacterial DNA by DNase I will convert the solution from a viscous gel into a watery yellowish solution with an off-white precipitate at the bottom of the tube. Mixing of the tube will form a homogeneous solution, but the IBs will again sediment to the bottom.

5. When digestion is complete, as evidenced by a watery, non-viscous consistency, transfer the solution (~20 ml) to a round bottom centrifuge tube and centrifuge in an SS34 rotor at 6500 rpm (10 000g) for 10 min. IBs form a large white/tan coloured pellet; the supernatant is discarded. (The pellet should be firm, and the supernatant can be poured off without danger of losing the pellet.) Each tube, corresponding to 250 ml of original culture, is sequentially washed with 10 ml each of no salt, low salt, high salt, and no salt buffers.

6. Add 10 ml of 0.5% TritonX-100 (TX-100), 1 mM EDTA to the tube and resuspend the IBs by pipetting. IBs are then pelleted by centrifugation as before.

7. Add 10 ml of a low salt solution (10 mM Tris pH 8, 5 mM EDTA, 0.15 KCl, 0.5% TX-100) to the pellet and resuspend the pellet by pipetting. Careful resuspension and disruption of clumps are essential to prepare inclusions free of major contaminants. Collect IBs by centrifugation.

8. Wash IBs with 10 ml of a high salt solution (10 mM Tris pH 8, 5 mM EDTA, 1.5 M KCl, 0.5% TX-100) and collect by centrifugation.

9. Wash IBs with 10 ml 0.5% TX-100, 1 mM EDTA and collect by centrifugation.

10. Resuspend each tube of IBs in 4 ml (8 ml total volume for 500 ml culture) of 20 mM Tris pH 8, 1 mM EDTA, 8 M urea. Pipette the solution up and down; the IBs should dissolve and yield a slightly yellow solution that is not viscous. If viscous, the DNase digestion was incomplete. Small brown clumps with a translucent halo around them are clumps of bacteria that were not resuspended.

C. Purification and preparation of spin labelled vimentin

Reagents and equipment

FPLC, e.g. Pharmacia

Source S column

Spin label compound (O-87500, (1-oxyl-2,2,5,5,-tetramethyl-Δ3-pyrroline-3-

methyl) methanethiosulfonate [MTSL], Toronto Research Chemicals, Toronto, Canada)

Superose column

TCEP (tris-(2-carboxylethyl) phosphine, Molecular Probes, Eugene, OR)

Procedure

1. Dissolve IBs (from 500 ml of bacterial culture) in 8 ml of 8 M urea (20 mM Tris pH 8, 1 mM EDTA, 8 M urea) and filter through a 0.2 micron filter (Pall Serum Acrodisc, Fisher Scientific).

2. Chromatograph 4 ml of inclusion body solution on a Hi Load 16/60 Superdex 200 column. The column is run with 20% buffer B, giving conditions of 20 mM Tris, pH 8, 1 mM EDTA, 0.2 M NaCl. Electrophorese column fractions on an SDS-PAG and visualize the proteins by Coomassie blue staining. Pool peak fractions.

3. Desalt the pooled vimentin peak by chromatography over a High-Prep26/10 desalting column.

4. Concentrate the desalted vimentin by chromatography over a Source 15 S column. The peak, typically 2.0 ml, is used for spin labelling.

5. Add TCEP (100 mM stock in H_2O) to a final concentration of 100 μM and incubate at room temperature for 30 min.

6. Spin label the reduced vimentin by addition of spin label to a final concentration of 500 μM; continue incubation for 1 h.

7. After spin labelling, add 8 ml of buffer A (8 M urea, 20 mM tris, 1 mM EDTA, pH 8.0) and chromatograph the spin labelled protein over the Source 15S as before. Collect the purified peak and store at $-80\,^{\circ}$C.

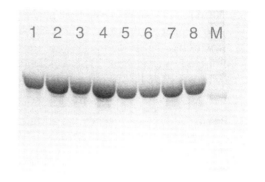

Figure 6.36 Representative spin-labelled vimentin samples. Samples of Source S fractions 21 and 22 of spin-labelled vimentin mutants cys[342], cys[345], cys[346] and cys[349] are shown following electrophoresis on SDS-PAGE and staining with Coomassie Blue. In each pair of lanes, fraction 21 is first, followed by fraction 22; Lanes 1, 2: vimentin cys[342]; lanes 3,4, vimentin cys[345]; lanes 5,6, vimentin cys[346]; lanes 7,8, vimentin cys[349]. Markers in lane M are Benchmark protein standards from Invitrogen. Vimentin migrates between the 50 and 60 kD bands

Figure 6.36 shows the results of the above purification/labelling scheme, for four separate vimentin mutants.

D.   Assembly of intermediate filaments

Equipment

Dialysis tubing and clips, e.g. Spectra/Por 6 regenerated cellulose, 10 000 molecular weight cut-off dialysis tubing (Fisher Scientific)

Procedure

1. Assemble purified vimentin in 8 M urea into filaments by dialysis against buffers without urea. Single-step dialysis can be performed using 20 mM Tris pH 7.5 and either 160 mM KCl or NaCl [6, 7]. Dialysis is performed at room temperature, overnight. For EPR studies, protein concentrations >1 mg/ml are used. For observation of filaments by electron microscopy, protein concentrations

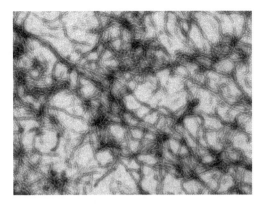

Figure 6.37 Intermediate filaments assembled from spin-labelled vimentin 342C. Spin-labelled vimentin was assembled by dialysis against 20 mM Tris pH 7.5, 160 mM NaCl, overnight followed by negative staining and visualization by EM

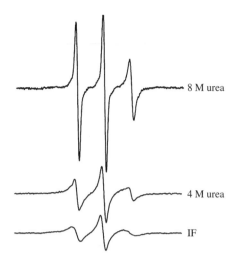

Figure 6.38 Normalized EPR spectra from spin-labelled vimentin cys^{342}. Spectra were collected from monomers (8 M urea), dimers (4 M urea) and filaments (20 mM Tris pH 7.5, 160 mM NaCl)

of 0.2–0.5 mg/ml are used. Figure 6.37 shows an example of filaments assembled from vimentin cys^{342}.

2. If EPR spectra are to be recorded at multiple steps during vimentin assembly, dialysis can be performed in a stepwise fashion [8]. Starting conditions are 20 mM Tris pH 8.0, 1 mM EDTA, 8 M urea (buffer A + 8 M urea). Vimentin dimers can be produced by dialysis against buffer A + 4 M urea. Vimentin tetramers can be produced by dialysis against 5 mM Tris pH 8 [9]. Filaments can be assembled from any of these intermediate steps by dialysis against assembly buffer, 20 mM Tris pH 7.5, 160 mM NaCl. Perform dialysis for 6–8 h, followed by overnight dialysis, for filament formation.

EPR

EPR measurements can be carried out in a JEOL X-band spectrometer fitted with a loop-gap resonator (Molecular Specialties, Inc., Milwaukee, WI), or equivalent [10, 11]. An aliquot of purified, spin-labelled protein is placed in a sealed quartz capillary (0.84 mm OD, Ruska Instrument Corp., Houston, TX) and inserted into the resonator. Spectra of samples at room temperature (20–22 °C) were obtained by a single 60 s scan over 100 G at a microwave power of 2 mW, a receiver gain of 250–400 and a modulation amplitude optimized to the natural line width of the individual spectrum. For intact filaments, spin label concentration is typically in the range of 50–70 μM. Intermediate filament samples tend to aggregate upon further concentration, so instrument sensitivity should be optimized to obtain good signal to noise, especially for specimens displaying broad line widths. However, protofilaments of IF oligomers formed in low ionic strength, can be concentrated using centrifugal devices providing excellent signal to noise ratio with even less sensitive instrumentation. For spectra obtained at −100 °C, the microwave power is reduced (100 μW) to avoid saturation. A representative example of EPR data is shown in Figure 6.38.

Acknowledgements

This research was supported by a UC Davis Health System Research Grant (JFH), a US Army Medical Research Acquisition Activity (grant number DAMD17-02-1-0664, JFH), National Institutes of Health Grant R01EY08747 (PGF) and core grant P30EY-12576 (JFH and PGF) and The March of Dimes Birth Defect Foundation (JV).

References

1. Strelkov, S., Herrmann, H., Parry, D., Steinert, P. and Aebi, U. (2000) *J. Invest. Dermatol.*, **114**, 779.
2. Strelkov, S. V., Herrmann, H., Geisler, N., Wedig, T., Zimbelmann, R., Aebi, U. and Burkhard, P. (2002) *EMBO (European Molecular Biology Organization) J.*, **21**, 1255–1266.
3. Hess, J. F., Voss, J. C. and FitzGerald, P. G. (2002) *J. Biol. Chem.*, **277**, 35 516–35 522.
4. Sambrook, J. and Russell, D. W. (2001) *Molecular Cloning: a Laboratory Manual*, 3rd edn (J. Sambrook, ed.), Cold Spring Harbor Press, Cold Spring Harbor, NY.
5. Nagai, K. and Thøgersen, H. C. (1987) *Method. Enzymol.*, **153**, 461–481.
6. Herrmann, H., Hofmann, I. and Franke, W. W. (1992) *J. Mol. Biol.*, **223**, 637–650.
7. Eckelt, A., Herrmann, H. and Franke, W. W. (1992) *Euro. J. Cell Biol.*, **58**, 319–330.
8. Carter, J. M., Hutcheson, A. M. and Quinlan, R. A. (1995) *Exp. Eye Res.*, **60**, 181–192.
9. Rogers, K. R., Herrmann, H. and Franke, W. W. (1996) *J. Struct. Biol.*, **117**, 55–69.
10. Froncisz, W. and Hyde, J. S. (1982) *J. Magn. Reson.*, **47**, 515–521.
11. Hubbell, W. L., Froncisz, W. and Hyde, J. S. (1987) *Rev. Sci. Instrum.*, **58**, 1879–1886.

Neurofilament assembly

Shin-ichi Hisanaga and Takahiro Sasaki

Introduction

Neurofilament (NF) is the neuron-specific intermediate filament (IF) [1, 2]. NF is the most abundant cytoskeleton in axon, essential for radial growth of axons. NF is composed of three subunits, NF-L (61 kDa), NF-M (95 kDa) and NF-H (110 kDa). Like other IF proteins, each subunit consists of three domains; the N-terminal head, central α-helical rod and C-terminal tail domain. The head domain is thought to be the region regulating the filament assembly–disassembly by phosphorylation. The rod domain is involved in the filament formation via coiled-coil interaction. The tail domain, which of NF-M and NF-H are highly phosphorylated and much longer than NF-L and other IF proteins, extrudes from the filament core and interacts with each other and with other cellular components [3, 4]. Therefore, NF-L is the basic subunit for filament formation, and NF-M and NF-H are suggested to be incorporated at the surface of filaments.

The assembly of NF has been extensively studied, but how these subunits are arranged in NF has not been discovered yet. Furthermore, abnormal metabolism of NF proteins has been reported recently with several neurodegenerative diseases. Accumulation of NF in the cell body and proximal axons of motor neurons is a well-known pathology of amyotrophic lateral sclerosis (ALS) [5]. The deletion and insertional mutations of the NF-H tail domain are discovered in sporadic ALS patients and are suggested to be a risk factor. Several point mutations in the NF-L gene are recently reported to be associated with Charcot-Marie-Tooth disease (CMT), the most common form of hereditary motor and sensory neuropathy [6, 7]. NF-L with CMT mutants (P8R and Q333P) is suggested to be an inability to form filaments properly and disturbs axonal transport in neurons [8, 9]. The point mutation of NF-M is also found in familial Parkinson's disease [10]. It will be important to determine the effects of these mutations on filament assembly.

The NF assembly has usually been assessed *in vitro* with purified proteins and in cultured cells by transfection. The NF-L purified in a denaturing solution containing $6 \sim 8$ M urea can assemble into long 10 nm filaments upon dialysis against a physiological solution (i.e. 0.15 M NaCl and several mM $MgCl_2$ at pH around 7) at $37\,^\circ$C. NF-M and NF-H need NF-L to form 10 nm filaments, although they themselves form small oligomeric aggregates. The *in vitro* reassembly system is used to examine the assembly conditions and processes, and capable of investigating the detailed filament structures by electron microscopy. It is also possible to examine disassembly of filaments by phosphorylation with several protein kinases including PKA, PKC, CaMKII and Rho kinase [11]. The expression of NF proteins in SW-13 cl.2/Vim⁻ cells which lack the cytoplasmic IF proteins is also often used to estimate the filament assembly in cellular conditions. This assay can be performed without purification of NF proteins. This method revealed that rodent NF proteins are obligate heteropolymers; NF-L required coexpression of either or both NF-M and NF-H for filament formation [12, 13]. We will describe here the assembly system methods *in vitro* with NF proteins purified from mammalian spinal cords and from *E. coli*-expressed NF-L and in SW-13 cl.2/Vim⁻ cells with transfection of cDNAs encoding NF protein.

Reagents

A. For assembly in vitro

PEM buffer (100 mM Pipes, pH 6.8, 1 mM EGTA, 2 mM MgCl₂, 1 mM DTT, 0.4 mM Pefabloc SC (Merck) ① and 10 μg/ml leupeptin)

PEMU buffer (PEM buffer containing 6 M urea)

PEMN buffer (PEM buffer containing 0.15 M NaCl)

Bovine or porcine spinal cords

Rat or mouse NF-L cDNA cloned into an *E. coli*-expression vector

B. For assembly in SW-13 cl.2/Vim⁻ cells

cDNAs of NF subunits cloned into human cytomegalovirus promoter- or rous sarcoma virus promoter-driven expression vectors

Transfection reagent, Fugene 6 (Roche) or Lipofectamine 2000 (Invitrogen)

All standard laboratory reagents should be of high purity and good quality deionized water used

Equipment

A. For assembly in vitro

Polytron homogenizer for spinal cords ② or sonicator for *E. coli*

DEAE cellulose (DE52, Whatman) column and Mono Q column (Amersham Biosciences)

Centricon-10 ③ (Millipore)

Centrifuge

Materials for preparation of negatively stained EM specimens

Transmission electron microscope

B. For assembly in SW-13 cl.2/Vim⁻ cells

Facilities for cell culturing

Antibody to NF proteins

Materials for preparation of immunofluorescent staining

Fluorescence microscope or laser scanning microscope

Procedure

I. NF assembly in vitro

A. Preparation of crude NF proteins

(a) NF proteins from spinal cords
1. Bring bovine or porcine spinal cords from a local slaughterhouse to the lab on ice as quickly as possible. ④
2. Remove meninges from the spinal cords and measure the weight.
3. Homogenize the spinal cords in an equal volume (w/v) of PEM buffer (100 mM Pipes, pH 6.8, 1 mM EGTA, 2 mM MgCl₂, 1 mM DTT, 0.4 mM Pefabloc SC (Merck) and 10 μg/ml leupeptin) with a polytron homogenizer.
4. Centrifuge the homogenate at 10 000g for 15 min at 4 °C.

5. Take the supernatant and recentrifuge it at 100 000g for 60 min at 4 °C.

6. Discard the supernatant and dissolve the pellet in PEM buffer containing 6 M urea (PEMU).

7. Centrifuge the suspension at 100 000g for 30 min at 4 °C.

8. Remove the supernatant as the crude NF proteins.

(b)   NF-L from *E. coli* ⑤

The expression of NF-L protein in *E. coli* is performed by a standard protocol using a T7 promoter-driven system. *E. coli* strain 'BL21(DE3) pLysS' carrying pET23-rat NF-L is induced to express NF-L protein by 0.5 mM IPTG for 3 h at 37 °C.

1. Collect *E. coli* by centrifugation at 8 000g and wash once with 0.85% NaCl.

2. Suspend cells with 5 ml of PEM buffer per 100 ml of culture.

3. Sonicate the suspension to disrupt cells with cooling.

4. Centrifugate the homogenate at 100 000g for 30 min at 4 °C.

5. Dissolve the pellet with 5 ml of PEMU per 100 ml of culture. ⑥

6. Stand for 1 h at room temperature.

7. Centrifugate the suspension at 100 000g for 30 min at 4 °C.

8. Recover the supernatant as the crude NF-L preparation.

B. Purification of NF proteins

1. Apply the crude NF proteins or crude NF-L to a DEAE-cellulose column equibrated with PEMU.

2. Wash the column with 3 ~ 5 column volumes of PEMU.

3. Elute the NF proteins with 5 column volumes of buffer with a linear gradient of NaCl from 0 to 0. 25 M. ⑦

4. Collect the NF-L- or NF-M-rich fractions and apply it to a Mono Q anion exchanger column equibrated with PEMU.

5. Wash the column with 5 column volumes of PEMU.

6. Elute the NF proteins with 10 column volumes of PEMU with a linear gradient of NaCl from 0 to 0.5 M.

7. Collect NF-L and NF-M separately, and concentrate to 2 ~ 3 mg/ml by Cetricon-10.

C. NF assembly

The reconstitution of NF is performed by dialysis of the NF protein solution (2 ~ 3 mg/ml) against the PEM buffer containing 0.15 M NaCl (PEMN) at 37 °C for more than 3 h. ⑧ For the assembly of filaments composed of NF-L/M or NF-L/H heteropolymers, their ratio should be 2 : 1 for both NF-L/NF-M and NF-L/NF-H. Assembly of filaments can be checked by pelleting them by centrifugation at 100 000g for 30 min, followed by SDS-PAGE. For EM observation, dilute the reassembled filaments to 0.1 mg/ml and process them for negative staining.

II. NF assembly in SW-13/cl.2 vim⁻ cells

SW-13/cl.2 vim⁻ cells, established by Dr R Evert (Univ. of Colorado Health Sciences Center, CO), were cultured in Dulbecco's modified Eagle's medium with 10% fetal bovine serum at 37 °C in 5% CO_2. SW-13/cl.2 vim⁻ cells can be transfected with calcium phosphate precipitation or lipofection-based reagents, although the latter method is easier. We successfully transfected plasmids using Fugene 6 (Roche) or Lipofectamine 2000 (Invitrogen) according to the manufacture's instructions. The human cytomegalovirus promoter or rous sarcoma virus promoter can be used to express the NF proteins in the cells. Expressed NF is visualized by the general immunofluorescence technique. Alternatively, self-fluorescence of enhanced green fluorescent protein (EGFP) fused to the NF protein is also available. EGFP-linked to the N-terminal end, but not C-terminal end, of NF subunits can assemble into filaments. Observation of NFs can be performed by a standard fluorescence microscope, but observation by a laser scanning microscope reveals more detailed structures.

Examples of representative data are shown in Figures 6.39 and 6.40.

Notes

① We use Pefabloc SC as a Serine-protease inhibitor in place of PMSF, because Pefabloc SC is water-soluble and is not inactivated in aqualous solutions.

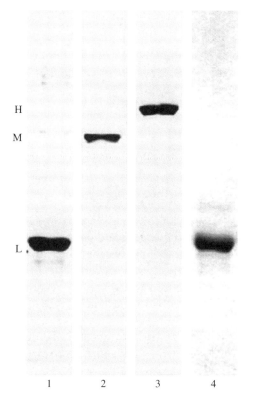

Figure 6.39   SDS-PAGE of NF proteins purified from bovine spinal cords and from *E. coli*. Lanes 1–3 are NF-L, NF-M and NF-H from bovine spinal cords, respectively and lane 4 is NF-L from *E. coli*

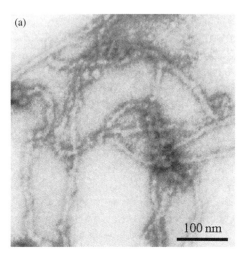

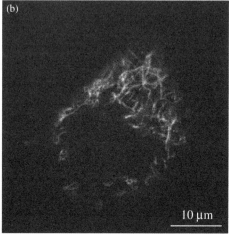

Figure 6.40   (a) A negative staining electron micrograph of NFs reassembled from porcine NF-L and (b) a laser scanning fluorescence micrograph of rat NF-L and EGFP-tagged rat NFH co-expressed in SW-13 cl.2/Vim⁻ cells

② A glass–Teflon homogenizer or juicer mixer can be used, although the spinal cords are more finely homogenized with a polytron.

③ Filter device for ultrafiltration.

④ Rat or mouse spinal cords can be used for preparation of NF. Rat or mouse spinal cords are easily ejected from an upper backbone by injecting water into the lower part of the backbone. However, the separation of NF-M and NF-L by column chromatographies is difficult in comparison with porcine or bovine NF. If you want to use rat or mouse NF-L, we recommend to express it in *E. coli* as described in section II).

⑤ Preparation of NF-M and NF-H from *E. coli* is not recommended, although purification of NF-H can be done. Not only is their expression low, but also a lot of truncated forms are generated particularly for NF-M.

⑥ NF-L is obtained as a inclusion body but is solubilized by 6 ~ 8 M urea.

⑦ NF-H is eluted earlier than NF-L and NF-M as a single band on the gel of SDS-PAGE, but NF-L and NF-M are eluted with a overlap in the order of NF-M and NF-L.

⑧ NF-L at lower concentrations (0.1 mg/ml) can assemble into filaments but higher concentrations are better to obtain longer filaments. The extended dialysis up to 24 h results in longer filaments.

References

1. Pant, H. C. and Veeranna (1995) *Biochem. Cell Biol.*, **73**, 575–592.
2. Julien, J. P. and Mushynski, W. E. (1998) *Prog. Nucleic Acid Res. Mol. Biol.*, **61**, 1–23.
3. Sasaki, T., Taoka, M., Ishiguro, K., Uchida, A., Saito, T., Isobe, T. and Hisanaga, S. (2002) *J. Biol. Chem.*, **277**, 36032–36039.
4. Julien, J. P. and Mushynski, W. E. (1983) *J. Biol. Chem.*, **258**, 4019–4025.
5. Carpenter, S. (1968) *Neurology*, **18**, 841–851.
6. Mersiyanova, I. V., Perepelov, A. V., Polyakov, A. V., Sitnikov, V. F., Dadali, E. L., Oparin, R. B., Petrin, A. N. and Evgrafov, O. V. (2000) *Am. J. Hum. Genet.*, **67**, 37–46.
7. De Jonghe, P., Mersivanova, I., Nelis, E., Del Favero, J., Martin, J. J., Van Broeckhoven, C., Evgrafov, O. and Timmerman, V. (2001) *Ann. Neurol.*, **49**, 245–249.
8. Perez-Olle, R., Leung, C. L. and Liem, R. K. (2002) *J. Cell Sci.*, **115**, 4937–4946.
9. Brownlees, J., Ackerley, S., Grierson, A. J., Jacobsen, N. J. O., Shea, K., Anderton, B. H., Leigh, P. N., Shaw, C. E. and Miller, C. C. J. (2002) *Hum. Mol. Gen.*, **11**, 2837–2844.
10. Lavedan, C., Buchholtz, S., Nussbaum, R. L., Albin, R. L. and Polymeropoulos, M. H. (2002) *Neurosci. Lett.*, **322**, 57–61.
11. Hisanaga, S., Gonda, Y., Inagaki, M., Ikai, A. and Hirokawa, N. (1990) *Cell Regul.*, **1**, 237–248.
12. Ching, G. Y. and Liem, R. K. (1993) *J. Cell Biol.*, **122**, 1323–1335.
13. Lee, M. K., Xu, Z., Wong, P. C. and Cleveland, D. W. (1993) *J. Cell Biol.*, **122**, 1337–1350.

PROTOCOL 6.36

α-Synuclein fibril formation induced by tubulin

Kenji Uéda and **Shin-ichi Hisanaga**

Introduction

Human α-synuclein was originally identified as the precursor protein of non-β-amyloid β component (NAC) in an amyloid-enriched fraction from Alzheimer's disease brains [1]. Later, three missense mutations in the α-synuclein gene were discovered in certain pedigrees with autosomal-dominant familial Parkinson's disease and dementia with Lewy bodies. Increasing evidence suggests that α-synuclein is a common pathogenic molecule in several neurodegenerative diseases, collectively known as synucleinopathies. α-Synuclein is a soluble neuronal protein of unknown function, but is the major filamentous component of pathological inclusions in those diseased brains. α-Synuclein has the property of forming fibrils by itself *in vitro*, and mutations of α-synuclein accelerate the fibril formation [2]. Furthermore, several factors are known to accelerate fibril formation *in vitro*. It is essential to understand the mechanism for filament formation in neurodegenerative diseases. Recently, it was demonstrated that tubulin seeds α-synuclein fibril formation under physiological conditions [3]. Here, the procedures for preparing proteins and fibril formation, and the detection of fibrils are described.

Reagents

Express human α-synuclein in *Escherichia coli* BL21(DE3)pLysS using an expression vector with isopropyl-1-thio-β-D-galactopyranoside (IPTG) and purified by HPLC methods [3, 4]. Briefly, a soluble fraction of *E. coli* expressing α-synuclein was fractionated by HPLC column chromatographies of anion exchange (DEAE 5PW, TOSO, Tokyo, Japan), hydrophobic (Phenyl TOYOPEARL, TOSO), and reversed phase (TSK-gel phenyl 5PW, TOSO) columns sequentially. From 1 l cell culture, $80 \sim 100$ mg of α-synuclein is obtained with a purity of more than 98%. ① Purified α-synuclein can be lyophilized and stored at $-80\,°C$. Microtubule proteins can be isolated from rat or porcine brains by two cycles of a temperature-dependent assembly and disassembly according to Shelanski *et al.* [5]; see also *Protocol 6.34*. Purify tubulin from the microtubule proteins by phosphocellulose (Whatman P11) column chromatography [6] and store at $-80\,°C$. All standard laboratory reagents should be of high purity and good quality. Milli Q water should be used.

Equipment

Eppendorf or other laboratory tubes

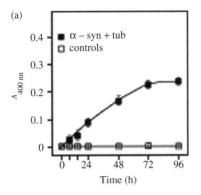
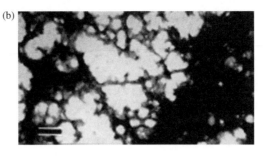

Figure 6.41 Fibril formation of α-synuclein in the presence of tubulin: (a) Fibril formation of α-synuclein monitored by quasi-elastic light scattering. α-Synuclein (300 μM) in 25 μl of PBS was incubated at 37 °C with 1 μM of tubulin, and 1 μl of the reaction mixture was taken and measured its scattering at 400 nm at various times. The fibrils began to form after 6 h of incubation, and this formation steadily increased and reached a plateau in 3 days. An open square represents four kinds of controls that were completely flat. Controls: (1) 300 μM α-synuclein; (2) 1 μM tubulin; (3) 300 μM α-synuclein+1 μM BSA; (4) 300 μM BSA + 1 μM tubulin. The value represents the average of three determinations. (b) A negative staining electron micrograph of representative α-synuclein fibrils. A number of long, 10-nm-wide filaments are formed. Bar = 100 nm

37 °C water bath

Materials for preparation of negatively stained EM specimens (see *Protocol 2.4*) Microcentrifuge

Spectrophotometer equipped with a ultra-micro amount sample cell (we use a Hitachi Gene Spec III, which has enabled ultra-micro volume quantitative analysis, using as little as 1 μl)

Transmission electron microscope

Procedure

1. Thaw α-synuclein and tubulin stored at −80 °C quickly in a water bath and keep them on ice. ②

2. Add ice-cold sterilized PBS, pH 7.4 to make 300 μM solution of α-synuclein with 1 μM of tubulin in 25 μl.

3. Incubate at 37 °C without agitation for 24 h (or longer, as required). Fibrils begin to form after 6 h of incubation.

4. Monitor fibril formation by light scattering at OD 400 nm using 1 μl of the reaction mixture with a Gene Spec III spectrophotometer. ③ ④ ⑤

5. Prepare EM specimens negatively stained with 2% uranyl acetate.

6. Assess the fibril formation on specimen grids by an EM.

An example of fibril formation by α-synuclein is shown in Figure 6.41.

Notes

① Purified recombinant α-synuclein is unstructured, heat-stable, highly soluble, and exists as a monomer in an aqueous solution.

② Tubulin is an unstable protein, losing its polymerization activity upon keeping it on ice for long time. Tubulin should be stored in aliquots to avoid repeated freezing/thawing and thawed freshly on each experimental day.

③ Take 1 µl aliquots from the incubation mixture into an ultra-micro sample cell without agitation at the time of monitoring.

④ α-Synuclein can undergo self-aggregation under certain conditions, such as increasing time lag, high temperature, high concentration and low pH.

⑤ Light scattering intensity indicates the total mass of fibrils formed.

References

1. Uéda, K., Fukushima, H., Masliah, E., Xia, Y., Iwai, A., Yoshimoto, M., Otero, D. A. C., Kondo, J., Ihara, Y. and Saitoh, T. (1993) *Proc. Nat. Acad. Sci. USA*, **90**, 11 282–11 286.

2. Ma, Q. L., Chan, P., Yoshii, M. and Uéda, K. (2003) *J. Alzheimer Dis.*, **5**, 139–148

3. Alim, M. A., Hossain, M. S., Arima, K., Takeda, K., Izumiyama, Y., Nakamura, M., Kaji, H., Shinoda, T., Hisanaga, S. and Uéda, K. (2002) *J. Biol. Chem.*, **277**, 2112–2117.

4. Hossain, S., Alim, A., Takeda, K., Kaji, H., Shinoda, T. and Uéda, K. (2001) *J. Alzheimer Dis.*, **3**, 577 584.

5. Shelanski, M. L., Gaskin, F. and Cantor, C. R. (1973) *Proc. Nat. Acad. Sci. USA*, **70**, 765–768.

6. Weingarten, M. D., Lockwood, A. H., Hwo, S. Y. and Kirschner, M. W. (1975) *Proc. Nat. Acad. Sci. USA*, **72**, 1858–1862.

Amyloid-β fibril formation *in vitro*

J. Robin Harris

Introduction

Fibrillogenesis of the amyloid-β peptide (Aβ) and inhibition of fibrillogenesis is of considerable interest for biomedical scientists and clinicians concerned with the development and prevention of Alzheimer's disease. The Aβ peptide, proteolytically cleaved *in vivo* from the soluble amyloid precursor protein, exists as a range of peptide length, between 39 and 44 amino acids. The 42 amino acid peptide ($A\beta_{1-42}$) is believed to be the predominant amyloid peptide incorporated in fibril form within Alzheimer senile plaques [1]. Although biophysical (e.g. spectroscopy and atomic force microscopy), and biochemical studies (e.g. thioflavin-T fluorescence, Congo red binding) have been used to assess Aβ fibrillogenesis [2], TEM is the most rapid and direct procedure, and it enables comment to be made on the progression of different oligomerization, protofibril and fibril states, and the interaction of fibrils with other proteins and physiological reagents such as cholesterol and sphingomyelin [3].

Reagents

Synthetic amyloid-β (Aβ) peptide (1–42) (human, mouse or rat sequence, and synthetic equivalents of Aβ mutants), is available from most biochemical suppliers, often as the trifluoroacetate salt. Store at −20 °C, or below.

Chemicals that potentiate (e.g. cholesterol, ganglioside GM1, metal ions) or inhibit (e.g. aspirin, nicotine, vitamin E) fibrillogenesis can also be incorporated within this system, as required.

All standard laboratory reagents should be of high purity and good quality deionized water used. ①

Equipment

Incubator (37 °C), with shaker

Microcentrifuge or other laboratory tubes

Microcentrifuge

Materials for preparation of negatively stained EM specimens (see *Protocol 2.4*)

Spectrometer/fluorescence spectrometer (if biochemical quantitation of fibrillogenesis is required)

Access to a transmission electron microscope (TEM)

Procedure

1. Warm the lyophilized Aβ peptide to room temperature before opening.

2. Add ice-cold deionized water (or buffer at defined pH) to produce a ~0.2 mM/~1 mg/ml solution of Aβ peptide. Keep on ice.

3. Disperse and solubilize the peptide by repeated gentle pipetting for ~2–5 min. Remove any insoluble aggregates by centrifugation, if necessary. ②

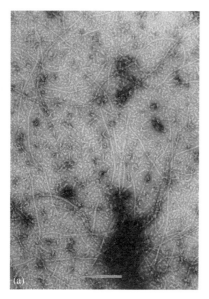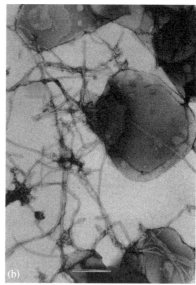

Figure 6.42 Example of human $A\beta_{1-42}$ peptide following 24 h incubation at 37 °C in the absence (a) and presence (b) of cholesterol microcrystals. Note that in (a) there is a background spread of oligomeric $A\beta$ particles, alongside short and longer A fibrils. In (b) all the oligomeric particles have polymerized to generate single and clustered fibrils which also bind to the cholesterol microcrystals. Negatively stained with 2% uranyl acetate. The scale bars indicate 200 nm (JRH, previously unpublished data)

4. Aliquot the $A\beta$ solution into several tubes containing an equal volume containing differing additives. ③ ④

5. Incubate at 37 °C with constant oscillation for 24 h (or longer, as required).

6. Prepare EM specimens negatively stained with 2% uranyl acetate (or other suitable negative stain [4].

7. Assess the formation of protofibrils and helical fibrils by studying specimen grids in a TEM.

8. $A\beta$ fibril formation can also be assessed by light microscopy/spectroscopy to determine Congo red binding or by thioflavin-T fluorescence, both of which indicate alpha-helix to beta-sheet conversion [2].

A representative example of $A\beta_{1-42}$ fibrillogenesis is shown in Figure 6.42.

Notes

This procedure will take approximately 24 h.

① Use HPLC grade, deoxygenated water. Metal ion contamination (e.g. Cu, Zn, Fe, Al) should be avoided.

② This will necessitate determination of the protein concentration of the supernatant.

③ That is, water/buffer alone, plus and appropriate concentration of a compound that may potentiate (cholesterol, sphingomyelin, gangliosides, multivalent metal ions) or inhibit $A\beta$ fibrillogenesis (e.g. vitamin E, $A\beta$ fragments, apolipoprotein E4, catalase, aspirin, nicotine, Congo red and other drugs) [3, 5].

④ In view of the potentially hazardous nature of amyloid peptides and the various additives or drugs used in this protocol, if injested, appropriate care should be taken throughout.

References

1. Rochenstein, E.,   Mallory, M.,   Mante, M., Sisk, A. and Masloaja, E. (2001) *J. Neurosci. Res.*, **66**, 573–582.

2. Nybo, M., Svegag, S.-E. and Nielsen, E. H. (1999) *Scan. J. Immunol.*, **49**, 219–223.

3. Harris, J. R. (2002) *Micron*, **33**, 609–626.

4. Harris, J. R. and Scheffler, D. (2002) *Micron*, **33**, 461–480.

5. Ono, K.,   Hasegawa, K.,   Yamada, M.   and Naiki, H. (2002) *Biol. Psychiatry*, **52**, 880–886.

Soluble Aβ₁₋₄₂ peptide induces tau hyperphosphorylation *in vitro*

Terrence Town and Jun Tan

Introduction

Alzheimer's disease (AD) has two key pathologies: the presence of neurofibrillary tangles (NFTs) and deposition of amyloid-β peptide (Aβ) as senile plaque. NFTs are predominately comprised of hyperphosphorylated tau protein, assembled primarily in the paired helical filament (PHF) conformation [1, 2]. While the mechanistic link between tau hyperphosphorylation and NFTs is not well understood, it is generally thought that increased phosphorylation promotes PHFs, resulting in disruption of the physiological role of tau in stabilizing neuronal microtubules that are necessary for axoplasmic flow [3]. Tau appears to be hyperphosphorylated at specific epitopes (particularly in the proline-rich region that contains several serine/threonine-proline motifs) in AD brain [2].

One key question concerning tau hyperphosphorylation is whether or not it is promoted by Aβ. It is now generally accepted that Aβ directly promotes increased levels of intracellular Ca^{2+} in neurons, thereby causing neuronal injury and apoptosis [4]. Recent evidence has established the signalling cascade that promotes neuronal cyclin-dependent kinase 5 (Cdk5) activation induced by p25, with the first event being increased intracellular Ca^{2+} levels, which then promotes calpain activation, which, in turn, goes on to directly

cleave p35 to p25 [5]. Excess p25 then complexes with Cdk5, thereby boosting its activity, an event that is associated with tau hyperphosphorylation at AD-specific phosphoepitopes and, not surprisingly, neurotoxicity [5, 6]. Further, as studies of human biopsies and aged canine and primate brains have consistently shown that dystrophic neurites precede deposition of Aβ in the formation of neuritic plaques [7–10], the putative role of soluble forms of Aβ peptide per se in promotion of tau phosphorylation is of interest. We have recently demonstrated that, in p35-overexpressing neuron-like cells, soluble Aβ is a potent activator of the Cdk5/p25 pathway, leading AD-like tau phosphorylation *in vitro* [11].

Reagents

1. The p35-overexpressing N2a cell line is required for these experiments and is available from our laboratory upon request. These cells are grown in Eagle's minimum essential media supplemented with 2 mM L-glutamine and Earle's balanced salt solution adjusted to contain 1.5 g/l sodium bicarbonate, 0.1 mM non-essential amino acids, 1.0 mM sodium pyruvate, 10% fetal bovine serum and 400 μg/ml of G418, and are induced to differentiate into neuron-like cells by serum withdrawal

and the addition of 0.3 mM dibutyryl cAMP for 48 h.

2. To examine phospho-tau and total tau levels, the following anti-tau antibodies are needed: AT8 (Innogenetics, Norcrosis, GA), AT270 (Innogenetics), pS199 (BioSource International, Inc. Camarillo, CA), and pS396 (BioSource International, Inc.) against phosphorylated forms of tau, and C-17 (Santa Cruz Biotechnology, Santa Cruz, CA) against total tau protein.

3. Synthetic human $A\beta_{1-42}$ peptide is available from QCB Biosource International (Camarillo, CA) and other suppliers.

4. To perform Western immunoblotting, the following reagents are needed in addition to the above-mentioned primary antibodies: buffers [cell lysis buffer containing phosphatase and protease inhibitors supplied as $10\times$ from NEB, Beverly, MA; Tris-buffered saline (TBS), Tris/glycine/SDS buffer, laemmli sample buffer, and blocking and antibody dilution buffer consisting of TBS with 5% w/v non-fat dry milk and 0.05% Tween-20, obtained from Bio-Rad]; 10% ready gels (Bio-Rad); Hybond PVDF membranes (Bio-Rad); appropriate secondary antibodies conjugated with horseradish peroxidase (HRP) and luminol reagent (Santa Cruz Biotechnology).

5. To perform immunocytochemistry, a cell permeabilizing solution of 0.05% v/v Triton X-100 diluted in complete medium described above is needed, and a phosphate-buffered saline (PBS) rinse solution, a fixing solution of 4% w/v paraformaldehyde in PBS, serum-free blocking solution (Dako, Carpinteria, CA), and the LSAB + immunochemistry kit (Dako, Carpinteria, CA) are needed.

Equipment

Immunoblot Apparatus

Fluor-S MultiImager with Quantity One™ software (Bio-Rad), for densitometric analysis of protein bands

Freezer $-80\,^{\circ}$C

Inverted light microscope, for tissue culture observation

Light microscope, preferably with a digitizing camera system

Microcentrifuge tubes

Microcentrifuge, $4\,^{\circ}$C

Protein gel electrophoresis system

Tissue culture CO_2 incubators, for growing cell cultures

Spectrophotometer

Procedure

Preparation of $A\beta_{1-42}$ peptide

Add Sigma water to human $A\beta_{1-42}$ peptide to a concentration of 250 μM immediately before all experiments. This stock solution can then be diluted to 1, 3 or 5 μM in cell culture media for tau phosphorylation experiments. Based on our recent data [11], more than 95% of the $A\beta_{1-42}$ employed in this experimental paradigm (including all doses studied, under both cell-free and cell-present conditions) remains in soluble form throughout the experiments.

Cell culture conditions

1. Plate p35-overexpressing N2a cells at 5×10^5 cells/well in six-well tissue culture plates (for Western immunoblotting analysis of phospho-tau), or plate on glass coverslip inserts in six-well tissue culture plates (for immunocytochemistry of phospho-tau) in complete media described above.

2. To differentiate these cells, subject p35-overexpressing N2a cells to serum

(a)

(b)

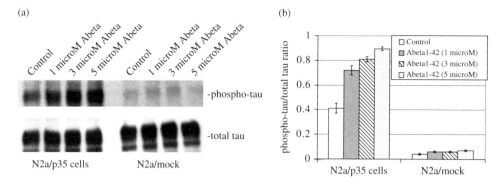

N2a/p35 cells   N2a/mock

Figure 6.43   Treatment of p35-overexpressing N2a cells with $A\beta_{1-42}$ results in increased tau phosphorylation by Western blot. p35 or mock-transfected N2a cells were differentiated and treated with a dose range of $A\beta_{1-42}$ (1, 3 or 5 μM $A\beta$) or went untreated (control) for 24 h. After incubation, phospho-tau epitopes were detected by Western blot and were normalized to total tau protein (as detected by anti-tau antibody C-17). (a) membranes were probed with AT8 (above) or C-17 (below), and representative blots illustrate an $A\beta_{1-42}$ dose–response relationship only in p35-transfected N2a cells. (b) densitometric analysis (ratio of phospho-tau to total tau signal ± 1 SEM, $n = 3$ for each condition presented) for pooled results from (A) reveals significant treatment effects of $A\beta_{1-42}$ dose on p35-transfected N2a cells ($P < 0.001$) but not on mock-transfected cells ($P > 0.05$)

withdrawal and give 0.3 mM dibutyryl cAMP for 48 h.

3. These cells then remain untreated (as a control), or are treated with a dose range of $A\beta_{1-42}$ (1, 3 or 5 μM) for 24 h.

Western immunoblotting analysis of tau hyperphosphorylation

1. Preparation of cell lysate:

   (a) Wash treated N2a cells twice in the six-well plate with ice-cold PBS and drain off the PBS.

   (b) Add ice-cold lysis buffer (containing 20 mM Tris, pH 7.5, 150 mM NaCl, 1 mM EDTA, 1 mM EGTA, 1% Triton X-100, 2.5 mM sodium pyrophosphate, 1 mM β-glycerolphosphate, 1 mM Na_3VO_4, 1 μg/ml leupetin and 1 mM PMSF) at a volume of 0.1 ml lysis buffer per 1×10^6 cells.

   (c) Scrape these cells off the plate with a plastic cell scraper and transfer the lysate into a microcentrifuge tube.

Gently rock the suspension on a rocker in the cold room for 15 min to lyse the cells.

   (d) Centrifuge the lysate at $14\,000g$ in a pre-cooled centrifuge (4 °C) for 15 min. Immediately transfer the supernatant to a fresh centrifuge tube.

   (e) Quantify the total protein content of supernatants for each sample using the Bio-Rad protein assay kit and aliquot the sample into several tubes and keep these tubes at −80 °C.

2. Western immunoblotting:

   (a) Perform SDS-polyacrylamide gel electrophoresis (SDS-PAGE) on cell lysates (50 μg of total protein of each sample) and transfer electrophoretically to Hybond PVDF membranes.

   (b) Wash the Hybond PVDF membrane twice with dH_2O and mark the transferred side of membrane with a pencil.

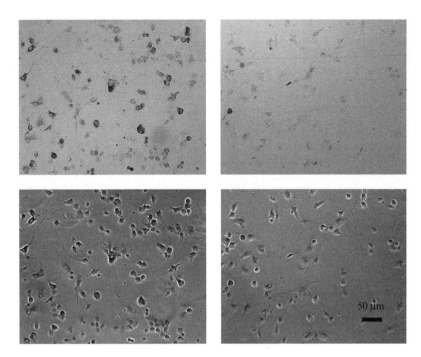

Figure 6.44 Increased tau phosphorylation *in situ* following treatment of p35-transfected N2a cells with Aβ$_{1-42}$. p35 or mock-transfected N2a cells were differentiated and incubated with Aβ$_{1-42}$ (3 μM) for 24 h. Tau phosphorylation was then detected using the AT270 antibody, as indicated by the brown reaction. Bright field (upper panels) and phase contrast (lower panels) micrographs showing phospho-tau in p35-transfected N2a cells in the absence (right panels) or presence (left panels) of treatment with 3 μM of Aβ$_{1-42}$

(c) Block the membrane in freshly prepared TBS containing 5% non-fat dry milk for 2 h at room temperature with agitation.

(d) Apply either anti-phospho-tau (AT8, 1 : 1000; AT270, 1 : 1000; pS199, 1 : 1500; pS396, 1 : 1500) or anti-total tau (C-17, 1 : 100) antibody diluted in fresh TBS containing 5% non-fat dry milk to the membrane and incubate for 2 h at room temperature with agitation.

(e) Wash the membrane three times for 5 min each with dH$_2$O.

(f) Incubate the membrane in the appropriate second antibody conjugated with HRP for 1 h at room temperature with agitation.

(g) Wash the membrane five times for 5 min each with dH$_2$O.

(h) Perform detection of proteins using the luminol reagent.

(i) Densitometric analysis is performed for the blots using the Fluor-S MultiImager with Quantity One™ software, and data are represented as a ratio of phospho-tau to total tau signal (see Figure 6.43).

Immunocytochemistry

1. Culture N2a cells at 5 × 10^5 cells/well on glass coverslips in six-well tissue culture plates and differentiate as described above. Twenty-four hours after treatment with a dose range of Aβ$_{1-42}$ as described above, fix

these cultured cells with ice-cold 4% paraformaldehyde in PBS (0.1 M, pH 7.4) for 10 min at 4 °C, and then permeabilize in 0.05% Triton X-100 diluted in complete medium.

2. After three rinses with ice-cold PBS, treat these cultured cells with endogenous peroxidase quenching (0.3% H_2O_2) and pre-block with serum-free protein blocking solution prior to primary antibody incubation.

3. Apply an anti-phospho-tau mouse monoclonal antibody (AT270, 1 : 100 dilution) to these cells overnight at 4 °C.

4. Perform immunocytochemistry using the LSAB + kit (utilizing an HRP-conjugated diaminobenzidine chromogen system).

5. Record morphology and phospho-tau immunoreactivity in these cells by brightfield microscopic examination, using a light microscope (see Figure 6.44).

References

1. Grundke-Iqbal, I., Iqbal, K., Tung, Y. C., Quinlan, M., Wisniewski, H. M. and Binder, L. I. (1986) Abnormal phosphorylation of the microtubule-associated protein tau (tau) in Alzheimer cytoskeletal pathology. *Proc. Nat. Acad. Sci. USA*, **83**, 4913–4917.

2. Friedhoff, P., von Bergen, M., Mandelkow, E. M. and Mandelkow, E. (2000) Structure of tau protein and assembly into paired helical filaments. *Biochem. Biophys. Acta*, **1502**, 122–132.

3. Drubin, D. G. and Kirschner, M. W. (1986) Tau protein function in living cells. *J Cell Biol.*, **103**, 2739–2746.

4. Mattson, M. P., Cheng, B., Davis, D., Bryant, K., Lieberburg, I. and Rydel, R. E. (1992) β-amyloid peptides destabilize calcium homeostasis and render human cortical neurons vulnerable to excitotoxicity. *J. Neurosci.*, **12**, 376–389.

5. Lee, M. S., Kwon, Y. T., Li, M., Peng, J., Friedlander, R. M. and Tsai, L. H. (2000) Neurotoxicity induces cleavage of p35 to p25 by calpain. *Nature*, **405**, 360–364.

6. Patrick, G. N., Zukerburg, L., Nikolic, M., de la Monte, S., Dikkes, P. and Tsai, L. H. (1999) Conversion of p35 to p25 deregulates Cdk5 activity and promotes neurodegeneration. *Nature*, **402**, 615–622.

7. Terry, R. D. and Wisniewski, H. M. (1970) The ultrastructure of the neurofibrillary tangle and the senile plaque. In: *CIBA Foundational Symposium on Alzheimer's Disease and Related Conditions* (G. E. W. Wolstenholme and M. O'Connor, eds), pp. 145–168. J&H Churchill, London.

8. Wisniewski, H. M., Johnson, A. B., Raine, C. S., Kay, W. J. and Terry, R. D. (1970) Senile plaques and cerebral amyloidosis in aged dogs. A histochemical and ultrastructural study. *Lab. Invest.*, **23**, 287–296.

9. Wisniewski, H. M., Ghetti, B. and Terry, R. D. (1973) Neuritic (senile) plaques and filamentous changes in aged rhesus monkeys. *J. Neuropath. Exp. Neurol.*, **32**, 566–584.

10. Martin, L. J., Pardo, C. A., Cork, L. C. and Price, D. L. (1994) Synaptic pathology and glial responses to neuronal injury precede the formation of senile plaques and amyloid deposits in the aging cerebral cortex. *Am. J. Pathol.*, **145**, 1358–1381.

11. Town, T., Zolton, J., Shaffner, R., Schnell, B., Crescentini, R., Wu, Y., Zeng, J., Delle Donne, A., Obregon, D., Tan, J. and Mullan, M. (2002) p35/Cdk5 pathway mediates soluble amyloid-β-peptide-induced tau phosphorylation *in vitro. J. Neurosci. Res.*, **69**, 362–372.

Anti-sense peptides

Nathaniel G. N. Milton

Introduction

Anti-sense peptide sequences are derived from the complementary strand of DNA encoding a given protein, read in the same open reading frame. The complementary strand DNA for each individual amino acid can be read in either the forward 3′-5′ or reverse 5′-3′ direction, adding degeneracy to the potential anti-sense peptide sequences. For a given peptide sequence it is possible to derive two anti-sense peptides from the DNA sequence. These sequences can then be used for synthesis of binding peptides and for screening databases for proteins with sequence similarity, which may act as binding proteins. Studies have shown that such anti-sense peptides bind with high affinity to the given protein and that they can also have sequence similarity to protein binding domains including receptor binding sites [1, 2]. Anti-sense peptide sequences can also be derived directly from the amino acid sequence of a protein, via reverse translation to produce a complementary DNA sequence. However, due to the degeneracy of the genetic code, there is typically more than one anti-sense sequence for any one protein sequence. The binding interactions between sense and anti-sense encoded peptides are mediated via hydropathic interactions [1]. Hydropathy profiles can also be used to derive anti-sense binding peptides [3].

The availability of vast protein and nucleotide sequence databases can be used both to derive anti-sense peptides and for sequence comparisons to identify potential protein–protein interactions. The simplicity of the anti-sense peptide sequence derivation plus availability of powerful sequence comparison programs to screen these databases makes this technique a cheap and rapid method for identification of potential protein–protein interactions, which can then form the basis for experimental studies. The use of this technique has been applied to identify novel Alzheimer's amyloid-ß peptide (Aß) binding proteins [4–6] and to characterize binding domains within known Aß binding proteins [7]. Such techniques are applicable to any protein or peptide and the resultant anti-sense sequences or protein binding domains can also be used in a similar manner to antibodies.

Reagents

Synthetic anti-sense peptides can be custom synthesized commercially or produced in-house. Biotin labelling of proteins can be carried out using commercially available kits [7]. Antibodies to target proteins can be obtained commercially or raised in house. Standard laboratory reagents should be of high purity and good quality deionized water used.

Equipment

Computer (PC or Macintosh) with access to Internet and Microsoft Word

ELISA plate reader with filters for absorbance at 405 and 450 nm

ELISA plate washer

Eppendorf or other laboratory tubes

Microtiter plates – NUNC Polysorb or Maxisorb for binding assays

Procedures

A. Derivation of anti-sense peptide sequences

The following method makes use of the Genbank nucleotide database available on the National Centre for Biotechnology Information website (http://www.ncbi.nlm.nih.gov/Entrez).

1. Access   http://www.ncbi.nlm.nih.gov/ Entrez website and perform a nucleotide sequence search for the protein being investigated. Copy the DNA sequence and paste into a Microsoft Word document. Delete all non-coding sections of DNA sequence to leave the coding sequence of interest. The resultant DNA sequence, typically starting with atg, should be converted to FASTA format by deletion of all spaces and numbers – use Microsoft Word Find and Replace options in the Edit toolbar. Alternatively a FASTA format DNA coding sequence can be imported from another source.

2. Convert the FASTA format sequence into a 'coding sequence' by insertion of spaces every three digits. This can be done using Find and Replace in Microsoft Word. In the Find What section, select 'Any Letter' three times. In the Replace With section select, 'Find What Text' followed by a single space. The resultant 'coding sequence' should be in the format 'atg gtc cag'. Save resultant 'coding sequence' text file.

3. Make a duplicate file of the coding sequence and convert each nucleotide triplicate to a 5′ to 3′ anti-sense amino acid based on Table 6.2 using Find and Replace. Where the anti-sense amino acid is a stop codon, replace the DNA triplet with X. Save the resultant 5′ to 3′ anti-sense peptide text file.

4. Make a duplicate file of the coding sequence and convert each nucleotide triplicate to a 3′ to 5′ anti-sense amino acid based on Table 6.2 using Find and Replace. Where the anti-sense amino acid is a stop codon, replace the DNA triplet with X. Save the resultant 3′ to 5′ anti-sense peptide text file.

B. Comparison of anti-sense peptide sequences with known proteins

The following methods make use of the BLAST program for sequence comparison with protein databases [8] available on the National Centre for Biotechnology Information website (http://www.ncbi.nlm.nih.gov/BLAST). Alternative sequence databases and comparison programs are available or can be obtained commercially. Protein binding can occur in both parallel and anti-parallel orientations. The BLAST search is restricted to comparisons of sequences in a parallel manner since it compares N to C terminally entered sequences. To overcome this problem it is possible to enter anti-sense peptide sequences in the C to N terminus orientation and perform BLAST searches. Therefore, four potential anti-sense sequence comparisons can be carried out.

1. Access the BLAST program on the Internet at http//www.ncbi.nlm.nih.gov/ BLAST.

2. Select Protein BLAST.

3. Then select Search for short nearly exact matches. This search uses a PAM30 matrix with Gap Costs of

Table 6.2 Identification of an alternative anti-sense amino acid that would recognize the respective sense amino acid

Amino acid	Coding DNA (5′-3′)	Anti-sense DNA (3′-5′)	Anti-sense 5′-3′ direction	Amino acid 3′-5′ direction
Ala (A)	GCA	CGT	Cys (C)	Arg (R)
	GCC	CGG	Gly (G)	Arg (R)
	GCG	CGC	Arg (R)	Arg (R)
	GCT	CGA	Ser (S)	Arg (R)
Arg (R)	AGA	TCT	Ser (S)	Ser (S)
	AGG	TCC	Pro (P)	Ser (S)
	CGA	GCT	Ser (S)	Ala (A)
	CGC	GCG	Ala (A)	Ala (A)
	CGG	GCC	Pro (P)	Ala (A)
	CGT	GCA	Thr (T)	Ala (A)
Asn (N)	AAC	TTG	Val (V)	Leu (L)
	AAT	TTA	Ile (I)	Leu (L)
Asp (D)	GAC	CTG	Val (V)	Leu (L)
	GAT	CTA	Ile (I)	Leu (L)
Cys (C)	TGC	ACG	Ala (A)	Thr (T)
	TGT	ACA	Thr (T)	Thr (T)
Glu (E)	GAA	CTT	Phe (F)	Leu (L)
	GAG	CTC	Leu (L)	Leu (L)
Gln (Q)	CAA	GTT	Leu (L)	Val (V)
	CAG	GTC	Leu (L)	Val (V)
Gly (G)	GGA	CCT	Ser (S)	Pro (P)
	GGC	CCG	Ala (A)	Pro (P)
	GGG	CCC	Pro (P)	Pro (P)
	GGT	CCA	Thr (T)	Pro (P)
His (H)	CAC	GTG	Val (V)	Val (V)
	CAT	GTA	Met (M)	Val (V)
Ile (I)	ATA	TAT	Tyr (Y)	Tyr (Y)
	ATC	TAG	Asp (D)	Stop (X)
	ATT	TAA	Asn (N)	Stop (X)
Leu (L)	CTA	GAT	Stop (X)	Asp (D)
	CTC	GAG	Glu (E)	Glu (E)
	CTG	GAC	Gln (Q)	Asp (D)
	CTT	GAA	Lys (K)	Glu (E)
	TTA	AAT	Stop (X)	Asn (N)
	TTG	AAC	Gln (Q)	Asn (N)
Lys (K)	AAA	TTT	Phe (F)	Phe (F)
	AAG	TTC	Leu (L)	Phe (F)
Met (M)	ATG	TAC	His (H)	Tyr (Y)
Phe (F)	TTC	AAG	Glu (E)	Lys (K)
	TTT	AAA	Lys (K)	Lys (K)
Pro (P)	CCA	GGT	Trp (W)	Gly (G)
	CCC	GGG	Gly (G)	Gly (G)
	CCG	GGC	Arg (R)	Gly (G)
	CCT	GGA	Arg (R)	Gly (G)

(continued overleaf)

Table 6.2   (*continued*)

Amino acid	Coding DNA (5'-3')	Anti-sense DNA (3'-5')	Anti-sense 5'-3' direction	Amino acid 3'-5' direction
Ser (S)	TCA	AGT	Stop (X)	Ser (S)
	TCC	AGG	Gly (G)	Arg (R)
	TCG	AGC	Arg (R)	Ser (S)
	TCT	AGA	Arg (R)	Arg (R)
	AGC	TCG	Ala (A)	Ser (S)
	AGT	TCA	Thr (T)	Ser (S)
Thr (T)	ACA	TGT	Cys (C)	Cys (C)
	ACC	TGG	Gly (G)	Trp (W)
	ACG	TGC	Arg (R)	Cys (C)
	ACT	TGA	Ser (S)	Stop (X)
Trp (W)	TGG	ACC	Pro (P)	Thr (T)
Tyr (Y)	TAC	ATG	Val (V)	Met (M)
	TAT	ATA	Ile (I)	Ile (I)
Val (V)	GTA	CAT	Tyr (Y)	His (H)
	GTC	CAG	Asp (D)	Gln (Q)
	GTG	CAC	His (H)	His (H)
	GTT	CAA	Asn (N)	Gln (Q)

9 for Existence and 1 for Extension. The Expect is set to 20 000. A Standard protein–protein BLAST [blastp] which uses the BLOSUM62 matrix with Gap Costs of 11 for Existence and 1 for Extension can also be used. For a Standard protein–protein BLAST [blastp] the Expect (statistical significance threshold for reporting matches against database sequences) can be reset to 100 or greater to account for the use of short peptide sequences in the search and increase the number of hits.

4. Alignments of peptides of <5 amino acids are considered non-significant under these conditions.

5. Sequences containing significant gaps (>10%) should also not be used since hydropathic binding interactions require a direct alignment between each sense protein residue and its complementary anti-sense peptide or binding domain residue.

6. The sequence information obtained within a BLAST search can be printed, copied and stored as required.

7. Using the options section it is possible to limit searches to sequences from different databases and select the species to be searched.

8. Within the BLAST search results access to the aligned sequences is also available and from this information about proteins interacting with the target protein can be obtained.

9. If a specific protein is known to bind the target protein a Pairwise BLAST with selection of BLAST 2 sequences can be performed. Select BLAST 2 sequences and the blastp (protein) comparison. Either the BLOSUM62 matrix with Gap Costs of 11 for Existence and 1 for Extension or the PAM30 matrix with Gap Costs of 9 for Existence and 1 for Extension can be selected.

10. The Expect (statistical significance threshold for reporting matches against database sequences) should be reset to 100 or greater to account for the use of short peptide sequences in the search and increase the number of hits.

11. Enter the anti-sense peptide sequence as sequence 1 and the interacting protein sequence as sequence 2. The interacting protein sequence identification either as a protein accession number or GI number can be entered rather than the full FASTA format protein sequence.

Choice of anti-sense peptide for synthesis

Anti-sense peptides can be synthesized based on the sequences obtained in Procedure A above. A good discussion of the choice of anti-sense peptide to synthesize is included in Bost and Blalock [1]. Regions of binding proteins identified in Procedure B can also be synthesized. Shorter anti-sense peptides will show less specificity, in general a 10-amino-acid peptide will be sufficient. The results from the comparison of anti-sense sequences with known proteins will help in determining the peptide to be synthesized and should help in deciding whether to use the $3'$ to $5'$ or $5'$ to $3'$ anti-sense peptide sequences. Information on structural features of a protein is particularly useful and a simple computer program to predict protein secondary characteristics is available [9]. If the sequence of interest contains an X, i.e. the non-coding DNA strand encodes a stop codon, then an alternative amino acid must be substituted. From Table 6.2 an alternative anti-sense amino acid that would recognize the respective sense amino acid can be identified. Where there is a choice then information from the BLAST comparisons may be of use and

an amino acid substitution could use a residue found in a protein with sequence similarity. The use of N or C terminal Cys residues to allow chemical linkage to carrier proteins, agarose or biotin labels should be considered. N-terminal acetylation and C terminal amidation can also be considered [1].

C. Direct binding of anti-sense peptides to target proteins

For this assay either a biotin labelled form of the target protein or an antibody specific for the target protein is required.

1. A 500 ml solution of carbonate buffer containing 15 mM Na_2CO_3, 35 mM $NaHCO_3$, 0.01% NaN_3 (w/v) pH 9.6 should be prepared (can be stored for <14 days at 4 °C).

2. Anti-sense peptides (1 μg/ml) should be prepared in carbonate buffer (10 ml/plate).

3. To each of the inner 60 wells on a NUNC Polysorb or Maxisorb 96-well ELISA plate add 100 μl of anti-sense peptide solution or carbonate buffer (for control non-specific binding determination). Cover plates with plate sealer (NUNC). Incubate plates at 4 °C overnight.

4. Remove plate sealer and wash plates three times with carbonate buffer (200 μl/well).

5. Prepare a 0.2% casein solution in carbonate buffer (20 ml/plate) and add 200 μl/well to each well. Incubate plates at room temperature for 1 h.

6. Prepare a 1 l solution of assay buffer containing 50 mM Tris-HCl, 0.01% NaN_3 (w/v), 0.1% BSA (Cohns Fraction V) (w/v), 0.1% Triton X-100 (v/v) pH 7.5.

7. Wash plates three times with assay buffer (200 µl/well).

8. Prepare a 5× stock solution of target protein (either biotin labelled or unlabelled as available) in assay buffer for addition to plates in the presence of competing anti-sense peptides or test compounds. Prepare competing anti-sense peptides and test compounds in assay buffer in polypropylene tubes at 5× required test concentration.

9. Prepare tubes containing target protein plus competing anti-sense peptides or test compounds and add 100 µl/well to coated ELISA plates. Incubate overnight at 4 °C.

10. Wash plates three times with 200 µl/ well assay buffer.

11. For biotin labelled target proteins prepare alkaline phosphatase polymer-streptavidin conjugate (1 : 1000 dilution of stock solution) in assay buffer. Add 100 µl/well to ELISA plates and incubate at 37 °C for 2 hs. Go to step 15.

12. For unlabelled target proteins prepare antibody at an appropriate dilution in assay buffer. Add 100 µl/well to ELISA plates and incubate at 37 °C for 2 h.

13. Wash plates three times with 200 µl/ well assay buffer.

14. Add 100 µl/well appropriate anti-IgG-alkaline phosphatase conjugate (1 : 2000 dilution of neat conjugate in assay buffer) and incubate for 2 h at room temperature.

15. Wash plates three times with 200 µl/ well assay buffer.

16. Prepare substrate containing 2.7 mM *p*-nitrophenylphosphate, 10 mM diethanolamine, 0.5 mM $MgCl_2$ pH 9.8. Add 100 µl/well and allow

sufficient colour development. Colour development can be stopped by addition of 50 µl/well 3 M NaOH if required.

17. Read absorbance at 405 nm using an ELISA plate reader.

References

1. Bost, K. L. and Blalock, J. E. (1989) Preparation and use of complementary peptides. *Methods Enzymol.*, **168**, 16–28.
2. Heal, J. R., Roberts, G. W., Raynes, J. G., Bhakoo, A. and Miller, A. D. (2002) Specific interactions between sense and complementary peptides: the basis for the proteomic code. *ChemBioChem.*, **3**, 136–151.
3. Villain, M., Jackson, P. L., Manion, M. K., Dong, W. J., Su, Z., Fassina, G., Johnson, T. M,. Sakai, T. T., Krishna, N. R. and Blalock, J. E. (2000) De novo design of peptides targeted to the EF hands of calmodulin. *J. Biol. Chem.*, **275**, 2676–2685.
4. Milton, N. G. N. (2001) Phosphorylation of amyloid-ß at the serine 26 residue by human cdc2 kinase. *NeuroReport*, **12**, 3839–3844.
5. Heal, J. R., Roberts, G. W., Christie, G. and Miller, A. D. (2002) Inhibition of ß-amyloid aggregation and neurotoxicity by complementary (antisense) peptides. *ChemBioChem* **3**, 86–92.
6. Milton, N. G. N. (2002) Peptides for use in the treatment of Alzheimer's disease. *Patent Publication*, WO 02/36614 A2.
7. Milton, N. G. N., Mayor, N. P. and Rawlinson, J. (2001) Identification of amyloid-ß binding sites using an antisense peptide approach. *NeuroReport*, **12**, 2561–2566.
8. Altschul, S. F., Madden T. L., Alejandro A. Schaeffer, A. A., Zhang, J., Zhang, Z., Miller, W. and Lipman D. J. (1997) Gapped BLAST and PSI-BLAST: a new generation of protein database search programs. *Nucleic Acids Res.*, **25**, 3389–3402.
9. Krchnak, V., Mach, O. and Maly, A. (1989) Computer prediction of B-cell determinants from protein amino acid sequences based on incidence of beta turns. *Methods Enzymol.*, **178**, 586–611.

PROTOCOL 6.40

Interactions between amyloid-ß and enzymes

Nathaniel G. N. Milton

Introduction

The neurotoxicity of the amyloid-ß peptide (Aß) is thought to be of major importance in the pathology of Alzheimer's disease [1] and compounds that prevent Aß neurotoxicity have therapeutic potential. The Aß peptide also interacts with a range of enzymes including catalase [2] and cyclin-dependent kinase-1 (CDK-1) [3]. Catalase enzyme activity is inhibited by Aß [2] and CDK-1 activity enhanced [3] allowing the use of these enzymes in functional bioassays to identify compounds that prevent actions of Aß. The binding characteristics of these enzymes [4, 5] can be used to identify compounds that interact with different regions of Aß and provide a method for rapid screening to identify Aß binding compounds. Microtiter plate binding and functional enzyme activity assays are rapid methods for identifying compounds that interact with Aß prior to screening for ability to influence Aß neurotoxicity. Compounds identified using these methods can be further screened for effects on Aß fibrillogenesis (see *Protocol 6.38*), Aß phosphorylation (see *Protocol 6.42*) or Aß neurotoxicity (see *Protocol 6.42*).

Reagents

Synthetic Aß peptides (commercially available)

N-terminal Biotin labelled Aß 1–40 and Aß 25–35 (commercially available)

Biotinylated H1 peptide (PKTPKKAKKL) available from Promega

Control peptides should be unrelated to Aß or scrambled sequence variants of Aß.

The reverse Aß 40-1 peptide interacts with the catalase Aß binding domain (Human catalase residues 400–409) and should not be used in binding assays or catalase inhibition assays. Purified or recombinant enzyme preparations are readily available commercially.

Anti-phosphoserine and phosphothreonine antibodies, secondary antibody alkaline phosphatase and HRP conjugates and TMB substrate systems are all commercially available. For enzyme activity assays preparations with known activity are required and should be stored at $-70\,^{\circ}C$; 30% hydrogen peroxide solutions are readily available.

The $[\gamma^{32}P]$-ATP with a specific activity of 3000 Ci/mmol 10 µCi/µl is commercially available and fresh batches should be obtained just prior to carrying out Procedure D (see below).

A liquid scintillation cocktail for non-aqueous samples is required; the choice should be appropriate for the laboratory disposal regulations and ß-counter available. All other standard laboratory reagents should be of high purity and good quality deionized water used.

Equipment

β-Counter

ELISA plate reader with filters for absorbance at 405 and 450 nm

ELISA plate washer

Eppendorf or other laboratory tubes

Freezer, $-70\,^{\circ}C$

Glass tubes

Incubator, $37\,^{\circ}C$

Microtiter plates – NUNC Polysorb or Maxisorb for binding assays

Microcentrifuge

Nitrogen gas supply

Scintillation tubes

Sonicator

Procedures

A. Preparation of unaggregated Aß

For Aß binding and enzyme activity bioassays monomeric unaggregated Aß peptide is required and the peptide should be prepared as follows. This method can be used for all Aß forms, including N-terminally biotinylated derivatives, and should also be used for appropriate control peptides.

1. Lyophilized Aß peptide should be resuspended in neat HPLC grade trifluoroacetic acid to produce a 1 mg/ml solution in a glass tube.

2. Sonicate the resuspended peptide for 15 min at $24\,^{\circ}C$ and if not completely dissolved add further trifluoroacetic acid.

3. Remove trifluoroacetic acid under a stream of nitrogen.

4. Resuspend peptide in hexafluoroisopropanol.

5. Remove hexafluoroisopropanol under a stream of nitrogen.

6. Repeat steps 4 and 5 two further times.

7. For storage resuspend peptide in hexafluoroisopropanol, aliquot into polypropylene tubes, remove hexafluoroisopropanol under a stream of nitrogen and store tubes at $-70\,^{\circ}C$.

8. Resuspend dried down peptide in appropriate buffer prior to use in binding and enzyme activity assays.

B. Binding of Aß to catalase and CDK-1

For Aß binding assays purified or recombinant enzyme preparations can be used. All Aß peptides should be prepared as above.

1. A 500 ml solution of carbonate buffer containing 15 mM Na_2CO_3, 35 mM $NaHCO_3$, 0.01% NaN_3 (w/v), pH 9.6 should be prepared (can be stored for <14 days at $4\,^{\circ}C$).

2. Catalase or CDK-1 (1 µg/ml) should be prepared in carbonate buffer (10 ml/plate).

3. To each of the inner 60 wells on a NUNC Polysorb or Maxisorb 96-well ELISA plate add 100 µl of enzyme solution or carbonate buffer (for control non-specific binding determination). Cover plates with plate sealer (NUNC). Incubate plates at $4\,^{\circ}C$ overnight.

4. Remove plate sealer and wash plates three times with carbonate buffer (200 µl/well).

5. Prepare a 0.2% casein solution in carbonate buffer (20 ml/plate) and add 200 µl/well to each well. Incubate plates at room temperature for 1 h.

6. Prepare a 1 l solution of assay buffer containing 50 mM Tris-HCl, 0.01% NaN_3 (w/v), 0.1% BSA (Cohns Fraction V) (w/v), 0.1% Triton X-100 (v/v), pH 7.5.

7. Wash plates three times with assay buffer (200 μl/well).

8. Prepare a 1 nM solution of biotinylated Aß peptide in assay buffer. This is a 5× stock for addition to plates in the presence of test compound and competing unlabelled Aß peptides. Prepare unlabelled Aß peptides and test compounds in assay buffer in polypropylene tubes at 5× required test concentration.

9. Prepare tubes containing biotinylated Aß, competing peptides and test compounds and assay buffer to give a final 1× solution. Add 100 μl/well to coated ELISA plates and incubate at 37 °C for 4 h.

10. Wash plates three times with assay buffer (200 μl/well).

11. Prepare alkaline phosphatase polymer-streptavidin conjugate (dilute stock solution to 0.5 μg/ml) in assay buffer. Add 100 μl/well to ELISA plates and incubate at 37 °C for 2 h.

12. Wash plates three times with assay buffer (200 μl/well).

13. Prepare substrate containing 2.7 mM p-nitrophenylphosphate, 10 mM diethanolamine, 0.5 mM $MgCl_2$, pH 9.8. Add 100 μl/well and allow sufficient colour development. Colour development can be stopped by addition of 50 μl/well 3 M NaOH if required.

14. Read absorbance at 405 nm using an ELISA plate reader.

C. Inhibition of catalase activity bioassay for Aß

Catalase EC 1.11.1.6 from human erythrocytes has been used for the development of this bioassay for Aß [2]. Purified catalase from other sources and immunoprecipitated catalase can also be used. If catalase from another source is used an Aß binding assay (Procedure B) should be carried out initially to confirm the presence of an Aß binding domain. It is essential to use monomeric Aß preparations as described above since aggregated material shows a reduced ability to inhibit catalase activity [2]. Many compounds and buffer components can influence catalase activity and all assays should include controls to ensure that only a direct action on Aß induced catalase inhibition is being observed. The catalase inhibitor 3-amino-triazole is a useful control material that acts via a similar mechanism to Aß [2, 6].

1. Prepare a 32.4 mM solution of ammonium molybdate (20 g/500 ml) and mix on a heated stirrer until a clear solution is obtained. This solution is stable for <14 days and can be stored at room temperature.

2. Freshly prepare a 100 mM phosphate buffer containing 88 mM Na_2HPO_4, 12 mM NaH_2PO_4, pH 7.4.

3. Prepare a 5% marvel solution in phosphate buffer (20 ml/plate) and add 200 μl/well to each well of a 96-well microtiter plate. Incubate at 37 °C for 1 h.

4. Wash plates three times with phosphate buffer (200 μl/well).

5. Prepare a 100 U/ml solution of catalase in phosphate buffer in a polypropylene tube. In 1.5 ml microcentrifuge tubes prepare dilutions of this top standard to give solutions of 50, 25, 12.5, 6.25, 3.13 and 1.56 U/ml. Also prepare a 50 U/ml solution for incubation with Aß and test compounds.

6. Prepare Aß peptides (10 μM stock) and test compounds in phosphate buffer.

7. Add Aß and test compounds to buffer to give a final Aß concentration of

4 µM in 1.5 ml microcentrifuge tubes. Add 10 µl of each test solution or 10 µl phosphate buffer for controls and standards to wells of a 96-well microtiter plate.

8. Add 10 µl of 50 U/ml catalase solution to each test well and 10 µl of standards to appropriate wells. Add 30 µl phosphate buffer to each well and incubate at 37 °C for 1 h.

9. Prepare substrate solution containing 15 µl (6.5 µmol) H_2O_2 in phosphate buffer.

10. Add 50 µl/well substrate to each well of the microtiter plate using a multi-channel pipette, wait 60 s and then add 100 µl/well 32.4 mM solution of ammonium molybdate. It is essential that the timing of the additions to each well is consistent so that the enzyme is exposed to H_2O_2 for the same time (60 s) in all wells.

11. Read absorbance at 405 nm using an ELISA plate reader.

12. Calculate enzyme activity in each test well from the standard curve.

D. Activation of CDK-1 bioassay for amyloid-ß

CDK-1 activity is dependent on its phosphorylation state and requires the presence of a cyclin component [3, 4]. Recombinant CDK-1/Cyclin-B1 preparations which are biologically active are available. Alternatively active CDK-1 can be immunoprecipitated from cell lysates. An ELISA-based protocol for determination of CDK-1 phosphorylation of either Histone H1 or CSH 103 peptide substrates can be used for comparison.

1. Prepare a Tris assay buffer containing 50 mM Tris-HCl, 10 mM $MgCl_2$,

1 mM EGTA, 2 mM DTT, 40 mM ß-glycerophosphate, 20 mM *p*-nitrophenylphosphate, 0.1 mM sodium vanadate, 50 µM ATP, pH 7.4.

2. Biotinylated H1 peptide (PKTPKKA-KKL) is used as substrate at a final concentration of 25 µM. Prepare a stock of 125 µM peptide in Tris assay buffer.

3. Test compounds and Aß peptides should be prepared in Tris assay buffer at 5× final concentration.

4. Prepare a 5× stock of [γ^{32}P]-ATP (with a specific activity of 3000 Ci/mmol) in Tris assay buffer.

5. Prepare a 200 U/ml solution of CDK-1/Cyclin-B1 in Tris assay buffer.

6. In 1.5 ml polypropylene tubes add 5 µl substrate, 5 µl test compounds, 5 µl [γ^{32}P]-ATP and 5 µl Aß or control peptide. Mix well and then add 5 µl/tube of CDK-1/Cyclin-B1 solution. Incubate for 10 min at 30 °C in a shaking water bath.

7. Prepare a 7.5 M guanidine HCl solution.

8. At the end of the incubation time add 12.5 µl guanidine HCl solution to each tube to terminate the reaction.

9. Spot a 15 µl aliquot of each sample onto a SAM2™ streptavidin membrane (Promega, UK) to isolate the biotinylated substrate.

10. Wash the membrane four times in 2 M NaCl.

11. Wash the membrane four times in 2 M NaCl containing 1% (v/v) H_3PO_4.

12. Wash the membrane twice in deionized H_2O.

13. Cut out spots from membrane and place in scintillation tubes.

14. Add an appropriate liquid scintillation cocktail suitable for non-aqueous samples and measure the radioactivity of [γ^{32}P] incorporated into the biotinylated substrate.

15. The enzyme activity of each sample should be determined from a standard curve.

E. ELISA-based activation of CDK-1 bioassay for amyloid-ß

1. For ELISA-based assay use carbonate buffer (Procedure B, step 1, above) to prepare 1 µg/ml solutions of either Histone H1 or CSH 103 peptide substrates. Coat and block ELISA plates as described in Procedure B, steps 3–5 (above).

2. Prepare an assay buffer containing 50 mM Tris-HCl, 10 mM MgCl$_2$, 1 mM EGTA, 2 mM DTT, 40 mM ß-glycerophosphate, 20 mM *p*-nitrophenylphosphate, 0.1 mM sodium vanadate, pH 7.4.

3. Prepare 250 µM ATP in assay buffer.

4. Test compounds and Aß peptides should be prepared in assay buffer at 5× final concentration.

5. Prepare a 200 U/ml solution of CDK-1/Cyclin-B1 in assay buffer.

6. To each well of a substrate coated ELISA plate add 20 µl ATP, 20 µl test compounds and 20 µl Aß peptides and 20 µl assay buffer and then add 20 µl/well of CDK-1/Cyclin-B1 solution. Incubate for 60 min at 30 °C.

7. Wash plates three times with 200 µl/well 50 mM Tris-HCl containing 0.1% Triton X-100, pH 7.5.

8. Add 100 µl/well anti-phosphoserine or anti-phosphothreonine antiserum (1 : 2500 dilution of neat antibody) and incubate for 2 h at room temperature.

9. Wash plates three times with 200 µl/well 50 mM Tris-HCl containing 0.1% Triton X-100, pH 7.5.

10. Add 100 µl/well anti-rabbit IgG-HRP conjugate (1 : 2000 dilution of neat conjugate) and incubate for 2 h at room temperature.

11. Wash plates three times with 200 µl/well 50 mM Tris-HCl containing 0.1% Triton X-100, pH 7.5.

12. Add 100 µl/well TMB substrate. After sufficient colour development add 50 µl/well 1 M phosphoric acid and measure absorbance at 450 nm using an ELISA plate reader.

References

1. Selkoe, D. J. (1999) Translating cell biology into therapeutic advances in Alzheimer's disease, *Nature*, **399** (Suppl.), A23–A31.
2. Milton, N. G. N. (1999) Amyloid-ß binds catalase with high affinity and inhibits hydrogen peroxide breakdown. *Biochem. J.*, **344**, 293–296.
3. Milton, N. G. N. (2002) The amyloid-ß peptide binds to cyclin B1 and increases human cyclin-dependent kinase-1 activity. *Neurosci. Lett.*, **322**, 131–133.
4. Milton, N. G. N. (2001) Phosphorylation of amyloid-ß at the serine 26 residue by human cdc2 kinase. *NeuroReport*, **12**, 3839–3844.
5. Milton, N. G. N., Mayor, N. P. and Rawlinson, J. (2001) Identification of amyloid-ß binding sites using an antisense peptide approach. *NeuroReport*, **12**, 2561–2566.
6. Milton, N. G. N. (2001) Inhibition of catalase activity with 3-amino-triazole enhances the cytotoxicity of the Alzheimer's amyloid-ß peptide. *NeuroToxicology*, **22**, 767–774.

Amyloid-ß phosphorylation

Nathaniel G. N. Milton

Introduction

Phosphorylation of the amyloid-ß peptide (Aß) is thought to be of major importance in the neurotoxicity of the Aß peptide [1]. The cyclin-dependent kinase-1 (CDK-1) enzyme is able to phosphorylate the Aß peptide at the serine 26 residue. Inhibition of CDK-1 and related kinases can prevent both neurotoxicity and phosphorylation of Aß [1, 2]. Mutation of the serine 26 residue to an alanine prevents Aß neurotoxicity, suggesting that the phosphorylated form may represent the neurotoxic intermediate. Observations that the phosphorylated forms of Aß show markedly increased cytotoxicity [2] further support this view. The presence of phosphorylated Aß in extracts from differentiated human teratocarcinoma cell line, Ntera 2/cl-D1 neurons (NT-2N) and Alzheimer's brain suggests that this form of Aß may be a legitimate target for therapeutic intervention. These results also suggest that measurement of phosphorylated Aß could have diagnostic potential as well as a use in screening for therapeutic compounds.

Reagents

Synthetic phosphorylated Aß peptides can be commercially synthesized as custom peptides. All Aß peptides should be prepared as described in *Protocol 6.41*, Procedure A. Full details of all other reagents required are detailed in *Protocol 6.41*.

Equipment

As described in *Protocol 6.40*. Cell culture facilities are required, including a sterile hood and 37 °C CO_2 incubator

Procedures

A. Culture of human NT-2N cells for neurotoxicity and phosphorylation of amyloid-ß

Human teratocarcinoma cell line, Ntera 2/cl-D1 (NT2), can be obtained from the American Type Culture Collection (ATCC, Rockville, MD, USA). Differentiation into neuronal (NT2N) cells should be carried out as follows:

1. NT2 cells are maintained in Dulbecco's modified high glucose Eagle's medium (DMEM), supplemented with 5% fetal bovine serum and 1% antibiotic mixture comprising penicillin–streptomycin–amphotericin, in a humidified atmosphere at 37 °C with 10% CO_2. Cells should be split 1 : 3 twice a week.

2. For differentiation 1×10^6 cells should be plated in a 75 cm^2 culture flask in DMEM, supplemented with 10% fetal bovine serum, 1% antibiotic mixture comprising penicillin–streptomycin–amphotericin and 10 µM all trans retinoic acid (10 mM stock retinoic acid in DMSO should be diluted into

media just prior to use). Media should be changed three times a week and cells maintained in retinoic acid for 5 weeks.

3. Cells should be replated at a 1:6 split ratio in DMEM supplemented with 10% fetal bovine serum and 1% antibiotic mixture comprising penicillin–streptomycin–amphotericin.

4. Change medium three times over a 1 week period.

5. After 1 week add medium further supplemented with the mitotic inhibitors cytosine arabinoside (1 μM), fluorodeoxyuridine (10 μM) and uridine (10 μM) to inhibit the division of non-neuronal cells.

6. Culture for 2 weeks in DMEM supplemented with mitotic inhibitors with three medium changes/week.

7. Prepare a 1 mg/ml stock solution of poly-D-lysine in sterile water. Sterilize using a 0.2 μm syringe filter (can be stored in aliquots at $-20\,^{\circ}$C). Dilute stock solution 1:100 in sterile distilled water and add diluted poly-D-lysine solution into culture dishes (for 96-well tissue culture plates add 125 μl/well; for 24-well culture plates add 1.25 ml/well). After 3 h remove poly-D-lysine solution and allow dishes to air-dry in a sterile cabinet.

8. Matrigel matrix should be thawed at 4 $^{\circ}$C overnight and diluted 1:25 in cold DMEM. Add Matrigel dropwise (for 96-well tissue culture plates add 1 drop/well; for 24-well culture plates add 5 drops/well) and spread evenly over surface. Allow to air-dry in a sterile cabinet. Store plates until use (stable for 2 months).

9. Repeat application of Matrigel on day of use.

10. After 2 weeks treat cells with trypsin for 2 min, mechanically dislodge cells by gently striking flasks and remove cell suspension.

11. Replate the cell suspension using culture plates pretreated with poly-D-lysine and Matrigel matrix. For neurotoxicity experiments 5×10^3 cells/100 μl culture medium should be plated in 96-well tissue culture plates. For Aβ phosphorylation determination 5×10^4 cells/1 ml culture medium should be plated in 24-well tissue culture plates.

12. Cells should be cultured in DMEM, supplemented with 10% fetal bovine serum and 1% antibiotic mixture comprising penicillin–streptomycin–amphotericin for a further 2 weeks prior to experimentation. Medium should be changed twice a week.

13. On day of experimentation test compounds can be added in fresh medium and neurotoxicity or Aβ phosphorylation determined after an appropriate incubation period.

14. The neurotoxicity of phosphorylated Aβ can be measured using MTT reduction or trypan blue dye exclusion as indicators of cell viability. For MTT reduction prepare a 5 mg/ml solution of MTT in DMEM, sonicate briefly and sterilize using a 0.2 μm syringe filter (can be stored in aliquots at $-20\,^{\circ}$C). After incubation of cells with test compounds add MTT to a final concentration of 0.4 mg/ml to each well and incubate cells for a further 6 h. Prepare cell lysis buffer, sufficient for 100 μl/well, containing 20% sodium dodecyl sulfate and 50% N,N-dimethylformamide pH 4.7. After incubation with MTT add cell lysis buffer (100 μl/well) and leave plate at 37 $^{\circ}$C overnight. Read absorbance

at 570 nm. For trypan blue exclusion after incubation of cells with test compounds dislodge cells from wells and add an equal volume of 0.4% trypan blue. Count the numbers of viable (non-stained) and dead cells using a haemacytometer.

B. Extraction of phosphorylated amyloid-ß from cells

The following procedure should be used to extract Aß peptides from cells:

1. Prepare a 100 ml extraction buffer containing 0.2% DEA, 50 mM NaCl and 0.1 mM sodium vanadate. Remove culture medium from cells in 24-well culture plates and add 0.5 ml extraction buffer to each well. Disrupt cells by scraping and repeated pippetting.

2. Transfer disrupted cell suspension to a 1.5 ml polypropylene microcentrifuge tube, sonicate for 15 min and then centrifuge for 10 min at 13 000 rpm to remove cell debris.

3. Transfer supernatant to a 1.5 ml polypropylene microcentrifuge tube and add to an equal volume of 1 M Tris containing 0.2% Triton X-100 pH 8.0. Add 100 µl of anti-Aß 15–30 antiserum (1 : 5000 dilution of neat antiserum), mix and incubate at 4 °C overnight.

4. After incubation add 100 µl of protein-A agarose and incubate for 2 h at room temperature.

5. Centrifuge for 10 min at 13 000 rpm to pellet protein-A agarose. Remove supernatant and resuspend protein-A agarose in 1 ml 50 mM Tris containing 0.1% Triton X-100, pH 7.5.

6. Repeat step 5 twice.

7. Prepare elution buffer containing 20% acetonitrile in 0.1% trifluoroacetic acid.

8. Remove supernatant and resuspend protein-A agarose in 0.5 ml elution buffer, mix for 10 min. Centrifuge for 10 min at 13 000 rpm to pellet protein-A agarose. Transfer supernatant to a glass tube and resuspend protein-A agarose in 0.5 ml elution buffer. Repeat this step twice and pool the supernatants.

9. Prepare a Sep-Pak C_{18} column on a 10 ml syringe. Pass 5 ml methanol followed by 5 ml 0.5 M acetic acid over the column. Pass 5 ml 20% acetonitrile in 0.1% trifluoroacetic acid over the column and add the sample. Wash the column with 10 ml 20% acetonitrile in 0.1% trifluoroacetic acid followed by 10 ml deionized water. Elute bound peptides with 2 ml 70% acetonitrile in 0.1% trifluoroacetic acid.

10. Purified peptides should be dried down under a stream of nitrogen, resuspended in 1 ml of trifluoroethanol and dried down under a stream of nitrogen. Dried down extracts can be stored at −70 °C prior to assay.

C. Immunoassay measurement of phosphorylated amyloid-ß

1. A 500 ml solution of carbonate buffer containing 15 mM Na_2CO_3, 35 mM $NaHCO_3$, 0.01% NaN_3 (w/v), pH 9.6 should be prepared (can be stored for < 14 days at 4 °C).

2. Rabbit anti-phosphoserine antibody (1 : 1000 dilution of neat antibody) should be prepared in carbonate buffer (10 ml/plate).

3. To each of the inner 60 wells on a NUNC Polysorb or Maxisorb 96-well ELISA plate add 100 µl of anti-phosphoserine antibody solution or carbonate buffer (for control non-specific binding determination). Cover

plates with plate sealer (NUNC). Incubate plates at 4 °C overnight.

4. Remove plate sealer and wash plates three times with carbonate buffer (200 µl/well).

5. Prepare a 5% marvel solution in carbonate buffer (20 ml/plate) and add 200 µl/well to each well. Incubate plates at room temperature for 1 h.

6. Prepare 500 ml of assay buffer–50 mM Tris-HCl containing 0.1% BSA (Cohn Fraction V), 0.1% Triton X-100, pH 7.5.

7. Wash plates three times with 200 µl/well assay buffer.

8. Prepare synthetic phosphorylated Aß peptide standards (0–10 nM) and samples in assay buffer, add 100 µl/well to anti-phosphoserine coated plates, seal plates with plate sealer and incubate at 4 °C overnight.

9. Wash plates three times with 200 µl/well assay buffer.

10. Add 100 µl/well mouse anti-Aß monoclonal antibody 6F3D (1 : 3000 dilution of neat antibody in assay buffer) and incubate for 2 h at room temperature.

11. Wash plates three times with 200 µl/well assay buffer.

12. Add 100 µl/well anti-mouse IgG-HRP conjugate (1 : 2000 dilution of neat conjugate in assay buffer) and incubate for 2 h at room temperature.

13. Wash plates three times with 200 µl/well assay buffer.

14. Add 100 µl/well TMB substrate (KPL). After sufficient colour development add 50 µl/well 1 M phosphoric acid and measure absorbance at 450 nm using an ELISA plate reader.

D. Phosphorylation of amyloid-ß by CDK-1

CDK-1 activity can be measured as described in *Protocol 6.41*, Procedure D, using biotinylated Aß 1–42, Aß 1–40 or Aß 25–35 in place of the biotinylated H1 peptide substrate. The following immunoassay method has the advantage of not using a radioactive label:

1. A 500 ml solution of carbonate buffer containing 15 mM Na_2CO_3, 35 mM $NaHCO_3$, 0.01% NaN_3 (w/v), pH 9.6 should be prepared (can be stored for < 14 days at 4 °C).

2. Streptavadin (1 µg/ml) should be prepared in carbonate buffer (10 ml/plate).

3. To each of the inner 60 wells on a NUNC Polysorb or Maxisorb 96-well ELISA plate add 100 µl of streptavidin solution or carbonate buffer (for control non-specific binding determination). Cover plates with plate sealer (NUNC). Incubate plates at 4 °C overnight.

4. Remove plate sealer and wash plates three times with carbonate buffer (200 µl/well).

5. Prepare a 0.2% casein solution in carbonate buffer (20 ml/plate) and add 200 µl/well to each well. Incubate plates at room temperature for 1 h.

6. Prepare 500 ml 50 mM Tris-HCl containing 0.1% Triton X-100, pH 7.5.

7. Wash plates three times with 200 µl/well 50 mM Tris-HCl containing 0.1% Triton X-100, pH 7.5.

8. Prepare an assay buffer containing 50 mM Tris-HCl, 10 mM $MgCl_2$, 1 mM EGTA, 2 mM DTT, 40 mM ß-glycerophosphate, 20 mM *p*-nitrophenylphosphate, 0.1 mM sodium vanadate, pH 7.4.

9. Prepare 250 µM ATP in assay buffer.

10. Biotinylated Aß peptides are used as substrates at a final concentration of 25 μM. Prepare a stock of 125 μM peptide in Tris assay buffer.

11. Test compounds should be prepared in assay buffer at 5 × final concentration.

12. Prepare a 200 U/ml solution of CDK-1/Cyclin-B1 in assay buffer.

13. In 1.5 ml polypropylene tubes mix 5 μl substrate, 5 μl ATP, 5 μl test compounds and 5 μl assay buffer and then add 5 μl/tube of CDK-1/Cyclin-B1 solution. Incubate for 30 min at 30 °C in a shaking water bath.

14. Add 200 μl 50 mM Tris-HCl containing 0.1% Triton X-100, pH 7.5 to each sample. Pipette 100 μl of each sample into duplicate wells of a streptavidin coated ELISA plate. Incubate for 4 h at 4 °C.

15. Wash plates three times with 200 μl/well 50 mM Tris-HCl containing 0.1% Triton X-100, pH 7.5.

16. Add 100 μl/well anti-phosphoserine antiserum (1 : 2500 dilution of neat

antibody) and incubate for 2 h at room temperature.

17. Wash plates three times with 200 μl/well 50 mM Tris-HCl containing 0.1% Triton X-100, pH 7.5.

18. Add 100 μl/well anti-rabbit IgG-HRP conjugate (1 : 2000 dilution of neat conjugate) and incubate for 2 h at room temperature.

19. Wash plates three times with 200 μl/well 50 mM Tris-HCl containing 0.1% Triton X-100, pH 7.5.

20. Add 100 μl/well TMB substrate (KPL). After sufficient colour development add 50 μl/well 1 M phosphoric acid and measure absorbance at 450 nm using an ELISA plate reader.

References

1. Milton, N. G. N. (2001) Phosphorylation of amyloid-ß at the serine 26 residue by human cdc2 kinase. *NeuroReport*, **12**, 3839–3844.
2. Milton, N. G. N. (2002) Peptides for use in the treatment of Alzheimer's disease. *Patent Publication*, WO 02/36614 A2.

Smitin–myosin II coassembly arrays *in vitro*

Richard Chi and **Thomas C. S. Keller III**

Introduction

Smitin is a large smooth muscle protein that colocalizes with smooth muscle myosin II filaments *in vivo* and associates with reconstituted smooth muscle myosin II filaments *in vitro* [1]. Although the unusually high molecular weight and molecular morphology of the long, fibrous smitin molecules resemble those properties of striated muscle titin, an understanding of the molecular relationship between smitin and titin awaits determination of smitin protein or gene sequence.

Smitin can be isolated from a variety of smooth muscle tissues including avian gizzard, which provides an excellent source of the protein. Published protocols designed for the purification of smooth muscle myosin II routinely yield preparations of myosin II contaminated with significant amounts of smitin, not readily detected with SDS-PAGE protocols commonly used to analyse such preparations. Likewise, the following protocol, which is designed to maximize isolation of smitin, yields smitin preparations contaminated with significant amounts of myosin II but little or no other proteins. This preparation is useful for investigating the properties of smitin–myosin II coassembly *in vitro* [1].

In physiological ionic strength conditions, myosin in the smitin–myosin preparations self-assembles into long, sidepolar filaments similar to those found in the contractile apparatus of smooth muscle cells *in vivo*. In the presence of smitin, these myosin II filaments aggregate into irregular arrays. In low ionic strength conditions, smooth muscle myosin self-assembles into small bipolar myosin II filaments, which in the presence of smitin align end to end and side by side into highly regimented arrays. The organization of these coassembly arrays resembles that of the coassembly arrays formed by cellular titin and myosin II *In vitro* and may represent the organization of smitin and smooth muscle myosin II in the stress fibre-like structures in proliferative, motile smooth muscle cells.

Reagents

Buffer A: 50 mM KCl, 2 mM $MgCl_2$, 1 mM EDTA, 1 mM EGTA, 0.5 mM DTT, 0.2 mM PMSF, ① 0.002 mM leupeptin, 10 mM imidazole-Cl (pH 7.0)

Buffer B: 0.6 M KCl, 4 mM ATP, 2 mM $MgCl_2$, 1 mM EDTA, 1 mM EGTA, 0.5 mM DTT, 0.2 mM PMSF, 0.002 mM leupeptin, 10 mM imidazole-Cl (pH 7.0)

Buffer C: 0.2 M KCl, 1 mM EGTA, 0.5 mM EDTA, 0.2 mM DTT, 10 mM imidazole-Cl (pH 7.5)

Buffer D: 0.6 M KCl, 0.2 mM DTT, 10 mM imidazole-Cl (pH 6.9) (used also with inclusion of 0.4 M potassium phosphate)

Buffer E: 50 mM KCl, 2 mM $MgCl_2$, 5 mM EDTA, 0.1 mM EGTA, 0.2 mM DTT, 10 mM imidazole-Cl (pH 7.0)

Buffer F: 150 mM KCl, 2 mM MgCl$_2$, 5 mM EDTA, 0.1 mM EGTA, 0.2 mM DTT, 10 mM imidazole-Cl (pH 7.0)

Equipment

Waring Blender

Preparative Superspeed Refrigerated Centrifuge with large capacity rotor (e.g. GSA fixed-angle rotor and 250 ml centrifuge bottles or equivalent)

Sephacryl S-1000 Superfine Matrix column (2.5 × 90 cm)

Hydroxylapatite Matrix column (2.5 × 13.5 cm)

Fraction collector

Procedure

Three fresh adult chicken gizzards will provide 20 g of smooth muscle tissue, the optimal amount of smooth muscle for this protocol. ②

1. Remove and discard the gizzard surface fat tissue. Dissect the smooth muscle tissue off the tough gizzard lining. Dice the smooth muscle tissue into pieces smaller than 1 cm^2 with a sharp scalpel or razor blade.

2. Mix the gizzard tissue with buffer A (50 ml/g of starting smooth muscle tissue) in a blender and homogenize with three bursts of 10 s each at maximum speed.

3. Pellet the tissue by centrifugation at 5000g for 15 min.

4. Repeat steps 2 and 3 twice.

5. Resuspend the pellet with buffer B (2 ml/g of starting smooth muscle tissue) and stir gently for 30 min in an ice bath.

6. Pellet the tissue by centrifugation at 15 000g for 30 min.

7. Decant the extract supernatant. Load 20 ml of the extract onto a Sephacryl S-1000 column (2.5 × 90 cm) equilibrated and eluted with degassed buffer C. ③ Smitin and myosin coelute near the void volume of the column (~240–280 ml).

8. Assay column fractions for smitin by SDS-PAGE. ④ Pool the peak smitin fractions, minimizing actin contamination.

9. To remove contaminating actin, α-actinin, and other proteins, apply pooled smitin–myosin II fractions to a hydroxyapatite column (2.5 × 13.5 cm) equilibrated with buffer D and eluted with a gradient of 0.0–0.4 M potassium phosphate in buffer D. Purified fractions containing smitin and myosin typically elute in fractions between 0.23 and 0.26 M phosphate.

10. Smitin and myosin in preparations containing both proteins coassemble when dialysed into low ionic strength (buffer E) or physiological ionic strength (buffer F) conditions lacking ATP for 6–48 h. ⑤

11. Smitin–myosin coassemblies can be pelleted by centrifugation at 10 000g for 5 min.

Notes

The protein purification steps of this procedure require approximately 2 working days to complete. Prolonged preparation time results in high actin contamination and irreversible smitin/myosin aggregation. All steps should be performed at 4 °C.

① PMSF is hazardous; handle with care. Make a stable 0.2 M PMSF stock solution in ethanol. Minimize precipitation of PMSF on dilution into aqueous solution, by slowly expelling the aliquot of PMSF stock solution from a pipette tip submersed in vigorously stirring buffer solution. PMSF has a half-life as short as 30 min in some aqueous solutions; add to the buffer solution immediately before use.

② Chill the gizzards on ice as soon as possible after acquisition. Keep the smooth muscle tissue cold throughout the initial steps of the protocol to minimize actin contamination of the purified smitin–myosin II preparation. For optimal purity of smitin–myosin II preparations, use the gizzards within 2 h of sacrifice. Cleaned and diced gizzard smooth muscle that has been rapidly frozen in liquid N_2 and stored at $-20\,^{\circ}C$ until use also provides a suitable source of smitin for purification.

③ Use of Sephacryl S-500 instead of S-1000 yields similar results. Minimize actin filament contamination of the smitin–myosin II Sephacryl S-1000 fractions by loading the column with a 50 ml front of 0.6 M KI (made in buffer C) immediately ahead of the protein load. For convenience, this column can be run overnight.

④ SDS-PAGE: To maximize protein concentration in gel samples of the column fractions, make gel samples using 8 × or 10 × Laemmli sample buffer. Heating the samples to 90 $^{\circ}C$ for 3 min minimizes the amount of myosin heavy chain trapped in the smitin band on the gel. Overheating can cause loss of full-length smitin from the sample. Samples can be analysed in large or mini gel formats. Using an acrylamide stock with a 75 : 1 ratio of acrylamide to bis-acrylamide to form the running and stacking gels enhances migration of smitin into the gel. Proteins with polypeptide molecular weights ranging from greater than 3 MDa to 15 kDa, including smitin, myosin heavy and light chains, and actin, can be fractionated in lanes on a 4–20% acrylamide gradient running gel with a 3% stacking gel. Allowing the gel to run for an additional 25–33% of the time it takes the dye front to reach the bottom of the gel increases migration of smitin into the gel and increases fractionation resolution of the proteins in the sample.

⑤ Prepare dialysis tubing according to the manufacturer's recommendations. Select the dimensions of the tubing to maximize the surface to volume ratio of the bag. The rate of coassembly formation is inversely related to the protein concentration of the sample. At high protein concentration, smitin–myosin coassembly aggregates may be visible in the dialysis bag as white flocculent material in less than 6 h.

References

1. Kim, K. and Keller, T. C. (2002) Smitin, a novel smooth muscle titin-like protein, interacts with myosin filaments *in vivo* and *in vitro*, *J. Cell Biol.*, **156**, 101–112.

Assembly/disassembly of myosin filaments in the presence of EF-hand calcium-binding protein S100A4 *in vitro*

Marina Kriajevska, Igor Bronstein and **Eugene Lukanidin**

Reagents

pQE30 expression vector (QIAGEN), enzymes for RT-PCR and cloning, *E. coli* M15 competent cells (QIAGEN), IPTG (1 M in H_2O), 50% slurry of Ni-NTA resin (QIAGEN), buffer A (6 M GuHCl, 0.1 M Na-phosphate, 0.01 M Tris-HCl pH 8.0), buffer B (8 M urea, 0.1 M Na-phosphate, 0.01 M Tris-HCl pH 8.0), buffer C (8 M urea, 0.1 M Na-phosphate, 0.01 M Tris-HCl pH 6.3), 1 M imidazole, 5 M NaCl, 1 M DTT, glycerol, 1 M MES pH 6.2, acetic acid glacial, 0.5 M EDTA, 1 M $CaCl_2$, Protein assay reagent (Pierce), Ezblue gel staining reagent (Sigma)

Procedures

A. Isolation of the recombinant C-terminal fragment of the heavy chain of non-muscle myosin IIA

1. Clone cDNA corresponding to the C-terminal fragment of the heavy chain of non-muscle myosin IIA in BamHI site of pQE30 bacterial expression vector ① and use it for the transformation of *E. coli* M15 cells.

2. Prepare overnight culture and use it for the preparation of large-scale culture (0.5 l) according to the QIAGEN protocol.

3. Perform IPTG induction (1 mM) of culture for 5 h. Harvest cells by centrifugation at 3000*g* for 20 min.

4. Store cell pellet at $-80\,°C$ or process immediately purification of the myosin fragment in denaturing conditions according to the QIAGEN recommendations.

5. Briefly, resuspend pellet in buffer A (5 ml/g wet weight) and stir cells at room temperature until lysis of cells is completed.

6. Centrifuge the lysate at 20 000*g* for 20 min and mix supernatant with 4 ml of 50% slurry of Ni-NTA resin, which was previously equilibrated in buffer A.

7. After 1 h of stirring at room temperature (RT), apply resin to the column and wash consequently with buffers A, B and C until the A_{280} of the flow-through is lower than 0.01 at each step.

8. Perform refolding of the myosin fragment with the 30 ml of the linear 6 M−1 M urea gradient in 0.5 M NaCl, 20% glycerol, 10 mM Tris pH 7.4.

9. Elute protein in 0.5 ml fractions using 5 ml of the elution buffer (1 M urea, 0.5 M NaCl, 10% glycerol, 10 mM Tris pH 7.4, 250 mM imidazole).

10. Combine pick fractions and dialyse against 0.6 M NaCl, 10 mM MES pH 6.2, 1 mM DTT.

11. Centrifuge sample after dialysis for 10 min at 20 000g. Aliquot sample, freeze in liquid N$_2$ and store at −80 °C.

B. Isolation of recombinant S100A4

1. Clone cDNA of S100A4 in the pQE30 expression vector (QIAGEN) in BamHI site (2) and use for the transformation of E. coli M15 cells.

2. Prepare 1 l of bacterial culture as described for the myosin preparation.

3. Resuspend cell pellet in the sonication buffer (50 mM Tris-HCl pH 7.5, 200 mM NaCl) at 4 volumes per gram of wet weight.

4. After three times freezing (liquid N$_2$) and thawing (RT), sonicate cell lysate on ice three times for 20 s following centrifugation for 20 min at 20 000g.

5. Mix supernatant with 6 ml of 50% slurry of Ni-NTA resin equilibrated by the sonication buffer.

6. Raise NaCl concentration in the mixture up to 1 M and rotate for 1 h at +4 °C.

7. Load resin into the column and wash at +4 °C with 50 mM Tris, pH 7.5 until the A_{280} of the flow-through is less than 0.01.

8. Elute the S100A4 with buffer containing 1 M NaCl, 100 mM acetic acid pH 4.0.

9. Dialyse sample against access of the buffer containing 10 mM MES pH 6.2, 50 mM NaCl, 0.5 mM EDTA, 1 mM DTT overnight at +4 °C and centrifuge at 20 000g for 10 min.

10. Aliquot the protein, freeze in liquid N$_2$ and store at −80 °C. ③

11. Measure the concentration of the protein by using the Protein assay reagent (Pierce).

C. Preparation of myosin filaments

1. Prepare 100 μl solution of myosin (10 μg of the protein, final concentration 0.1 mg/ml) in 0.6 M NaCl, 10 mM MES pH 6.2, 1 mM DTT, 1 mM CaCl$_2$ at RT.

2. Dialyse for 5 h at RT against 200 ml of the buffer containing 10 mM MES pH 6.2, 1 mM DTT, 1 mM CaCl$_2$, 50 mM NaCl. Myosin filaments are formed during dialysis.

3. Precipitate filaments by centrifugation for 10 min at 20 000g and save pellet for the next step.

D. S100A4 stimulates disassembly of preformed myosin filaments

1. Add 50 μl of the S100A4 (1 mg/ml) in low ionic strength buffer (10 mM MES pH 6.2, 1 mM DTT, 1 mM CaCl$_2$, 50 mM NaCl) or 50 μl of high ionic strength buffer (10 mM MES pH 6.2, 1 mM DTT, 1 mM CaCl$_2$, 600 mM NaCl) without S100A4 to the myosin pellet (10 μg).

2. Incubate tubes at RT for 1 h with gentle agitation.

3. Centrifuge 10 min at 20 000g and analyse supernatants by running the 15% SDS-PAGE. Almost 80% of the C-terminal fragment becomes soluble in the presence of S100A4, while the high ionic strength buffer is able to solubilize 40% of the precipitated myosin fragment.

E. S100A4 prevents aggregation of myosin filaments in the presence of calcium

1. Prepare 100 µl mixture of myosin (10 µg with final concentration 0.1 mg/ml) and S100A4 (50 µg) in a buffer (0.6 M NaCl, 10 mM MES pH 6.2, 1 mM DTT) containing 1 mM $CaCl_2$ or 5 mM EDTA.

2. Dialyse proteins for 5 h at RT against 200 ml of buffer, 10 mM MES pH 6.2, 1 mM DTT, 1 mM $CaCl_2$ or 5 mM EDTA and different NaCl concentrations, 50, 100, 200, 300 and 400 mM.

3. Centrifuge samples for 10 min at 20 000g.

4. Analyse supernatant in the presence of the myosin fragment by running a 15% SDS-PAGE and staining with Ezblue gel staining reagent (Sigma). ④

Notes

① Synthesis of the cDNA corresponding to the C-terminal 202 aa fragment of the heavy chain of the human non-muscle myosin IIA: prepare total RNA from human platelets and perform RT-PCR using specific myosin primers, forward primer, CGGGATCCCAGAT CAACGCCGAC (+5303 to +5317), and reverse primer, CGGGATCCGG CTTATTCGGCAGG (+5894 to +5908).

② Synthesis of the human S100A4 cDNA: prepare total RNA from human platelets and perform RT-PCR with specific S100A4 primers, forward primer, CGGGATCCATGGC GTGCCCTCTG (+65 to +79) and reverse primer, CGGGATCCGAG TTTTCATTTCTTCCTGGG (+356 to +376).

③ Fresh prepared S100A4 keep at $-80\,°C$ in order to prevent aggregation of the protein.

④ In the presence of S100A4 and calcium, the myosin fragment remains in a soluble form in low ionic buffer.

Acknowledgements

This work was supported by grants from the Danish Cancer Society, Dansk Kræftforsknings Fond, Novo Nordisk Fonden and Yorkshire Cancer Research.

Collagen fibril assembly *in vitro*

David F. Holmes and Karl E. Kadler

Introduction

It has long been known that native-like collagen fibrils can be produced *in vitro* by self-assembly by incubation of extracted and purified type I collagen molecules in warm, neutral solution conditions [1, 2]. The individual fibrils have a polarized arrangement of molecules with an axial D-periodicity of 67 nm and are of near-uniform diameter. Collagen fibrils are stabilized in tissue by the formation of intermolecular cross-links involving specific lysine/hydroxylysine residues and aldimine derivatives of these. Collagen is commonly extracted from young tissues (typically calf skin or rat tail tendon) using acetic acid solution [3] which breaks these intermolecular cross-links. Alternatively collagen can be released by pepsin treatment which cleaves the telopeptides (terminal non-triple-helical domains) where cross-link sites are located. Fibril assembly is, however, critically dependent on the intactness of the telopeptides [2, 4]. Partial loss can lead to a changed morphology and loss of polarity. Extensive loss can completely inhibit fibril formation [4].

The fibrils in the final reconstituted gel of type I collagen have a fairly broad diameter distribution, compared with those in tissue, and this is dependent on the solution conditions at pH 7.4 (temperature and ionic strength, as well as ion type). Optimal conditions for the formation of compact, uniform-diameter, native-like fibrils have generally involved a temperature of 34 °C or less and a phosphate buffer [5, 6].

Different assembly pathways have been observed in the reconstitution of collagen fibrils [5, 7]. The transfer route used to raise pH, temperature and ionic strength from the initial conditions (cold, acid solution) to the reconstitution conditions (warm, neutral solution) has been found to be a major determinant of the assembly pathway [5]. Warming before neutralization ('warm start initiation') leads to abundant short 'early fibrils' with tapered tips. Similar 'early fibrils' are also observed during collagen fibril assembly in tissue [8]. In contrast, neutralization followed by warming ('neutral start initiation') leads to the early accumulation of filaments [5, 7].

The procedure described here is based on a set of 'optimal' reconstitution conditions: 200 µg/ml collagen, I 0.2, pH 7.4 phosphate buffer and a temperature of 34 °C. These conditions are established using the 'warm start' initiation (Figure 6.45(a)). The kinetics of assembly can be monitored by turbidimetry (Figure 6.45(b)) and the assembly products can be studied by transmission electron microscopy (Figure 6.45(c)).

Reagents

Extracted and purified type I collagen is available from biochemical suppliers. ① ②

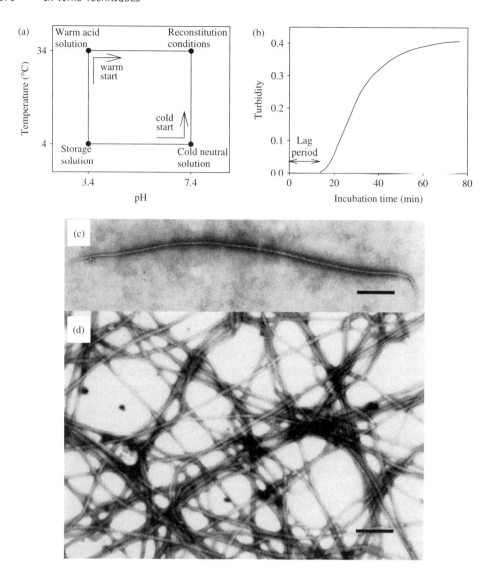

Figure 6.45 Reconstitution of type I collagen fibrils at 34 °C, I 0.2, pH 7.4. (a) Different initiation routes are possible in changing from cold acid to warm neutral conditions and this can affect both the assembly pathway and kinetics of fibril formation. (b) Typical turbidimetric curve (obtained at 313 nm) after 'warm start' initiation with I 0.2, pH 7.4 phosphate buffer at 34 °C. The turbidity is proportional to the extent of fibril assembly. (c) Typical 'early fibril' with tapered tips, found throughout the lag and early growth phases of reconstitution (scale bar = 300 nm). (d) Typical final gel of D-periodic fibrils (D = 65 nm in the dried and negatively stained preparation) with no fibril tips apparent (scale bar = 300 nm)

Suitable acetic-acid-soluble type I collagen can alternatively be extracted and purified by well-established procedures [3] with modifications to ensure minimal loss of the telopeptides [4]. Laboratory reagents should be of high purity.

Equipment

Temperature-controlled water bath

Spectrophotometer with temperature-controlled cell holder

Transmission electron microscope

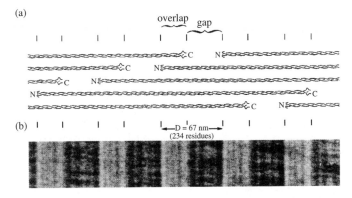

Figure 6.46   Axial structure of type I collagen fibril. (a) Schematic diagram to show the axial arrangement of collagen molecules in the fibrils. Short non-triple domains ('telopeptides') are located at the ends of the long triple helical (300 nm) domain. (b) Characteristic negative stain pattern (using 1% sodium phosphotungstate, pH 7) of native-type D-periodic collagen fibril reconstituted from acetic-acid-soluble type I collagen. This is shown at the same magnification and aligned with the schematic axial structure in (a). The gap/overlap structure of the fibril and the fine stain excluding telopeptide regions at the gap/overlap junctions are apparent

Procedure

1. Start with a collagen solution in 10 mM acetic acid solution (using dialysis if necessary) at 4 °C.

2. Dilute ③ the collagen to ~400 μg/ml with 10 mM acetic acid solution and keep at 4 °C.

3. Warm equal volumes of collagen solution and double strength phosphate buffer (124 mM Na_2HPO_4, 29.2 mM KH_2PO_4) separately to 34 °C for about 5 min. This yields final solution conditions of I 0.2, pH 7.4.

4. Mix the two volumes (acid collagen solution and double strength buffer) to start fibril reconstitution and incubate at 34 °C for several hours. ④

5. A uniform, cloudy gel will form. This can be pulled from the tube with forceps.

6. Prepare TEM specimens by wiping a small portion of gel over a carbon-filmed grid, washing with several drops of water and negatively staining with 1–2% sodium phosphotungstate, pH 7.0 (or other suitable stain, e.g. 1–2% uranyl acetate).

7. Examine the grid sample by TEM. The fibrils should be compact with a polarized, D-periodic stain pattern and uniform diameter along individual fibrils (Figures 6.45(d) and 6.46). Fibril diameters are typically in the range 30–90 nm.

Notes

① Suppliers include BD Biosciences, USA and Elastin Products Company, USA.

② The fibril reconstitution procedure described here is appropriate for acetic-acid-extracted collagen rather than pepsin-extracted.

③ Collagen concentration can be determined by a direct colorimetric assay ('Sircol' kit available from Biocolor Ltd, Northern Ireland).

④ The kinetics of assembly can be simply monitored by turbidimetry measurements using a spectrophotometer and recording the absorbance with time (see Figure 6.46(b)).

References

1. Gross, J. and Kirk, D. (1958) *J. Biol. Chem.*, **233**, 355–360.
2. Kadler, K. E., Holmes, D. F., Trotter, J. A. and Chapman J. A. (1996) *Biochem. J.*, **316**, 1–11.
3. Jackson, D. S. and Cleary, E. G. (1967) *Methods Biochem. Anal.*, **15**, 25–76.
4. Capaldi, M. J. and Chapman, J. A. (1982) *Biopolymers*, **21**, 2291–2313.
5. Holmes, D. F., Capaldi, M. J. and Chapman, J. A. (1986) *Int. J. Biol. Macromol.*, **8**, 161–166.
6. Williams, B. R., Gelman, R. A., Poppke, D. C. and Piez, K. A. (1978) *J. Biol. Chem.*, **253**, 6578–6585.
7. Gelman, R. A., Willaims, B. R. and Piez, K. A. (1979) *J. Biol. Chem.*, **254**, 180–186.
8. Birk, D. E., Hahn, R. A., Linsenmeyer, C. Y. and Zycband, E. I. (1996) *Matrix Biol.*, **15**, 111–118.

7

Selected Reference Data for Cell and Molecular Biology

David Rickwood

Chemical safety information

ABTS	May be harmful by inhalation, ingestion or skin absorption. Causes eye and skin irritation and irritating to mucous membrane and upper respiratory tract
Acetic acid	Harmful if swallowed, inhaled or absorbed through skin. Material extremely destructive of tissues of mucous membranes, upper respiratory tract, eyes and skin. Inhalation may be fatal
Acetone	Highly flammable. Use in well-ventilated area away from sources of ignition
Acridine orange	Harmful: possible risk of irreversible effects through inhalation, in contact with skin and if swallowed
Aminoethylbenzenesulfonyl fluoride (AEBSF)	Irritant to eyes, respiratory system and skin
8-aminonaphthalene-1, 3, 6-trisulfonic acid (ANTS)	Irritant to eyes, respiratory system and skin
Aminopterin:	Can be fatal if inhaled, swallowed or absorbed through skin. Teratogen
Bacto-tryptone	Irritant. Avoid contact with eyes and skin
Barbital	Harmful if swallowed. Causes irritation. May cause allergic skin reaction. Effects of overexposure include sedation, hypnosis and coma depending on amount ingested
Bleach	All cleaning solutions are potentially dangerous. Avoid contact with skin. Keep only small quantities on the open bench, within the confines of a spillage tray

(continued overleaf)

Cell Biology Protocols. Edited by J. Robin Harris, John Graham, David Rickwood
© 2006 John Wiley & Sons, Ltd

Borate buffer	May be harmful by inhalation, ingestion and skin absorption. Causes irritation. Vapour causes irritation to eyes, mucous membranes and upper respiratory tract
Boric acid H_3BO_3	Harmful by inhalation, in contact with skin and if swallowed. Irritating to eyes, respiratory system and skin. Possible teratogen. Reproductive hazard
Bradford Reagent	Causes burns, toxic by inhalation and if swallowed. Avoid contact with skin. In case of contact with eyes, rinse immediately with plenty of water and seek medical advice
5-bromo-2′-deoxyuridine (BrdU)	Experimental teratogen. May be harmful if swallowed, inhaled or absorbed through the skin
Calcein AM	May be harmful if inhaled, swallowed or absorbed through the skin. Toxicology not fully investigated
$CaCl_2$	Toxic; irritating to eyes, respiratory system and skin
Carbenicillin	Harmful
Cedarwood oil	Irritating to eyes, respiratory system and skin
Chloramine T	May be harmful by inhalation, ingestion and skin absorption. Causes eye and skin irritation. Causes irritation to mucous membranes and upper respiratory tract
Chloroform	Chloroform should be handled with care, and disposed of by means appropriate for organic solvents
Chloros	Irritating to eyes and respiratory system. Contact with acid liberates toxic chlorine gas
Chromic acid	Sulfuric acid is highly corrosive, and the making of chromic acid is a very hazardous procedure. Chromates are potential carcinogens. Protective clothing, including heavy-duty rubber gloves, must be worn while making chromic acid, and the operation carried out in a sink inside a fume cupboard. Make only small quantities
Cisplatin	Very toxic if inhaled, swallowed, or absorbed through the skin. Many platinum compounds are sensitizers, so even low exposure may be harmful if taking place repeatedly or over an extended period. Experimental carcinogen, teratogen. IARC probable human carcinogen. May cause reproductive damage. Human mutagenic data. May cause severe allergic reaction, which may get worse with continuing exposure
Citric acid	May be harmful by inhalation, ingestion or skin absorption. Causes eye and skin irritation and irritating to mucous membrane and upper respiratory tract
Colcemid	Toxic by inhalation, and if swallowed, and contact with skin. Possible risk of irreversible effects
Cryogen/liquid nitrogen	The handling and storage of liquid nitrogen must be performed strictly within laboratory safety regulations. The wearing of protective glasses should be obligatory and protective gloves when carrying any polystyrene container filled with liquid nitrogen

$CuSO_4$	Harmful if swallowed. Irritating to eyes and skin. Do not breathe dust
$CuSO_4.5H_2O$	Harmful if swallowed. Irritating to eyes and skin. Eye contact: irrigate thoroughly with water for at least 10 min. If discomfort persists obtain medical attention. Inhalation: remove from exposure, rest and keep warm. In severe cases obtain medical attention. Skin contact: wash off skin with water. Remove contaminated clothing and wash before reuse. In severe cases, obtain medical attention. Ingestion: wash out mouth thoroughly with water and give plenty of water to drink. Obtain medical attention
Cycloheximide	Very toxic if swallowed, inhaled or absorbed through the skin
Cyclosporin A	Potent immunosuppressant; may cause irritation to eyes, skin, mucous membranes. Harmful if swallowed. May cause irritation to the eyes, respiratory system and skin
Cytochalasin B	Very toxic by inhalation, in contact with skin and if swallowed
Diaminobenzidine tetrahydrochloride DAB	Carcinogenic
Diethanolamine	Harmful if swallowed, inhaled or absorbed through skin. Extremely destructive of tissues of mucous membranes, upper respiratory tract, eyes and skin. Inhalation may be fatal
Diethyl ether (ether)	Extremely flammable: flash point $-45\,°C$. Use in well-ventilated area away from sources of ignition. Forms explosive peroxides on prolonged storage or on exposure to light. Store only the minimum quantity necessary and always use a grade containing a stabilizer unless contrary to the chemistry employed in a particular protocol. Harmful by inhalation, in contact with skin and if swallowed
Digitonin	Harmful
Dimethyl sulfoxide (DMSO)	Harmful if swallowed. Irritating to eyes and skin. Readily absorbed through skin. Avoid breathing vapour. Wear protective clothing. May degrade under storage: keep only the minimum quantity necessary, under nitrogen if possible
DPX mountant	Contains xylene. Flammable, harmful by inhalation and in contact with skin. Irritating to skin. R-phrases: R1O-20/21-38. S-phrases: S25-36/37
EDTA	Toxic; irritating to eyes, respiratory system and skin
Eosin	Harmful by inhalation, in contact with skin, and if swallowed. Irritating
Ethane	Because of risk of explosion, the reliquification of gaseous ethane, taken from a cylinder of liquid ethane, must be carefully performed within a negative pressure fume extraction hood. Also, following the preparation of specimen grids, the small liquid ethane container must be left to evaporate within the fume hood
Ethanol	Highly flammable

(*continued overleaf*)

Ethanolamine	Harmful if swallowed, inhaled or absorbed through skin. Material extremely destructive of tissues of mucous membranes, upper respiratory tract, eyes and skin. Inhalation may be fatal
Ethidium bromide	Harmful. Ethidium bromide is a strong mutagen and can induce cancer. Always use gloves and clean up equipment carefully!
Ethyl acetate	Ethyl acetate should be handled with care, and disposed of by means appropriate for organic solvents
FITC	Harmful by inhalation, skin absorption or ingestion
Fluorescein isothiocyanate (FITC)	May be harmful by inhalation, ingestion or skin absorption. Causes eye and skin irritation. Irritating to mucous membranes and upper respiratory tract. Repeated exposure may cause asthma and allergic reactions. Prolonged exposure can cause nausea, dizziness, headache and lung irritation
Formaldehyde (liquid and vapour)	Toxic by inhalation, ingestion and by skin contact (causes sensitization). Possible risk of irreversible effects. Wear gloves and eye protection
Freund's adjuvant	Potentially harmful by inhalation, ingestion or skin absorption. Potent inflammatory agent if introduced intradermally or into eyes. In the case of skin or eye contact flush with copious amounts of water for at least 15 min
Giemsa staining solution	Contains methanol. Highly flammable. Toxic by inhalation and if swallowed. R-phrases: R11-23/25. S-phrases: S7-16-24-45
Glutaraldehyde	Glutaraldehyde is harmful if inhaled or if allowed to come into contact with the skin. Inhalation of the vapour may cause irritation to the mucous membranes, upper respiratory tract, eyes and skin. Sensitization with allergic, respiratory and skin reactions may occur. Handle in the fume cupboard, and wear gloves
Glycine	May be harmful if swallowed, inhaled or absorbed through skin. May cause irritation
Guanidinium hydrochloride	Harmful
HCl	Very toxic, causes burns
Hexylene glycol	Irritating to eyes and skin
Hydrochloric acid	May be fatal if inhaled, swallowed or absorbed through skin. Causes burns. Material extremely destructive of tissues of upper respiratory tract, eyes and skin. Inhalation may be fatal
Hydrofluoric acid	Safety precautions must be taken when working with these strong acids, particularly hydrofluoric acid, for which an antidote cream should be available in case of burns. Hydrofluoric acid must only be handled in the fume cupboard, using protective gloves and clothing
Hydrogen peroxide	Strong oxidizing agent

^{125}I-iodine	Radioiodine is a γ-emitter and is extremely hazardous. It is essential to consult your Radiation Protection Officer for a full list of precautions before using it. Minimize the risk of exposure to it by working in a designated fume hood with a lead shield between you and the source. Minimize also the time of exposure. Wear double gloves and lab coat. Prevent aerosol formation and monitor each step of the procedure and waste material with a monitor suitable for detecting γ-rays
KCl	Irritating to eyes, respiratory system and skin
KH$_2$PO$_4$	Irritating to eyes and skin. Eye contact: irrigate thoroughly with water. If discomfort persists obtain medical attention. Inhalation: remove from exposure. Skin contact: wash off skin with water. Ingestion: wash out mouth thoroughly with water and give plenty of water to drink. In severe cases obtain medical attention
KI	Irritating to eyes and skin. Eye contact: irrigate thoroughly with water. If discomfort persists obtain medical attention. Inhalation: remove from exposure. Skin contact: wash off skin with water. Ingestion: wash out mouth thoroughly with water and give plenty of water to drink. In severe cases obtain medical attention
KNO$_3$	Flammable. Contact with combustible material may cause fire. Extinguishing media: carbon dioxide, dry chemical powder or appropriate foam
KOH	Causes severe burns (see NaOH for details)
Leishman's staining solution	Contains methanol. Highly flammable. Toxic by inhalation and if swallowed. R-phrases: R11-23/25. S-phrases: S7-16-24-44
Liquid nitrogen	Causes low temperature burns ($-196\,^\circ$C) as liquid and as gas. Wear insulated gloves and goggles or face mask. Wear clothing which gives good skin and foot protection. Always handle in a well-ventilated area to prevent build-up of high nitrogen levels, which will reduce the concentration of available oxygen and cause asphyxiation
Lysozyme	Harmful
Mayer's haematoxylin	Harmful and irritating to eyes, respiratory system and skin
Methanol	Used as a fixative and a solvent for stains – highly flammable, toxic by inhalation and if swallowed
MgCl$_2$	Irritating to eyes, respiratory system and skin
MgSO$_4$.H$_2$O	Irritating to eyes and skin. Eye contact: irrigate thoroughly with water. If discomfort persists obtain medical attention. Inhalation: remove from exposure. Skin contact: wash off skin with water. Ingestion: wash out mouth thoroughly with water and give plenty of water to drink. In severe cases obtain medical attention

(continued overleaf)

$MnSO_4.H_2O$	Harmful: danger of serious damage to health by prolonged exposure through inhalation and if swallowed. Do not breathe dust. In case of contact with eyes, rinse immediately with plenty of water for 15 min and seek medical advice. In case of contact, immediately wash skin with soap and copious amounts of water. If inhaled, remove to fresh air. If not breathing give artificial respiration. If breathing is difficult, give oxygen. If swallowed, wash out mouth with water provided person is conscious
$Na_2EDTA.2H_2O$	Harmful by inhalation, in contact with skin and if swallowed. Irritating to eyes, respiratory system and skin. In case of contact with eyes, rinse immediately with plenty of water for 15 min and seek medical advice.
Na_2HPO_4	Irritating to eyes, respiratory system and skin
$Na_2MoO_4.2H_2O$	Harmful by inhalation, in contact with skin and if swallowed. Irritating to eyes, respiratory system and skin. In case of contact with eyes, rinse immediately with plenty of water for 15 min and seek medical advice. In case of contact, immediately take off all contaminated clothing and wash skin with soap and copious amounts of water. If inhaled, remove to fresh air. If not breathing give artificial respiration. If breathing is difficult, give oxygen. If swallowed, wash out mouth with water provided person is conscious
NaOH	Very corrosive, causes severe burns. Eye contact: rinse immediately with plenty of water for 15 min and seek medical advice. Skin contact: immediately wash skin with soap and copious amounts of water. Ingestion: if the chemical has been confined to the mouth give large quantities of water as a mouthwash. Ensure the mouth wash is not swallowed. If the chemical has been swallowed, give about 250 ml of water to dilute it in the stomach. In severe cases, obtain medical attention
N-ethylmaleimide (NEM)	Toxic; irritating to eyes, respiratory system and skin
NH_4NO_3	Flammable: Contact with combustible material may cause fire. Harmful by inhalation, in contact with skin and if swallowed. Irritating to eyes, respiratory system and skin. In case of contact with eyes, rinse immediately with plenty of water for 15 min and seek medical advice. In case of contact, immediately take off all contaminated clothing and wash skin with soap and copious amounts of water. If inhaled, remove to fresh air. If not breathing, give artificial respiration. If breathing is difficult, give oxygen. If swallowed, wash out mouth with water provided person is conscious
Nicotinic acid	Avoid contact with skin and eyes
Nitric acid	Corrosive. Safety precautions must be taken when working with strong acids. Nitric acid must only be handled in the fume cupboard, using protective gloves and clothing

o-phenylene diamine (OPD):	Harmful if swallowed, inhaled or absorbed through skin. Can cause irritation to eyes, skin mucous membranes and upper respiratory tract. Carcinogen
Osmium tetroxide	Osmium tetroxide is extremely hazardous. Very toxic. Flammable. Extremely harmful to eyes, respiratory system and skin. Readily volatilizes at room temperature. May produce irreversible effects. Danger of cumulative effect Always handle in the fume cupboard using protective gloves. Dispense over a safety tray to contain any spillage. Contaminated glassware and used osmium tetroxide fixative should be neutralized in 10% stannous chloride for 24–48 h, and then rinsed thoroughly with water
Paraformaldehyde	Paraformaldehyde is harmful if inhaled or if allowed to come into contact with the skin. Inhalation of the vapour may cause irritation to the mucous membranes, upper respiratory tract, eyes and skin. Sensitization with allergic, respiratory and skin reactions may occur. Handle in the fume cupboard, and wear gloves
Penicillin G	Harmful
Periodic acid	Contact with combustible material may cause fire; causes burns
Phenylmethylsulfonyl fluoride (PMSF)	Very toxic by inhalation, in contact with skin and if swallowed; causes burns, contact with water liberates extremely flammable gases
Potassium iodide	Harmful if swallowed, inhaled or absorbed through skin. Causes eye and skin irritation. Irritable to mucous membranes and upper respiratory tract. May cause allergic, respiratory and skin reactions.
Potassium permanganate	Harmful oxidizing agent
Presept	Harmful if swallowed
Propylene oxide	Harmful. Highly flammable. Irritating to eyes, respiratory system and skin. May form an explosive vapour mixture. Wear suitable protective clothing, gloves and face protection
Pyridoxine-HCl	Avoid contact with skin and eyes
Schiff's reagent	Causes burns; may cause cancer
Sodium azide	May be fatal if inhaled, swallowed or absorbed through skin. May cause eye and skin irritation
Sodium cacodylate	Toxic. Danger of arsenic poisoning. Poisonous by inhalation or ingestion. Avoid contact with skin and eyes. Do not breathe dust. Wear suitable protective clothing, gloves and face protection when preparing solutions and in their use. Danger of cumulative effect
Sodium dodecylsulfate	Harmful
Sodium metabisulfite	May be harmful by inhalation, ingestion and skin absorption. Causes eye and skin irritation. Persons with allergies and/or asthma may exhibit hypersensitivity to sulfites
Sodium tetrathionate	Irritating to eyes, respiratory system and skin

(continued overleaf)

Sulfuric acid	Highly corrosive. Avoid contact with skin and clothes. Carry out the operation over a sink, and use only small quantities. Wear protective gloves
Taab resin	Harmful. Can cause inflammation of the skin and mucous membranes. Can irritate respiratory system. Do not swallow or inhale. Avoid contact with skin and eyes. Over long exposure can produce skin sensitization
Taxol	Highly toxic cytotoxic agent
Tetramethylrhodamine	Irritant to eyes and skin
Thiamine-HCl	Avoid contact with skin and eyes
Tris	Irritating to eyes, respiratory system and skin
Trifluoracetate	Highly poisonous inhibitor
Triton X-100	Toxic; irritating to eyes, respiratory system and skin
Uranyl acetate	Very toxic. Radioactive. Irritating to eyes, respiratory system and skin. Do not breathe dust. Wear suitable protective clothing, gloves and face protection when preparing solutions and in their use. Danger of cumulative effect
Virkon	Irritant
Xylene	Toxic; carcinogen; highly flammable – handle in fume cupboard
$ZnSO_4.7H_2O$	Irritating to eyes and skin. Eye contact: irrigate thoroughly with water for at least 10 min. If discomfort persists obtain medical attention. Inhalation: remove from exposure, rest and keep warm. In severe cases obtain medical attention

Centrifugation Data

Calculation of centrifugal force

In centrifugation it is important to differentiate between the speed of centrifugation (RPM) and the centrifugal force (RCF or g) since these are often confused. The centrifugal force generated by a centrifuge can be easily calculated from the equation:

$$RCF = 11.18 \times R \times (RPM/1000)^2$$

where R is the distance from the centre of rotation in centimetres, that is the centrifugal force increases as the particles move down the centrifuge tube. As a general rule, the greater the centrifugal force the shorter the separation time. However, centrifugation also generates hydrostatic forces within the solution and so excessive centrifugal forces can disrupt some biological particles such as ribosomes.

Applications of centrifuge rotors

Type of rotor	Type of separation			
	Pelleting	Rate-zonal	Isopycnic (organelles)	Isopycnic macromolecules
Fixed-angle	Excellent but pellet on side	Poor	Poor	Excellent
Swinging bucket	Inefficient but small pellet	Excellent	Excellent	Acceptable
Vertical tube	Do not use	Good	Good	Excellent
Zonal rotor	Do not use	Excellent	Excellent	Acceptable

Calculation of pelleting times

Pelleting is the separation of particulate and non-particulate material and it is one of the simplest, most frequently used centrifugation techniques typically as part of a procedure for harvesting cells or the isolation of precipitated material. As a very approximate guide to the conditions needed to sediment various biological particles the following can be used:

Eukaryotic cells	$200g$ for 10 min
Nuclei	$1000g$ for 10 min
Mitochondria	$10\,000g$ for 10 min
Microsomes	$100\,000g$ for 60 min

As a general rule differential centrifugation produces enriched fractions rather than purified fractions, for example, the 'nuclear pellet' obtained by differential centrifugation almost always contains mitochondrial material that has co-sedimented with the nuclei. Similarly, the 'mitochondrial pellet' always contains material from lysosomes and peroxisomes. Hence differential centrifugation is usually carried out as an early step in the purification of subcellular components, often prior to further purification involving the use of rate-zonal or, more frequently, isopycnic gradient centrifugation.

The time to pellet particles depends on the k-factor of the rotor; the smaller the k-factor of a rotor the more efficient it is for pelleting particles. The time in hours for pelleting particles of known size (s-value) such as ribosomes can be calculated from the formula:

$$T = k\text{-factor}/s\text{-value}$$

But this formula does assume that sedimentation is taking place in a liquid with the same density and viscosity as water; in sucrose solutions the time required is significantly greater.

Care of centrifugation equipment

One of the major problems when using centrifuges is the corrosion of centrifuge rotors, particularly those made of aluminium alloys. Rotors made of aluminium alloys are very susceptible to severe corrosion even when left to soak in water overnight. Solutions left in aluminium rotors can cause internal corrosion of the metal alloy. After use always rinse, drain and dry rotors to avoid corrosion and the build-up of contamination. Always follow the manufacturers' recommendations regarding the care of centrifuge rotors. If at all possible use titanium or carbon composite rotors that are not affected by corrosion by aqueous solutions.

Radioisotope Data

Radioisotope	Half-life	Type of radiation	Shielding required	Comments
^{3}H	12.3 yr	β	None	
^{14}C	5760 yr	β	None	
^{22}Na	2.6 yr	β	Yes	
^{32}P	14.3 days	β	Yes, 10 mm Perspex	
^{33}P	25.4 days	β	Yes, 10 mm Perspex	Safer than ^{32}P
^{35}S	87.1 days	β	None	Volatile
^{51}Cr	28 days	EC γ	Yes, lead shielding	
^{59}Fe	46.3 days	βγ	Yes, lead shielding	
^{75}Se	121 days	EC γ	Yes, lead shielding	
^{125}I	60 days	γ	Yes, lead shielding	Volatile
^{131}I	8.1	β, γ	Yes, lead shielding	Volatile

Nuclease inhibitors

Nuclease inhibitors are included in gel electrophoresis procedures to prevent nucleic acid degradation during electrophoresis. Table 7.1 lists the main nuclease inhibitors used in many laboratory procedures. The RNase inhibitor diethyl pyrocarbonate (DEPC) is used in the water and autoclaved used to make up all gel solutions. DEPC-treated water should be autoclaved prior to use to prevent DEPC interfering with any further analyses. Inhibitors are also often used during the isolation of RNA and DNA to prevent degradation by the nucleases present in the host cells.

General manipulation procedures should also be observed to prevent nuclease action:

1. Wear plastic surgical gloves during all manipulations;

2. Autoclave or filter-sterilize all solutions;

3. Wipe surfaces clean prior to starting experiments;

4. Avoid leaning over or breathing into samples;

5. Autoclave or bake at 200 °C all utensils prior to experiments.

Table 7.1 Nuclease inhibitors and their working concentrations

Agent	Active concentration	Method of nuclease inactivation	Comments
Aurintricarboxylic acid (ATA)	10 μM	Complexes to a wide range of nucleases	
Bentonite	3 mg/ml	Inactivation by adsorbing to nucleases	
Diethyl pyrocarbonate (DEPC)	0.1%	Alkylates proteins disrupting protein structure	Toxic
Dithiothreitol (DTT)	1 mM	Reduces disulfide bonds, denaturing proteins	
EDTA	1–10 mM	Chelates divalent cations needed for ribonuclease activity	
Guanidine hydrochloride	8 M	Inactivates ribonucleases	Toxic
Guanidium thiocyanate	4 M	Inactivates ribonucleases	Toxic, strongest agent for ribonuclease inactivation
Heparin	0.5 mg/ml	Binds to basic ribonucleases	
8-Hydroxyquinoline	0.1% (w/v)	Inactivates ribonucleases	Very toxic
Macaloid	0.015% (w/v)	Adsorbs to ribonucleases	
2-Mercaptoethanol	0.1–0.25 M	Reduces disulfide bonds, denaturing proteins	Very toxic
Phenol/chloroform	50% (v/v)	Denatures ribonucleases	Toxic
Polyvinyl sulfate (PVS)	1–10 μg/ml	Complexes to basic nucleases	
Proteinase K	100–200 μg/ml	Hydrolysis of proteins	
Sodium dodecyl sulfate	0.1–1%	Disrupts protein structures	
Ribonucleoside vanadyl complex	10 mM	Binds to active site of ribonuclease	

Index

Page numbers followed by f indicate figures; page numbers followed by t indicate tables.

Aβ$_{1-42}$ peptide
  preparation of, 349
Abbé refractometer, 102
Acetone, 218
Achromat, 5
Acid phosphatase, 94
Acrylamide, 371
Adherins, 52
Agarose, 22
Agarose encapsulation
  for cell suspensions, 39
Agarose gel, 39
Alcian blue-coated mica, 30
Alcian blue-treated mica, 32
Alkaline phosphatase, 97, 260
  non-specific, 141
Alkaline phosphodiesterase, 97
  assay for, 143
Alkylation damage, 256
Alzheimer's disease (AD), 348
Amphibian Ringer's solution, 242
Amphotericin, 364
Amylases, 96
Amyloid-ß
  and enzymes, interactions between,
    359, 363
  fibrillogenesis, 359
  phosphorylation of, 364, 368
  phosphorylation by cyclin-dependent
    kinase-1 (CDK-1), 367, 368
  neurotoxicity, 359
  peptide, synthesis of 359, 364, 368
  phosphorylated
    extraction from cells, 366
    immunoassay measurement of,
      366, 367

Amyloid precursor protein, 345
Amyotrophic lateral sclerosis (ALS), 337
Anode buffer, 182
Anti-apoptotic proteins, 272
Anti-BrdU antibody, 212, 216
Anti-sense amino acid
  identification of, 355t
Anti-sense peptides, 353, 358
  comparison with known proteins, 354, 357
  derivation of sequences, 354
  direct binding to target proteins, 357, 358
  synthesis of 353, 354, 358
Anticoagulant, 80
Antigenic epitopes, 23
Antigenicity, 23, 38, 70
Antineuraminidase antibody, 283
Apical and basolateral domains
  fractionation from Caco-2 cells, 163, 164
  fractionation from MDCK cells, 165, 166
Apoptosis, 259
  immunocytochemical methods to study, 260,
    262
  inhibition of, 261, 262
Asialoglycoprotein
  endocytosis by rat liver, 191, 192, 193
Aspartic acid, 256
Assembly promotion, 329
A-type lamins, 226
Autoradiograms, 229
Autoradiography, 229
Avidin, 23
Axial resolution, 14

Ball-bearing homogenizer, 89, 99, 110, 164, 165,
  172, 174, 176, 188, 189

Cell Biology Protocols.   Edited by J. Robin Harris, John Graham, David Rickwood
© 2006 John Wiley & Sons, Ltd

Basolateral and bile canalicular plasma
    membrane domains
    separation from mammalian liver, 160, 161
Bath sonicator, 160
Beam-splitting prism, 11
Beckman fraction recovery system, 105, 129,
    172, 189, 192, 195
Beckman Optiseal tube, 129, 166
BEEM capsule, 48, 49
Bench centrifuge, 54, 58, 76
Biological specimens
    metal shadowing of, 35, 36
Biotin labelling, 353
Biotin-streptavidin binding affinity, 43
Block staining, 37, 48
Blocking buffer, 210
Blot overlay assay, 226, 229
Blotted proteins
    detection by enhanced chemiluminescence
        (ECL), 183, 184
Body tube, 2
Bongkrekic acid, 266
Bovine brain cytosol (BBC), 293–295
Bovine serum albumin (BSA), 29, 294
Bright field microscopy, 1, 6, 12
    setting microscope for, 7
Buoyant density gradient purification, 91
Buoyant density iodixanol gradients, 156

Cacodylate buffer, 45
Cadherins, 52
Calcein AM, 264
Capillary action, 68
Carbon coating, 48
Carbon films, 29
Carbon rod sharpener, 35
Carbon thickness monitor, 26
Carbon-coated mica, 26
Carbon-formvar
    preparation of, 25, 26
Carbon-formvar films, 29
Caspase inhibitor, 259
Caspase inhibitor ZVAD, 263f
Caspases, 259
Catalase, 95, 255, 359
Catalase assay, 123
    duration of, 123
    equipments for, 123
    procedure for, 123
    reagents for, 123
Cathode buffer, 182
Caveolae
    isolation of, 170, 171
Caveolin, 171

Cell
    analysis of, 51, 53
    concentration, calculation of, 68
    counting of, 65, 66
    counts, quantitation of, 67, 69, 70
    cultures, use of, 69, 70
    dead, removal using isopycnic centrifugation,
        65, 66
    freezing of, 74, 75
    isolation of, 51, 53
    isolation from body fluids, 61, 62
    membrane, 51
    recovery from effusions, 63
    recovery from monolayer cultures, 71, 72, 73
    thawing, 76, 77, 79
    viability, 65, 66
        calculation of, 68
        determination of, 65
Ceramides, 283
Charcot-Marie-tooth disease (CMT), 337
Chase period, 215
Chemiluminescence, 182
Chlorophyll, 97
Chlorophyll a/b complex
    reconstitution into liposomes, 300, 304
Chloroplast chlorophyll
    measurement of, 147
Chloroplast integrity
    assessment of, 148
Chloroplasts, 97
    functional assays, 97
    isolation from green leaves 145, 146
    isolation from pea seedlings, 145, 146
    isolation of, 97
Chromatic beam splitter, 13, 14
Chromatin, 220
Chromatin assembly, 207f
Chromosomes,endogenous
    nuclear assembly around, 232f
Circular dichroism (CD), 304
Cisplatin, 250
Cisplatin nanocapsules, 250
    characterization of, 252, 253
    delivery of contents, 253
    electron micrograph of, 252f
    encapsulation efficiency, 252
    principle, 251
    shape of, 252, 253
    size of, 252, 253
    synthesis, procedure for, 251, 252
Citrate, 80, 82, 83
Class II cabinet, 71, 76
Classical nuclear matrix method, 46, 48
Clathrin, 286, 288

Clathrin-coated vesicles (CCVs)
  purification from rat brains, 286, 287
CLSM, *See* Confocal laser scanning microscopy
    (CLSM)
Cold trypsinization, 58, 59
Collagen, 60
Collagen fibril
  in vitro assembly of, 375, 376, 377, 378
Collagenase, 60
  concentration used for tissue disaggregation,
      61*t*
  types of, 61*t*
Collagenase type Ia, 60
Collagenase type IV, 60
Colloidal gold, 23
  conjugate, 43, 46
Colorimetric assay, 157
Complementary strand, 353
Compound microscope
  components of, 2, 6
Condenser, 2, 3
Condenser-iris diaphragm, 2
Confocal laser scanning microscope (CLSM),
    14, 242
Confocal microscopy, 14, 16, 154
  image collection, factors affecting, 16*t*
Continuous carbon films
  preparation of, 25, 26
Continuous cultures, 70
  contamination, signs of, 70
Continuous linear gradients, 156
Coomassie blue, 181, 332
Coverslip, 4, 8, 67
Creatine phosphokinase, 204
Cristae, 255
Critical point drying, 48
Cross-contamination, 70, 72
Cross-link stabilized nuclear matrix method, 46
Cross-linking agents, 18
Cryoelectron microscope, 22, 34
Cryoholder system, 34
Cryomicroscopy, 23
Cryopreservation, 72, 73
  critical factors in, 72
Cryoprotectant, 72
Cryotransfer, 34
Cultured cells
  purification of mitochondria in Nycodenz
      gradient, 114, 115
Cyclin-dependent kinase-1 (CDK-1), 359, 364
Cyclosporin A, 266
Cysteine proteases, 259
Cytochalasin, 233
Cytochrome *c*

measuring release in isolated mitochondria,
    271
Cytochrome *c*, 255
Cytomatrix
  of keratinocyte, 323*f*
Cytomatrix proteins
  detection by immunogold embedment-free
      electron microscopy, 317, 325
    cell extraction 319, 320
    equipments for, 319
    preparation for 320, 321
    reagents for, 318, 319
    staining with antibodies, 321, 322
Cytoskeletal buffer, 220
Cytoskeleton, 96
  imaging by embedment-free electron
      microscopy, 44, 45, 46, 50
Cytosolic thioredoxin peroxidase I
    (cTPxI), 255
Cytosolic thioredoxin peroxidase II, 255
Cytotoxic effect, 253
Cytotoxicity, 66, 253, 364

Dark field microscopy, 11, 12
Dehydration, 47
Dense sucrose solution (DSS), 139
Densitometric scanning, 229
Density gradient centrifugation, 91, 92
Density gradient fractionation, 96
Deoxyribonuclease I, 102
Depurination
  of DNA, 212, 218
Desmosomes, 96
Detergent-resistant membranes (DRMs), 155
DEVD, 259
Dewar flask, 76
Dextran, 96, 135
4'-6-diamidino-2-phenylindole (DAPI), 210, 211,
    224, 330
Diaminobenzidine (DAB), 183
Diatrizoate, 78
Diethylene glycol distearate (DGD), 44, 48, 318
Differential cell fractionation, 314
Differential centrifugation, 90, 91
  equipment for, 91
  problems in, 90
Differential interference contrast (DIC), 1
  microscopy, 11
Digestion buffer, 222
Diluent, 74
Dimethyldichlorosilane, 100
Dimethylsulfoxide (DMSO), 72, 74, 76
Dioleoyl-phoshatidylglycerol (DOPG), 252
Dioleoyl-phosphatidic acid (DOPA), 252

Dioleoyl-phosphatidylethanolamine
    (DOPE), 252
Dispase, 60
    concentration used for tissue
        disaggregation, 61t
    types of, 61t
DNA
    double labelling with halogenated thymidine
        analogues, 214, 215, 216
        safety precautions, 216
    single labelling with halogenated thymidine
        analogues, 209, 210, 212, 213
    fingerprinting, 70
    proteins and halogen-dU-substituted
        simultaneous immunostaining of, 217, 218,
        219
DNase I, 103, 106
DNases, 249
Dounce homogenization, 166, 176
Dounce homogenizer, 89, 91, 99, 109–112, 139,
    160, 172, 174, 188
Droplet immunolabelling, 32
Droplet negative staining procedure, 27, 28
    equipments for, 27
    preparation of
        duration, 28
    procedure for, 27, 28
    reagents for, 27
Dual-beam spectrophotometer, 131
Dye exclusion method, 66
Dystrophic neurites, 348

EDTA, 80, 82, 83, 192
EGTA, 88
Electroblotting, 171
Electron microscopy (EM), 21, 24, 317
    methods of, 22
    preparative equipments, 24
    reagents used for, 24
    resinless section, 44
    sections for, staining of, 40, 41
    whole mount, 44
Electron-dense marker, 23
Electrostatic interactions, 251
Embedment-free electron microscopy (EFEM),
    44, 50, 317
Endocytic intermediates
    electron micrographs of, 288, 292
    reconstitution on lipid monolayer, 288, 289,
        291, 292
Endocytosis, 158, 159, 253, 288
    analysis of, 159
        iodixanol gradients for, 189t
    ligand labelling, 158

Endoplasmic reticulum (ER), 51, 95, 96
    analysis in sedimentation velocity iodixanol
        density gradients, 177, 178, 178t, 179
    analysis using iodixanol density gradients,
        174, 175, 176
    buoyant density separation of, variations in
        gradient conditions, 175t
    chemical markers for, 95, 96
    enzyme markers for, 95, 96
    rough, 95, 96
    separation in preformed sucrose gradient, 127,
        128
    separation in self-generated iodixanol
        gradient, 129, 130
    smooth, 95, 96
    subfractions
        analysis using sucrose density gradients,
            172, 173
Endoplasmic reticulum-golgi-plasma membrane
    pathway
    membrane compartments in, analysis of, 156,
        157
        buoyant density iodixanol gradients, 156
        by electroblotting, 157
        by SDS-PAGE, 157
        sedimentation velocity iodixanol gradients,
            157
        sucrose-D₂O gradients, 156
Endosome-lysosome events
    analysis in mammalian liver, 194, 195, 196
Endosomes, 95, 96, 159, 195
    early, 190
    recycling, 190
Enhanced chemiluminescence (EC), 314
Envelope suspension buffer (ESB), 102
Enzyme activity, 60
Epifluorescence, 13
    setting microscope for, 14, 17
Epifluorescence microscope, 13f, 210, 215, 218
    dichroic mirror block in, 13f
EPR, 335
ER-golgi intermediate compartment (ERGIC),
    96, 156
    analysis using iodixanol density gradients,
        174, 175, 176
Erythrocyte ghost pellet, 137
Erythrocyte ghosts, human
    purification of, 137, 138
Erythromycin, 66
Ethanol, 89
Euchromatin, 44
EVON chopstick electrodes, 297
Extraction buffer (EB), 104, 221
Eyepiece, 1, 2, 5

FASTA format, 354
Ferricyanide, 97
β fibril formation, 345, 347
    equipments for, 345
    procedure for, 345, 346
    reagents for, 345
Fibrillogenesis, 345
Aβ fibrillogenesis, 345
Ficoll, 195
Field iris diaphragm, 2
Filters, 2
Fiske-Subbarow reducing agent, 132, 141
FITC-dextran, 296
Fixatives, 18t
Flameless atomic absorption spectrometry, 252
Fluorescence microscopy, 1, 12, 13
Fluorescence unit, 120
Fluorimeter, 120
Fluorinert, 176, 179, 192, 195
Fluorochrome dye, 12
Fluorochromes, 218
Fluorometric assay, 157
Flutec-blue, 192, 195
Formaldehyde, 18, 217
Formaldehyde fixative, 222
Formvar, 28
Formvar film, 46, 49
Fractin, 261
Fractin staining, 261
Freeze-fracture, 23
Fume extraction hood, 33
Functional enzyme activity assays, 359

β-Galactosidase, 94
    fluorometric assay, 120
    spectrophotometric assay, 119
Galactosyl transferase, 96
Galactosylation, 96
Ganglioside
    flip-flop translocation in lipid bilayers, 283
Gelatinous pellet, 102
Gibbs-Donnan equilibrium, 78
Glucose-6-phosphatase, 95
Glucose-6-phosphatase assay, 132
ß-D glucosidase, 96
Glutaraldehyde, 18, 24, 36
Glutaraldehyde fixation, 32, 39
Glutaraldehyde fixative, 45, 47
Glutathione, 258
Glutathione-dependent peroxidase, 255
Glycerol, 72, 181
Glycolipids, 281
    asymmetric incorporation into liposomal
        membranes, 280, 281, 283

asymmetric incorporation into membranes,
    280, 285
Glycolysis, 255
Glycosylphosphatidylinositol (GPI), 280, 283
GM3
    neuraminidase cleavage of, 282f
GM3 ganglioside
    sphingomyelinase induced flip-flop of, 284f
Golgi
    analysis using iodixanol density gradients
        duration of, 175
Golgi complex, 51
Golgi membrane, 96
    analysis in sedimentation velocity iodixanol
        density gradients, 177, 178, 178t, 179
    analysis using iodixanol density gradients,
        174, 175, 176
    binding of vimentin to, 310f
    buoyant density separation of, variations in
        gradient conditions, 175t
    enzyme marker of, 96
    isolation from liver, 134, 135
    purification of, 308
    tubule formation from, 293, 294, 295
    purified
        vimentin binding to, biochemical assay for,
            307, 308, 309, 312
trans-Golgi network (TGN), 95, 96, 154
    analysis in sedimentation velocity iodixanol
        density gradients, 177, 178, 179
Golgi stack, negatively stained, 294f
Golgi subfractions
    analysis using sucrose density gradients, 172,
        173
Golgi-vimentin interaction, 307, 311f, 312
    immunofluorescence analysis of, 309, 310, 312
Gradient master, 121, 127, 170, 178
GST-synaptobrevin, 275

Haemocytometer chamber, 69f
Halogenated nucleosides, 210
Hanks balanced salt solution (HBSS)
    without calcium and magnesium, 54, 57, 58,
        71, 74
Hela cells, 204
Hematoxylin, 49
Heparin, 80, 82, 83, 104
Hepatocyte, reconstituted
    micrograph of, 279f
Hepatocytes, 158
Hepes-buffered saline (HBS), 80, 82, 83, 163
Heterochromatin, 44, 93, 103
Heteropolymers, 338
Hexylene glycol buffer, 100

[³H]-inulin, 296
Hoechst 33258, 70
Holey carbon support films, 33
    preparation of, 25, 26
Homogenization, 88, 90
Homogenization buffer (HB), 172
Homogenization medium (HM), 89t, 98, 110,
    121
Horseradish peroxidase (HRP), 159
Hotplate/magnetic stirrer, 56, 58
Human Aβ₁₋₄₂ peptide, 346f
Hydropathic interactions, 353
Hydroperoxide tolerance
    determination of, 256, 257
Hydrophilicity, 28
Hydrostatic pressure, 91, 93
Hyperosmotic medium, 90
Hypoosmotic disruption, 97
Hypoosmotic medium, 90, 99
Hypotonic lysis, 96
Hypotonicity, 64

IF oligomers, 335
Immunoblot analysis, 275
Immunoblotting, 231
Immunocytochemistry, 351, 352
Immunodepletion, 236
    column preparation for, 235, 236
    of proteins, 234
Immunofluorescence, 12, 238, 239
Immunofluorescence microscopy, 231, 317
Immunofluorescent signal, 213, 216
Immunofluorescent staining
    indirect, 209
Immunolabelling, 23, 36
    post-embedding indirect, 42, 43
Immunolocalization, 46
Immunonegative staining, 26, 29, 30
Immunostaining, 18
Inclusion bodies (IBs)
    isolation of, 332, 333
    purification of, 332, 333
Interference filters, 13
Intermediate filament (IF), 307, 331, 335f, 337
    assembly of, 334, 335
Interpupillary distance, 7
Inverted phase contrast microscope, 71, 74
Iodinates density gradient media
    molecular structure of, 77f
Iodixanol, 122, 143, 154, 181, 185, 189, 191
Iodixanol gradient, 155
Iodixanol working solution (IWS), 124
p-Iodonitrotetrazolium violet, 94, 116
Ischaemia, 259

Isoenzymes, 70
Isoosmotic medium, 78, 89
Isopycnic centrifugation, 65, 105

JEOL X-band spectrometer, 335
Jurkat cells, 90

Karyotypes, 70
Keratinocytes, 296
Kohler illumination, 6, 11
    setting of, 8

Labconco auto densi-flow device, 129, 169, 172,
    176, 189, 192, 195
Laemmli gel electrophoresis system, 157
Laminopathies, 226
Lateral displacement, 11
Lead citrate, 41, 43
Levamisole, 141
LHCIIb, See Light-harvesting chlorophyll (Chl)
    a/b protein complex of photosystem II
    (LHCIIb)
LHCIIb complex
    fluorescence emission spectra of, 302f
LHCIIb liposomes
    fluorescence emission spectra, 303f
LHCIIb proteoliposomes, 304
    circular dichroic absorption spectra of, 302f
    fluorescence emission spectra of, 302f
Light microscope 1, 2t, 6
    care of, 16
    limit of resolution, 2
    maintenance of, 16
    parts of, 4f
    problems in, 16
Light mitochondrial pellet (LMP), 114, 117
Light path
    conventional vs confocal microscopy, 15f
Light source, 2
Light-harvesting chlorophyll (Chl) a/b protein
    complex of photosystem II (LHCIIb), 300
    function of, 300
Lipid flip-flop movement
    detection of, 283
Lipid rafts
    isolation of, 167, 168t, 169
        duration of, 168
        equipments for, 167
        by gradient separation, 167
        procedure for, 167, 168
        from post-nuclear supernatant, 167
        reagents for, 167
        from total cell lysate, 167

Lipophilicity, 250
Liposomes, 158, 250, 251, 280, 281
    GPI incorporation into the outer monolayer of, 282*f*
Liver endoplasmic reticulum (ER)
    invitro reconstitution of, 277, 278, 279
London resin, 38
Lowicryl resin, 38
Lymphocyte separation medium, 80
Lymphocytes, 90
Lymphoprep, 65, 78, 80
Lyophilization, 252
Lysine/hydroxylysine residues, 375
Lysosomes, 51, 91, 93, 94, 109, 114, 190, 192, 194
    enzyme markers of, 94
    isolation in Nycodenz gradient, 117, 118
    isolation of, 94
    marker of, 94
    purification of, 94

Madin Darby canine kidney (MDCK) cells, 296
Magnification system, 1
Mammalian liver
    fractionation of light mitochondrial fraction, 125*f*
    isolation of nuclei from, 98, 99
    purification of mitochondria in Nycodenz gradient, 114, 115
Mammalian peroxisomes
    isolation in iodixanol gradient, 121, 122
Mannitol, 108
Membrane proteins
    SDS-PAGE of, 180, 181
Membrane vesicles
    separation from cytosolic proteins, 157, 158
Membranes and cytosolic fractions
    separation from bacteria, 185, 186
    separation from mammalian cells, 185, 187
Metal shadowing, 23
Metaphase chromosomes
    isolation of, 100, 101
Methacrylate resin, 42
Methylamine tungstate, 28
Methylamine vanadate, 28
Metrizamide, 100
Metrizoate, 77
Mg-ATPase, 103
Microscope, 1
    components of
        body, 2
        condenser, 3
        eyepiece, 5
        lamp, 2

    objective, 4
    stage, 4
    illumination system of, 2
    magnification system of, 2
Microscopy
    techniques of, 6
Microsomes, 154
Microtiter plate binding, 359
Microtubule assembly promoters (MTAPs), 326
Microtubule organizing centre, 310
Microtubule peroxisome interaction, 313, 315, 315*f*
Microtubule-associated proteins (MAPs), 330
Millicell electrical resistance system (ERS), 296
Mitochondria, 51
    enzyme markers of, 93, 94
    functional marker of, 93, 94
    isolated
        protein import into, 272
    isolation of, 93, 94
Mitochondrial fraction, light
    preparation from cultured cells, 110, 111
    preparation from tissues, 110, 111
Mitochondrial inhibitors, 258
Mitochondrial membrane potential, 264
Mitochondrial permeability transition
    $\Delta\Psi$m loss determination
        in cells using confocal laser imaging, 264, 265, 266, 267
        in isolated mitochondria using confocal laser imaging, 264, 267
    $\Delta\Psi$m loss measurement
        in cells on FACS, 270
        in isolated mitochondria on FACS, 270
        in isolated mitochondria using fluorometer, 268, 269
    PT loss determination
        in cells using confocal laser imaging, 264, 265, 266, 267
        in isolated mitochondria using confocal laser imaging, 264, 265, 266, 267
    PT loss measurement
        in cells on FACS, 270
        in isolated mitochondria on FACS, 270
        in isolated mitochondria using fluorometer, 268, 269
Mitochondrial thioredoxin peroxidase I, 255
Mitotic lysates
    preparation of, 230, 231
Molar extinction coefficient
    p-nitrophenol, 143
    of reduced cytochrome *c*, 131
    of reduced INT, 116
Monolayer cultures, 71

MTT reduction, 365
Mycoplasma, 70
Myosin filaments
   aggregation, prevention by S100A4, 374
   assembly/disassembly in presence of S100A4, 372, 374
   disassembly, S100A4 stimulated, 373
   preparation of, 373
Myosin II filaments, 369

Na$^+$/K$^+$-ATPase, ouabain-sensitive
   assay for, 144
N-acetyl-glucosamine, 94, 96
NADPH-cytochrome c reductase, 95
   determination of, 95
NADPH-cytochrome c reductase assay, 131
Nanocapsules, 250, 253
Nanoprecipitate, 251
Negative staining, 22, 28
Negative staining-carbon film technique, 31, 32
Neogalactosylalbumin, 192
Neubauer chamber, 67
Neubauer haemocytometer, 74
Neuraminidase, 281, 283
Neuroblastoma cells, 178
Neurodegenerative diseases, 259
Neurofibrillary tangles (NFTs), 348
Neurofilament (NF), 337
Neurofilament (NF) assembly, 337, 338, 340, 340f, 341
   in SW-13/cl.2 vim$^-$ cells, 339, 340
Neurofilament (NF) proteins
   purification of, 339
   SDS-PAGE of, 340f
Neuronal apoptosis
   in vitro assessment of, 259, 260, 262, 263
Neurotoxicity, 250, 359, 364
Neutral protease, 60
Newton's rings, 67
Nigrosin, 66
Nitrocellulose, 182
Nitrocellulose membrane, 297
Nitrogen cavitation, 99, 101, 163
Nitrophenol, 94
Nocodazole, 233
Nomarski optics, 266
Non-β-amyloid β component (NAC), 342
Non-immune serum, 29
Non-nucleolar matrix proteins, 107
Nosepiece, 2
Nuclear assembly, 238
   and immunofluorescence, 237, 238, 239
   steps in, 237

Nuclear components
   isolation of, 92, 93
Nuclear envelope (NE), 226, 237
   isolation of, 102, 103
Nuclear ghosts, 102, 103
Nuclear lamina, 226
Nuclear matrix
   classical, 223, 224
   cross-link stabilized, 223
   imaging by embedment-free electron microscopy, 44, 50
      equipments for, 46
      procedure for, 46, 50
      reagents for, 44, 45
      using resinless sections, 48, 50
      solutions for, 45
      using whole mount of cells, 46, 48
   of keratinocyte, 323f
   preparation of, 106
   ultrastructure of, 221f
   uncovering in cultured cells, 220, 223, 224
Nuclear matrix-lamin interactions
   blot overlay assay, 226, 228, 229
   nuclear reassembly assay, 230, 232, 233
Nuclear pore complex (NPC), 237, 240
   isolation of, 104, 105
Nuclear transport
   optical single transporter recording (OSTR) measurements of, 244, 245, 245f, 246f, 247
Nuclei
   pelleting of, 92
   purification of, 92
Nucleocytoplasmic transport, 241f
   measurement of, 242f
   measurement using Xenopus oocyte nuclei, 240, 243
Nucleolar decondensation, 103
Nucleoli, 106
   preparation of, 107
Nucleoplasmic lamin binding proteins, 226
Nucleosomal histones, 224
Nucleosome assembly
   coupled to DNA repair synthesis, 204, 205, 206, 208
Nucleosomes, 204
5'-nucleotidase, 97
   assay for, 141
Nucleotide excision repair (NER) pathway, 204
Nucleotide precursor, 210
Nycodenz, 78, 101, 117, 143, 154, 181
Nycoprep, 78, 80, 117

Objective lens, 1, 2, 6*f*
  properties of, 5*t*
Observation methods
  selection of, 3*f*
Octylglucopyranosid (OGP), 306
Ocular lens, 1
Oil immersion microscope
  focusing procedure, 9
Oil immersion objective, 8, 9
Optical single transporter recording (OSTR), 244
OptiPrep, 78, 80, 82, 83, 98, 121, 165, 170, 186
OptiPrep diluent (OD), 124
Optiseal tube, 192
Orcinol reagent, 133
Organelles
  analysis in performed iodixanol gradient, 124,
    125, 126
  fractionation in iodixanol gradient, 125*t*
  functional analysis of, 88, 97
  isolation of, 88, 97
Osmic acid, 24
Osmium fixative, 45
Osmium tetroxide, 37, 38, 45
Osmotic shock, 73
Osmotic stress, 256
Ouabain, 144
Oxidative phosphorylation, 89, 255

p20 antibody, 261
p20 staining, 261
p35-overexpressing N2a cells, 349
Palmitoylation, 274
Parafilm, 27
Pellet
  contamination, prevention of, 90
Pelleting, 90, 92, 119
Penetratin, 259
Penicillin, 364
Peptide J, 262
  toxicity on cortical neuron cultures, 262*f*
Percoll, 92, 94
Perichromatin fibrils, 220
Peripheral blood mononuclear cells (PBMC)
  purification on density barrier, 80, 81
  purification using barrier flotation technique,
    83, 84, 86
  purification using mixer technique, 82
Peripheral blood mononuclear cells (PBMC),
    human
  isolation of, 77, 79
    using barrier flotation technique, 78, 79
    using density barrier, 77, 78
    using mixer technique, 78
Permeability transition (PT), 264

Permeability transition (PT) inhibitor, 266
Peroxidase, 103
Peroxide, 123, 258
Peroxiredoxins, 255
Peroxisome-microtubule binding assay, 313,
    314, 314*f*, 315
Peroxisomes, 89, 91, 93–95, 192, 313
  enzyme markers of, 95
Phase contrast microscope, 9, 10*f*, 61, 102, 106,
    176, 134
  limitations of, 9
  setting microscope for, 11, 13
Phenol red, 61
Phenylmethylsulfonylfluoride (PMSF), 233, 370
Phosphate buffered saline (PBS), 29, 39, 42,
    167, 183
Phosphatidylcholine, 281, 284
Phosphatidylethanolamine, 284
Phospholipase A_2 enzyme
  Ca^{2+}-independent, 293
Phospholipase C, 285
Phospholipids, 288
Phosphorylated lamina proteins, 231
Phosphotungstic acid (PTA), 293
Photobleaching, 15
Photomultiplier (PMT), 14
Photosystem 2
  reconstitution into liposomes, 305, 306
    biobeads pretreatment, 306
    equipments for, 306
    liposome preparation, 305
    procedure for, 306
    reagents for, 306
Plan Achromat, 5
Plan Apochromat, 5
Plasma membrane, 96, 97
  analysis using enzyme markers, 97
Plasma membrane domains, 155, 156
Plasma membrane sheets
  isolation from rat liver, 139, 140
Plasmid DNA, 206
Plastoquinone, 305
Platinum shadowing, 291
Plunge-freezing apparatus, 33
Poly-L-lysine coating, 18, 19
Polyacrylic acid, 291
Polymorphonuclear leukocytes (PMNs), 78, 82
Polysaccharide, 78
Polystyrene box, 75
Polysucrose, 195
Polysucrose gradients, 195
Polysucrose-Nycodenz gradient, 194
Polytron, 145
Polytron homogenizer, 89, 134, 160

Polyvinylidine difluoride, 157
Ponceau S stain, 228, 229
Post-embedding immunogold cytochemistry
    on COLO 20 cells, 324f
Post-nuclear supernatant (PNS), 165
Potassium ferricyanide, 148
Potter-Elvehjem homogenizer, 89, 98, 108, 110,
    111
Pregnant mare's serum gonadotropin (PMSG),
    235
Propidium iodide, 249
Protease inhibitor, 112
Protein assay kit, 309
Proteins
    immunogold localization of, 46
Proteoliposomes, 158
Pulmonary oedema, 233
Pulse period, 215

Quartz crystal, 26

Radioactivity, 233
Rat liver
    isolation of heavy mitochondrial fraction
        from, 108, 109
    microsomes, fractionation in self-generated
        iodixanol gradient, 130f
    sinusoidal domain
        isolation using antibody-bound beads, 162
Red blood cells (RBC)
    removal by snap lysis, 64
    removal using isopycnic centrifugation, 65, 66
Refractive index, 11
Refractometer, 103, 139
Renaturation, 229
Resin, 38
Resin embedding, 37, 38
Resinless section electron microscopy, 44
Resolution, 1, 2
Ribonucleoprotein (RNP), 220
RNA analysis, 133
RNA splicing, 220
Rough ER suspension medium (RSM), 127

S100A4, recombinant
    isolation of, 373
Saccharomyces cerevisiae, 255
Scintillant, 136
SDS-PAGE, 231, 275
SDS-PAGE gel, 181

Sectioning
    of frozen tissues, 19
    of paraformaldehyde-fixed, paraffin-embedded
        tissues, 19
Sedimentation velocity iodixanol
    gradients, 157
Semi-dry blotting, 182
Semi-dry Western blotting, 157
Seminiferous tubules, 297
Serine-protease inhibitor, 340
Sertoli cells, 296, 297
    assembly of TJ-permeability barrier, 298f
    electron micrograph of, 297f
Silicotungstate, 28
Smitin, 369
    isolation of, 369
Smitin-myosin II coassembly arrays, 369, 371
    synthesis of , 369, 370
SNARE complexes, ternary
    formation of, 274, 274f, 275, 276
Sodium cacodylate, 45
Sodium citrate, 40
Sodium phosphotungstate, 28
Soluble NSF attachment protein receptor
    (SNARE) proteins, 274
Sonication, 156
Sonicator, 171
Specimen
    fixation, 18, 19
    metal shadowing, 23
    preparation of, 16, 17
    staining of, 16, 17
    tissue sections, preparation of, 19
Spectrophotometer, 119, 123, 377
Sphingolipid, 155
Sphingomyelin, 280, 284
Sphingomyelinase, 280, 285
Stains, 19, 19t
Stereomicroscope, 240
Streptavidin, 23
Streptolysin O (SLO), 248
    cell permeabilization with, 248, 249
Streptomycin, 364
Subcellular membranes
    fractionation of, 154, 159
        methods, 154, 155
Subconfluent monolayer, 46, 48
Succinate dehydrogenase, 93
Succinate-INT reductase assay, 116
Sulfydryl-activatable toxin, 248

Supercoiling assay, 208
Supernatant, 63, 69
Suspension buffer (SB), 106
Swinging-bucket rotor, 90, 92
Syntaxin, 274
α-Synuclein fibril formation
    tubulin induced, 342, 343, 343*f*, 344
Synucleinopathies, 342

Tau hyperphosphorylation, 348
    Aβ$_{1-42}$ peptide induced, 348, 349, 352
    western immunoblotting analysis of, 350, 351
Tau phosphorylation,351*f*
Taxol, 326, 329
TdT mediated dUTP nick end labelling
    (TUNEL), 260
    staining, 260, 261
Teflon block, 289, 289*f*
Tetramethylrhodamine (TMRM), 265
Thermanox coverslip, 48, 49
Thin section electron microscopy, 106
Threonine, 256
Thymidine, 215
Tight junction
    assembly of, 296, 299
        cell cultures for, 296, 297
        procedure for transepithelial electrical
            resistance (TER)measurement, 297,
            299
Tissue
    disaggregation using collagenase, 60, 62
    disaggregation using dispase, 60, 62
    mechanical disaggregation of, 52, 53
    disaggregation by mechanical
        mincing/chopping, 54, 55
    methods for disaggregating, 53*t*
    red blood cell removal during disaggregation,
        63
    disaggregation by warm trypsinization, 56, 57
Tissue culture hood, 46
Tissue processing, 37, 38
Titanium oxysulfate, 95, 123
Titin, 369
Topoisomerase, 208
Transblotted proteins, 228
Transepithelial electrical resistance (TER), 296
Transfection, 250, 337
Transferrin
    endocytosis in transfected MDCK cells, 188,
        190

Transitional endoplasmic reticulum
    (tER), 278
Transitional solvent, 49
Trehalose, 23
Trichloroacetic acid (TCA), 132
Tricine-buffered saline, 80, 82, 83
Triton X-100, 106
Trypan blue, 66–68, 249
Trypan blue dye exclusion, 365
Trypsin, 56
Trypsin inhibitors, 56, 57, 59
Trypsinization, 57, 59, 224
    cold, 56, 58, 59
    warm, 56
Tubulation buffer, 294
Tubulin, 326, 343
Tubulin assembly
    monitored on absorption spectrometer, 328*f*
    induced by microtubule assembly promoters
        (MTAPs), 326, 330
    induced by taxol, 326, 327, 330
    monitored by fluorescence, 329*f*
Turbidimetry, 375
Type I collagen fibril
    axial structure of, 375, 378
    reconstitution of, 376*f*

UDP-galactose galactosyl transferase
    assay of, 136
Ultra-Turrax, 145
Ultrasonication, 104, 105
Ultrasonicator, 104, 107
Uranyl acetate, 24, 40, 43
Uranyl acetate stain, 291
Uranyl formate, 28

v-SNARE synaptobrevin, 274
Vacuum coating unit, 31
VAMP seev-SNARE synaptobrevin
Vectashield solution, 239
Vernier scale, 4
VersaDoc imager, 184
Vesiculation, 155
Vimentin, 307
    production of, 331, 332
    spin labelled, 334*f*
        EPR spectra from, 335*f*
        preparation of, 333, 334
        purification of, 333, 334
Vitrification, 22, 23

Vitrified specimens
    negatively stained, 33, 34
    unstained, 33, 34

Waring blender, 145
Western immunoblotting, 349
Whole mount electron microscopy, 44
Working distance, 5

Xenopus laevis
    egg extracts, preparation of, 234, 235, 236

Yeast
    mitochondria, purification in Nycodenz
        gradient, 112, 113
    protective role of antioxidant proteins,
        determination of, 255, 258
    scheme viability of, 257f
    synthetic medium for, 256
Yeast peptone dextrose (YPD), 258
Yeast spheroplast, 112

ZVAD, 259